Chest Blunt Trauma
A Modern Approach to a Multidisciplir

Chest Blunt Trauma
A Modern Approach to a Multidisciplinary Disease

Edited by

Piergiorgio Solli
Unit of Thoracic Surgery, Royal Papworth Hospital NHS Foundation Trust, Cambridge, United Kingdom

Marco Scarci
Department of Thoracic Surgery, Imperial College NHS Healthcare Trust, London, United Kingdom; National Heart and Lung Institute, Imperial College, London, United Kingdom

Academic Press is an imprint of Elsevier
125 London Wall, London EC2Y 5AS, United Kingdom
525 B Street, Suite 1650, San Diego, CA 92101, United States
50 Hampshire Street, 5th Floor, Cambridge, MA 02139, United States

Copyright © 2025 Elsevier Inc. All rights are reserved, including those for text and data mining, AI training, and similar technologies.

For accessibility purposes, images in electronic versions of this book are accompanied by alt text descriptions provided by Elsevier. For more information, see https://www.elsevier.com/about/accessibility.

Publisher's note: Elsevier takes a neutral position with respect to territorial disputes or jurisdictional claims in its published content, including in maps and institutional affiliations.

No part of this publication may be reproduced or transmitted in any form or by any means, electronic or mechanical, including photocopying, recording, or any information storage and retrieval system, without permission in writing from the publisher. Details on how to seek permission, further information about the Publisher's permissions policies and our arrangements with organizations such as the Copyright Clearance Center and the Copyright Licensing Agency, can be found at our website: www.elsevier.com/permissions.

This book and the individual contributions contained in it are protected under copyright by the Publisher (other than as may be noted herein).

Notices
Knowledge and best practice in this field are constantly changing. As new research and experience broaden our understanding, changes in research methods, professional practices, or medical treatment may become necessary.

Practitioners and researchers must always rely on their own experience and knowledge in evaluating and using any information, methods, compounds, or experiments described herein. In using such information or methods they should be mindful of their own safety and the safety of others, including parties for whom they have a professional responsibility.

To the fullest extent of the law, neither the Publisher nor the authors, contributors, or editors, assume any liability for any injury and/or damage to persons or property as a matter of products liability, negligence or otherwise, or from any use or operation of any methods, products, instructions, or ideas contained in the material herein.

ISBN: 978-0-443-13866-9

For Information on all Academic Press publications
visit our website at https://www.elsevier.com/books-and-journals

Publisher: Stacy Masucci
Acquisitions Editor: Tracey Lange
Editorial Project Manager: Kristi Anderson
Production Project Manager: Omer Mukthar
Cover Designer: Miles Hitchen

Typeset by MPS Limited, Chennai, India

Contents

List of contributors — xiii

1. **Lung contusion, flail chest, acute respiratory distress syndrome, and respiratory failure after blunt chest trauma**
   Nora Mayer, Pietro Bertoglio, Savvas Lampridis, Marco Scarci and Fabrizio Minervini

Blunt chest trauma	1
Lung contusion	1
Epidemiology/pathophysiology	1
Lung laceration	2
Diagnostic	3
Therapy	4
Flail chest	4
Diagnostic and definition	4
Therapy	5
Conservative management	5
Pain management	5
Adjunctive management	6
Non-invasive/ventilation	6
Indications for rib stabilization	6
ARDS and respiratory failure	8
Epidemiology/pathophysiology	8
Risk factors	9
Diagnostic	9
Differential diagnosis	10
Therapy	10
Adjunct therapies	11
Recovery and long-term results	11
Research/future therapy options	12
Conclusion	12
References	12

2. **Anesthesiological aspects in the chest trauma patient**
   Emiliano Gamberini, Maria Maddalena Bitondo, Emanuele Russo, Domenico Pietro Santonastaso and Vanni Agnoletti

Initial management of chest blunt trauma	19

	Tracheobronchial injury and airway management	20
	Tension pneumothorax	23
	Open pneumothorax	24
	Hemothorax	24
	Cardiac tamponade	24
	Extracorporeal membrane oxygenation	25
	Anesthesia for surgical treatment	26
	Analgesia in blunt chest trauma	27
	Regional anesthesia	29
	Thoracic epidural analgesia	29
	Thoracic paravertebral block	29
	Serratus anterior plane block	30
	Erector spinae plane block	30
	Intercostal nerve block	31
	References	31
3.	**Scoring systems of blunt thoracic trauma and rib fractures** *Jennifer E. Baker, Kevin N. Harrell and Fredric M. Pieracci*	
	Introduction	37
	General properties of scoring systems	37
	Ideal characteristics of a chest wall scoring systems	40
	Thoracic chest wall taxonomy	41
	Current chest wall scoring systems	42
	Chest wall Organ Injury Scale	42
	Thoracic trauma severity score	45
	Rib fracture score	46
	Chest trauma score	46
	RibScore	46
	Sequential clinical assessment of respiratory function score	47
	Chest wall injury scoring systems and surgical stabilization	48
	Future directions	49
	Conclusions	50
	References	50
4.	**Taxonomy of rib fractures and blunt chest wall injury** *John G. Edwards*	
	AO Foundation/Orthopaedic Trauma Association fracture and dislocation classification compendium	56
	Chest Wall Injury Society Taxonomy	57
	CWIS/ASER working group	60
	Chest Injuries International Database	62
	Other rib fracture classifications	62
	SMuRFS	64
	Costal cartilage	64
	Injuries to the costal margin	65

	Injuries to the sternum	66
	Summary	67
	References	67

5. Rib fractures management algorithm
Didier Lardinois

Introduction	71
Management in flail chest patients	72
Summary	73
Management in no flail chest patients	76
Summary	79
Management of rib fracture nonunion	80
Summary	82
When is rib fixation in non-flail patients usually not necessary?	82
References	82

6. Early rib fixation: the evidence, the trials
Namariq Abbaker, Ahmed Hamada and May Al-Sahaf

Introduction	87
Evidence synthesis: guidelines and recommendations for early rib fixation	88
Trial analysis and discussion: understanding early surgical stabilisation	90
Movement toward surgical intervention	90
Indications for surgical approach	93
Timing to surgical intervention	94
Patient selection and allocation challenges	96
Multi-disciplinary approach	96
Quality of life	97
Cost-effectiveness	98
Summary	99
References	99

7. Modern approach to rib fixation: surgical techniques
Federico Raveglia, Riccardo Orlandi, Federica Danuzzo, Ugo Cioffi and Marco Scarci

Introduction	103
Minimal invasive surgical rib fixation	104
Internal rib fixation	105
Intrathoracic rib fixation	106
Full VATS approach	106
Partial VATS approach	107
The application of 3D-printing technology in rib fracture fixation	110
3D printing technology	110

	Preoperative planning with 3D printing models	110
	Customized implants, surgical guides and templates, minimally invasive SSRF	111
	Potential for training and education	111
	Future directions and challenges	112
	Evolution of the surgical approaches	112
	Surgical approaches	112
	Video-assisted thoracoscopic hybrid approaches	118
	References	119
8.	**Postoperative complications**	
	Rachel Chubsey, Savannah Gysling and Edward J. Caruana	
	Introduction	123
	Early procedure-related complications	125
	Bleeding	125
	Pulmonary complications	125
	Pleural effusion	125
	Pneumothorax	126
	Ventilatory failure	126
	Wound infection	126
	Neuromuscular weakness	127
	Hardware infection and failure	127
	Hardware infection	128
	Mechanical hardware failure	129
	Long-term complications	130
	Fracture nonunion or malunion	130
	Chronic pain and irritation	131
	References	131
9.	**Rib malunion**	
	Athanasios Kleontas and Kostas Papagiannopoulos	
	Definition	137
	Types	137
	Incidence	139
	Risk factors	140
	Main symptom	140
	Diagnostic approach	141
	Differential diagnosis	142
	Treatment options	143
	Surgical treatment	144
	Indications for surgical intervention	144
	Surgical options	145
	Associated complications	149
	Prognosis	150
	Prevention strategies	150
	References	151

10. Slipping rib syndrome
Athanasios Kleontas and Kostas Papagiannopoulos

Definition	153
History	154
Types	154
Forms	154
Demographics	155
Incidence	155
Etiology	155
Pathogenesis	155
Main symptom	156
Diagnostic approach flow chart	156
Physical examination	157
Differential diagnosis	159
Diagnostic examinations	159
Treatment options flow chart	160
Surgical treatment	161
Indications for surgical intervention	162
Surgical options	162
Patient monitoring	165
Prevention strategies	165
References	166

11. Role of VATS in nonpenetrating chest trauma
Jury Brandolini, Fabrizio Minervini and Pietro Bertoglio

Introduction	169
The development of VATS surgery and its use in chest trauma	169
Benefits, features, and timing of VATS surgery in blunt trauma patients	170
VATS for posttraumatic haemothorax	172
VATS for posttraumatic empyema	173
VATS for posttraumatic hernias	174
VATS for posttraumatic chest wall injuries	176
Conclusions	176
References	177

12. Hydropneumothorax: why, when, how
Savvas Lampridis, Fabrizio Minervini and Marco Scarci

Introduction	181
Historical background	181
Terminology	182
Epidemiology	183
Pathophysiology	184
Hemothorax	184

Pneumothorax	185
Primary evaluation	186
Imaging investigations	187
Plain radiography	187
Ultrasonography	187
Ultrasonography versus radiography	188
Computed tomography	189
Management	191
Massive hemothorax	191
Hemothorax	192
Retained hemothorax	192
Pneumothorax	193
Prognosis	195
Conclusions	196
References	196

13. Sternal fracture
Korkut Bostanci and Zeynep Bilgi

Epidemiology	203
Mechanism of injury and implications	203
Clinical presentation and diagnosis	204
Management	206
References	210

14. VATS for delayed hemothorax
Jacie Jiaqi Law and Giuseppe Aresu

Introduction	213
Definition, epidemiology, and classification of delayed hemothorax	213
An overview of existing DHTX literature	215
Clinical and radiological predictive factors of delayed hemothorax	215
The Quebec clinical decision rule	220
Complications of delayed hemothorax	221
The diagnostic and therapeutic role of VATS in delayed hemothorax	222
The role of uniportal VATS in blunt chest trauma	224
Conclusion	225
References	226

15. Chest drains: which, where, when
Tanzil Rujeedawa, Ahmed Mohamed Osman and Adam Peryt

Introduction	233
Which	234
Chest drain components	234
When	235

Indications of chest tube insertion and drainage	235
Precautions to chest tube insertion	237
Where	238
Chest tube insertion	238
Chest tube removal	240
Potential pitfalls and complications	241
Summary	241
References	242

16. Chest tube placement: options and guidelines
Kunal Bhakhri and Anuj Wali

Introduction	245
Indications	245
Chest tubes for pneumothorax	246
Do you need a chest drain for pneumothorax?	246
Positive pressure ventilation	248
Chest tubes for haemothorax	249
Does any volume of blood loss mandate surgery?	249
Do you need a chest drain for "small" haemothorax?	250
Management of retained haemothorax	251
Chest tube management	252
Suction versus no suction	252
Do you need prophylactic antibiotics?	252
Summary	253
References	253

17. Diaphragmatic rupture and herniation following blunt trauma: minimally invasive versus open approach
Duaa Ali Faruqi, Waad Attafi, M. Yousuf Salmasi and Nizar Asadi

Introduction	257
Anatomy of the diaphragm	258
Physiology of the diaphragm	259
Mechanism of diaphragmatic injury	260
Blunt diaphragmatic trauma	260
Penetrating trauma	261
Clinical presentation and diaphragmatic rupture	261
Acute presentation	261
Delayed presentation	261
Diagnostic imaging	262
Chest X-ray	262
Ultrasound	263
Computed tomography scan	263
Commonly associated injuries	265
Mechanisms of repair	267
Laparoscopic approach	267

Other minimally invasive approaches	273
Open approach	275
Primary repair versus mesh	277
Primary repair	277
Mesh	277
References	278

18. Traumatic pericardial effusion and cardiac tamponade
Filippo Antonacci, Ahmed Mohamed Osman and Piergiorgio Solli

Introduction	283
Clinical findings	284
Investigations	285
Treatment	286
Conservative approach	286
Pericardiocentesis	288
Discussion	289
References	291

19. Airway and esophageal injuries
Sahar Hasanzade and Marco Scarci

Airway and esophageal injuries	293
Introduction	293
Clinical presentation	293
Mechanisms of trauma	293
Diagnosis	294
Treatment	294
Operative management	297
Complications	298
Airway injuries	299
Mechanism	299
Clinical presentation	299
Evaluation	301
Management	302
Complications	304
References	304

Index	307

List of contributors

Namariq Abbaker Imperial College NHS Healthcare Trust and National Heart and Lung Institute, Division of Thoracic Surgery, London, United Kingdom

Vanni Agnoletti Anesthesia and Intensive Care Unit, "M. Bufalini" Hospital, Local Health Authority of Romagna, Cesena (FC), Italy

May Al-Sahaf Imperial College NHS Healthcare Trust and National Heart and Lung Institute, Division of Thoracic Surgery, London, United Kingdom

Filippo Antonacci Unit of Thoracic Surgery, IRCCS Az Univ Ospedaliera Policlinico S. Orsola, Bologna, Italy

Giuseppe Aresu Department of Cardiothoracic Surgery, Royal Papworth Hospital, Cambridge, United Kingdom; Department of Cardiothoracic Surgery, Royal Victoria Hospital, Belfast, United Kingdom

Nizar Asadi Department of Thoracic Surgery, Harefield Hospital, London, United Kingdom

Waad Attafi Faculty of Medicine, Kings College London, London, United Kingdom

Jennifer E. Baker Department of Surgery, Division of Trauma, Denver Health Medical Center, Denver, CO, United States

Pietro Bertoglio Department of Thoracic Surgery, Azienda Ospedaliero-Universitaria di Bologna, Bologna, Italy; Division of Thoracic Surgery, IRCSS Azienda Ospedaliero-Universitaria, Bologna, Italy; Alma Mater Studiorum, University of Bologna, Bologna, Italy

Kunal Bhakhri Department of Thoracic Surgery, University College Hospital, London, United Kingdom

Zeynep Bilgi Department of Thoracic Surgery, Medeniyet University School of Medicine, Istanbul, Turkey

Maria Maddalena Bitondo Anesthesia and Intensive Care Unit, "Infermi" Hospital, Local Health Authority of Romagna, Rimini (RN), Italy

Korkut Bostanci Department of Thoracic Surgery, Marmara University School of Medicine, Istanbul, Turkey

Jury Brandolini Department of Thoracic Surgery, Azienda Ospedaliero-Universitaria di Bologna, Bologna, Italy

Edward J. Caruana Glenfield Hospital, Leicester, United Kingdom

Rachel Chubsey Nottingham City Hospital, Nottingham, United Kingdom

Ugo Cioffi Department of Surgery, University of Milan, Milan, Italy

Federica Danuzzo San Gerardo Hospital, Thoracic Surgery Department, Monza, Italy

John G. Edwards Department of Cardiothoracic Surgery, Sheffield Teaching Hospitals NHS Foundation Trust, Northern General Hospital, Sheffield, United Kingdom

Duaa Ali Faruqi Faculty of Medicine, Imperial College London, London, United Kingdom; Faculty of Medicine, Kings College London, London, United Kingdom

Emiliano Gamberini Anesthesia and Intensive Care Unit, "Infermi" Hospital, Local Health Authority of Romagna, Rimini (RN), Italy

Savannah Gysling Queen Elizabeth Hospital Birmingham, Birmingham, United Kingdom

Ahmed Hamada Imperial College NHS Healthcare Trust and National Heart and Lung Institute, Division of Thoracic Surgery, London, United Kingdom

Kevin N. Harrell Department of Surgery, Division of Trauma, Denver Health Medical Center, Denver, CO, United States

Sahar Hasanzade Department of Thoracic Surgery, Imperial College NHS Healthcare Trust, London, United Kingdom

Athanasios Kleontas Department of Thoracic Surgery, European Interbalkan Medical Centre, Thessaloniki, Northern Prefecture, Greece

Savvas Lampridis Department of Cardiothoracic Surgery, Hammersmith Hospital, Imperial College Healthcare NHS Trust, London, United Kingdom; Department of Thoracic Surgery, Imperial College NHS Healthcare Trust, London, United Kingdom

Didier Lardinois Department of Thoracic Surgery, University Hospital Basel, Basel, Switzerland

Jacie Jiaqi Law Department of Cardiothoracic Surgery, Royal Papworth Hospital, Cambridge, United Kingdom; Department of Cardiothoracic Surgery, Royal Victoria Hospital, Belfast, United Kingdom

Nora Mayer Department of Thoracic Surgery, University Hospital Bern, Bern, Switzerland

Fabrizio Minervini Division of Thoracic Surgery, Cantonal Hospital Lucerne, Lucerne, Switzerland

Riccardo Orlandi San Gerardo Hospital, Thoracic Surgery Department, Monza, Italy

Ahmed Mohamed Osman Department of Thoracic Surgery, Nottingham University Hospital, Nottingham, United Kingdom

Kostas Papagiannopoulos Department of Thoracic Surgery, St. James' University Hospital, Leeds, West Yorkshire, United Kingdom

Adam Peryt Department of Thoracic Surgery, Royal Papworth Hospital, Cambridge, United Kingdom

Fredric M. Pieracci Department of Surgery, Division of Trauma, Denver Health Medical Center, Denver, CO, United States

Federico Raveglia San Gerardo Hospital, Thoracic Surgery Department, Monza, Italy

Tanzil Rujeedawa School of Clinical Medicine, University of Cambridge, Cambridge, United Kingdom

Emanuele Russo Anesthesia and Intensive Care Unit, "M. Bufalini" Hospital, Local Health Authority of Romagna, Cesena (FC), Italy

M. Yousuf Salmasi Faculty of Medicine, Imperial College London, London, United Kingdom; Department of Thoracic Surgery, Harefield Hospital, London, United Kingdom

Domenico Pietro Santonastaso Anesthesia and Intensive Care Unit, "M. Bufalini" Hospital, Local Health Authority of Romagna, Cesena (FC), Italy

Marco Scarci Department of Thoracic Surgery, Imperial College NHS Healthcare Trust, London, United Kingdom; National Heart and Lung Institute, Imperial College, London, United Kingdom

Piergiorgio Solli Unit of Thoracic Surgery, Royal Papworth Hospital NHS Foundation Trust, Cambridge, United Kingdom

Anuj Wali Department of Thoracic Surgery, University College Hospital, London, United Kingdom

Chapter 1

Lung contusion, flail chest, acute respiratory distress syndrome, and respiratory failure after blunt chest trauma

Nora Mayer[1], Pietro Bertoglio[2,3], Savvas Lampridis[4], Marco Scarci[4,5] and Fabrizio Minervini[6]

[1]Department of Thoracic Surgery, University Hospital Bern, Bern, Switzerland, [2]Division of Thoracic Surgery, IRCSS Azienda Ospedaliero-Universitaria, Bologna, Italy, [3]Alma Mater Studiorum, University of Bologna, Bologna, Italy, [4]Department of Thoracic Surgery, Imperial College NHS Healthcare Trust, London, United Kingdom, [5]National Heart and Lung Institute, Imperial College, London, United Kingdom, [6]Division of Thoracic Surgery, Cantonal Hospital Lucerne, Lucerne, Switzerland

Blunt chest trauma

Blunt thoracic trauma occurs in nearly two-thirds of polytrauma patients and has a mortality of 20%–25% (Dogrul et al., 2020). Scoring systems like the Thoracic Trauma Score (TTS) (Pape et al., 2000) and Pulmonary Contusion Score (PCS) have been developed to evaluate patients with blunt chest trauma (Gupta et al., 2021). These scores are selectively used for the assessment of thoracic trauma, rather than the well-known Injury Severity Score (ISS) (Baker et al., 1974) or the Trauma Injury Severity Score (TISS) (Boyd et al., 1987), generally applied in polytrauma patients.

Lung contusion

Epidemiology/pathophysiology

Pulmonary contusion is present in up to 80% of the patients suffering from blunt chest trauma with mortality rates up to 20% in severely injured adults (Hosseini et al., 2015; Simon et al., 2012).

Pathophysiological, pulmonary contusion is defined by alveolar edema and destruction of the alveolar septum and pulmonary interstitium. Hemorrhage into the alveoli triggers the activation of pro-inflammatory cascades, leading to accumulation of cytokines and chemokines as well as migration of neutrophil granulocytes (Rendeki & Molnár, 2019) and ultimately increased permeability of the alveolar-capillary membranes and microvascular leak leading to loss of the physiological structure and function of the injured lung parenchyma (Rendeki & Molnár, 2019). The resulting pulmonary edema leads to ventilation and perfusion mismatch, intrapulmonary shunting, and increased airway pressure (Fig. 1.1). As a result, lung compliance is reduced with decreased functional residual air volume (Seitz et al., 2010).

The clinical manifestations of lung contusion may be insidious and potentially remain unrecognized until complications occur (Rendeki & Molnár, 2019). In the case of major pulmonary contusion, the resulting tissue hypoxia clinically manifests as dyspnea, tachypnoea, and consequently tachycardia. Hence, patients with pulmonary contusions are at high risk of developing subsequent adverse events, including pneumonia and acute respiratory distress syndrome (ARDS) (Bakowitz et al., 2012; Cohn & Dubose, 2010), making early diagnosis of pulmonary contusion crucial. While the presence of multiple rib fractures has been shown to enlengthen hospital stay (Flagel et al., 2005), pulmonary contusion and accompanied lung injury were correlated with higher death/mortality rates (Kollmorgen et al., 1994).

Lung laceration

Lung laceration, the maximum form of pulmonary parenchymal injury, is defined as alveolar rupture and parenchymal shearing injury contrary to lung contusion (Peters et al., 2010). Up to 12% of blunt thoracic trauma patients suffer from lung lacerations (Elmali et al., 2007). In CT imaging, lung lacerations can be classified into four groups according to their appearance

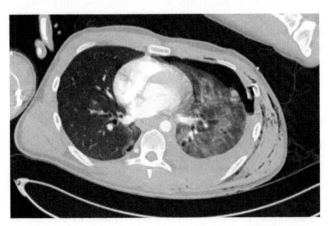

FIGURE 1.1 Left Pulmonary contusion in a 44-year-old patient with blunt chest trauma.

and causative trauma mechanism (Wagner et al., 1988). As per the definition of "laceration", they more commonly appear in penetrating rather than blunt chest trauma (Carboni Bisso et al., 2021; Elmali et al., 2007). As for pulmonary contusion, lung laceration might be underdiagnosed in plain chest radiographs. On CT, lacerations present as ovoid cystic lesions/pneumatoceles either filled with blood (intrapulmonary hematoma/hematopneumocele) or air/fluid air levels. Complications following lung laceration include pneumothorax, which is the most common complication and reason for pneumothorax in blunt chest trauma, followed by the formation of pulmonary abscess, bronchopleural fistula, or air embolism. In extreme cases with the connection of the circulatory bloodstream and airways, patients might even drown from their own blood (Dogrul et al., 2020). Treatment should be tailored according to the extent of the laceration ranging from conservative management over chest drain insertion to anatomical resection of the affected part of the lung (Mason et al., 2010).

Diagnostic

While plain chest radiograph can lead to suspicion of pulmonary contusion, computed tomography (CT) chest remains the gold standard for diagnosis and quantification of the extent of pulmonary contusion with high sensitivity and high diagnostic accuracy for additional thoracic injuries (Deunk et al., 2010; Gayzik et al., 2007; Palas et al., 2014) Fig. 1.2. The sole use of plain chest radiographs was found to often lead to underdiagnoses of pulmonary contusion, especially at the time of admission (Pape et al., 2000). The typical radiological findings for pulmonary contusion usually become visible after about 48–72 hours post-trauma (Rendeki & Molnár, 2019).

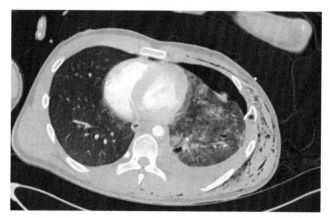

FIGURE 1.2 Lung laceration in the left lower lobe in a 19-year-old polytraumatized patient after a motorcycle accident.

The screening performance of ultrasonographic B-lines in the detection of lung contusion following blunt chest trauma as a cost-effective and easily made available bedside screening method has shown high sensitivity and specificity for lung contusion and showed high sensitivity for pulmonary contusion when more than 6 B-lines could be detected (Abbasi et al., 2018). Lung ultrasound in summary showed high accuracy in diagnosing pulmonary contusion and correlated well with CT chest in quantifying lung contusion in polytrauma patients (Sayed et al., 2022; Staub et al., 2018).

Therapy

According to the Eastern Association for the Surgery of Trauma (EAST) guidelines, the cornerstones of the treatment of pulmonary contusion are aggressive chest physiotherapy, meticulous fluid resuscitation to a reasonable extent, and sufficient analgesia while preventing mechanical ventilation in cases without respiratory failure (Simon et al., 2012). Supplemental oxygen can be provided (Cohn & Dubose, 2010). In cases of severe injury, intensive monitoring is essential for ensuring adequate oxygenation and organ perfusion. A pulmonary artery catheter may be useful to avoid fluid overload during resuscitation (Simon et al., 2012).

Patients should only ultimately be intubated with hypoxic respiratory failure failing to maintain adequate oxygenation with non-invasive positive-pressure ventilation. When required, mechanical ventilation should be tailored to optimize oxygenation while minimizing the potential for secondary lung injury.

Experimental mice studies recently revealed the reduction of the inflammatory response associated with lung contusion with 1.3% hydrogen inhalation and might therefore be considered an adjunct in the conservative treatment of pulmonary contusion (Ageta et al., 2023). Hydrogen inhalation therapy is applicable to humans and available for biomedical applications (Ono et al., 2017).

Flail chest

Diagnostic and definition

A flail chest is defined as fractures of two or more consecutive ribs or costal cartilages in at least two or more places, creating a floating chest wall segment, which moves paradoxically throughout the respiratory cycle, moving inward with inspiration and outward with exhalation. Flail chest is a clinical diagnosis confirmed by radiological findings (Ludwig & Koryllos, 2017; Peek et al., 2022). Awake patients may present with respiratory distress and severe chest pain while intubated patients become conspicuous experiencing ventilatory and oxygenation difficulties. While chest radiography is a valuable tool for the detection of pneumothorax and hemothorax in thoracic trauma, CT chest provides reliable information on costochondral junctions, sternal and rib fractures, and underlying pulmonary contusion (Schnyder & Wintermark, 2000).

Several risk factors associated with increased morbidity and mortality, including age, pre-existing comorbidities, higher numbers of rib fractures, and patients with concomitant injuries were identified (Battle et al., 2012). The RibScore, a novel radiographic scoring system to predict adverse outcomes in thoracic trauma patients, was found to linearly correlate with pneumonia, tracheostomy, and respiratory failure and consists of six variables: (1) six or more rib fractures, (2) bilateral fractures, (3) flail chest, (4) three or more severely (bicortical) displaced fractures, (5) first rib fracture, and (6) at least one fracture in all three anatomic areas (anterior, lateral, and posterior) (Chapman et al., 2016).

Therapy

Outcome prediction for patients with flail chest injuries with or without pulmonary contusion depends on accompanying injuries. Age, however, was a significant independent factor for an increase in mortality in patients with flail chest (Battle et al., 2012). The mortality of flail chest patients ranges between 10%—20% depending on additional injuries like extra-thoracic organ injuries and shock (Bastos et al., 2008). The incidence of pneumonia, ARDS, and sepsis were 21%, 14%, and 7%, respectively in a large study by Dehghan et al. (2014).

Conservative management

Pain management

The presence of pain impairs inspiration and expiration leading to a reduction of tidal volumes and suppression of the cough reflex which ultimately leads to atelectasis and pneumonia. Sufficient pain management is a key factor in the conservative management of flail chest patients to prevent respiratory failure. Opioids are still considered first-line therapy for acute rib fracture pain in combination with non-opioid analgesia to reduce potential addiction and misuse (Pharaon et al., 2015). The epidural catheter is the preferred mode of analgesia and superior to other regional analgesia strategies like intrapleural, paraspinal, or intercostal block as long as epidural analgesia is not excluded by any contraindications (Pain management for blunt thoracic trauma: A joint practice, 2023). Epidural analgesia was associated with a shorter duration of mechanical ventilation and a reduced rate of hospital-acquired pneumonia (Bulger et al., 2004).

Adjunctive management

Management of flail chest includes immediate assessment of the airway, breathing, and circulation and, for stable patients, monitoring of the respiratory status, pain control, lung physiotherapy, early mobilization, and adequate nutrition.

Non-invasive/ventilation

As a flail chest is accompanied by pulmonary contusion in most of the cases resulting from massive traumatic impaction on both the chest wall and the parenchyma, non-invasive and invasive ventilation may be indicated according to the same criteria as described for patients with lung contusion as described in the previous paragraph (Prunet et al., 2019). Moreover, mechanical ventilation, also called "internal pneumatic splinting" supports a more physical ventilation cycle in the presence of impaired chest wall mechanics (Pharaon et al., 2015).

Indications for rib stabilization

Open reduction and internal fixation of rib fractures in flail chest has not yet been widely established but has been shown to reduce intensive care unit (ICU) stay and mechanical ventilation, complications, and overall length of hospital stay (LOS) in hospital (Coughlin et al., 2016; Ingoe et al., 2019). Most recent studies could show statistically significant benefits in terms of above mentioned as well as the prevention of pneumonia, tracheostomy, and reduced mortality in carefully selected patients (Sawyer et al., 2022). While the stabilization of multiple rib fractures is still controversial, there is a strong indication to surgically treat flail chest in carefully selected patients. Evidence with adequate selection criteria for the most suitable patients for surgical management of flail chest however is sparse. Efforts have been made by different expert groups to establish an algorithm for appropriate patient selection but because of the heterogeneity of the studies, no universal guidelines have yet been established (Sawyer et al., 2022).

The aim of surgical stabilization is to restore the integrity and stability of the chest wall and ensure sufficient ventilation. Several different methods of rib fixation as well as sternal fixation in combined injuries have been described (Fokin et al., 2020; Sawyer et al., 2022). If surgery is considered, outcomes have been more favorable in early stabilization within 72 hours after initial trauma (Simmonds et al., 2023).

Hemothorax and hemopneumothorax occur in almost one out of three trauma patients with thoracic injuries (Management of posttraumatic retained hemothorax, 2019). Together with the stabilization of the ribs, a beneficial wash out of the hemothorax to prevent empyema can be done within the same procedure Figs. 1.3 and 1.4.

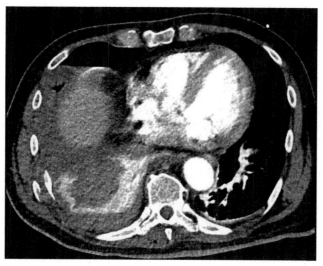

FIGURE 1.3 CT scan of a patient with multiple rib fractures and hemothorax.

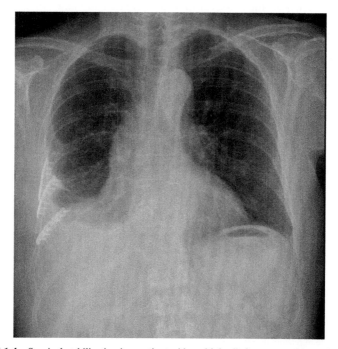

FIGURE 1.4 Surgical stabilization in a patient with multiple rib fractures and hemothorax.

ARDS and respiratory failure

Epidemiology/pathophysiology

ARDS is a syndrome with onset occurring within the first week after an event leading to pulmonary edema and hypoxemia which is associated with both high morbidity and mortality (Simmonds et al., 2023). ARDS can result from multiple etiologies and occurs in 17% after lung contusion and acute lung injury (ALI), however in up to 78% of patients with polytrauma. If more than 20% of the lung volume is injured, ARDS develops in 82% of the patients, whereas if affected lung volume is less than 20%, only 22% of the patients develop ARDS (A new method to analyze lung compliance when the pressure-volume relationship is nonlinear - PubMed Internet, 2023; Prunet et al., 2019).

ALI is defined as a PaO_2/FiO_2 ratio <300 mmHg (40 kPa) and ARDS is a sub-group defined based on more severely impaired oxygenation with a PaO_2/FiO_2 ratio <200 mmHg (26.7 kPa). In addition, ARDS is defined by bilateral pulmonary opacities on chest radiographs and respiratory failure not entirely explainable by heart failure or fluid overload.

ARDS was described at the American-European Consensus Conference (AECC) in 1994 (Bernard et al., 1994) and further categorized into three severity groups according to the Berlin modification of the American-European Consensus (2012) divided into mild, moderate, and severe ARDS according to the degree of hypoxemia, associated with increased mortality of 27%, 32%, and 45%, respectively (ARDS Definition Task Force et al., 2012) (Table 1.1).

The histopathological features of ARDS are not represented in the clinical classification of the AECC. More than 50% of the patients with clinical diagnosis of moderate to severe ARDS showed diffuse alveolar damage with hyaline membranes at post-mortem examination (Diffuse alveolar damage, 2023). Shear

TABLE 1.1 Berlin definition of the ARDS.

Categorization of ARDS severity	Three levels of hypoxemia, defined as $PaO_2: FiO_2$ ratio £ 300 mmHg	Mortality	Median duration of mechanical ventilation in survivors
Mild	$PaO_2:FiO_2$ 201–300 mmHg	27% (95% CI, 24%–30%)	5d (IQR 2–11)
Moderate	$PaO_2:FiO_2$ 101–200 mmHg	32% (95% CI, 29%–34%)	7d (IQR 4–14)
Severe	$PaO_2:FiO_2$ £ 100 mmHg	45% (95% CI, 42%–48%)	9d (IQR 5–17)

ARDS, Acute respiratory distress syndrome.

forces can appear between the injured, atelectatic parenchyma and the uninjured lung regions, resulting in atelectrauma. As described in lung contusion, immune reactions on a molecular level are consequently leading to the release of cytokines and a systemic inflammatory response syndrome, contributing to the development of multisystem organ failure (Tremblay & Slutsky, 1998).

Risk factors

The main risk factors for ARDS in trauma patients attributed to the initial trauma itself is pulmonary contusion followed by secondary risks that arise in the posttraumatic phase during the stay in the intensive care unit like pneumonia, aspiration of gastric contents, hemorrhagic shock as well as modifiable measures like mass transfusion of blood products and the amount of crystalloid resuscitation within the first 24 h of injury given (Tran et al., 2022). In addition, ARDS development was associated with fractures of the femur, humerus, and pelvis as well as chest injuries including rib/sternal fracture and hemo-/pneumothorax (Navarrete-Navarro et al., 2006).

Diagnostic

The typical imaging findings of ARDS are bilateral pulmonary infiltrates on chest radiograph and bilateral ground-glass opacities or consolidation on CT chest, mainly distributed in the dependent parenchymal areas Fig. 1.5.

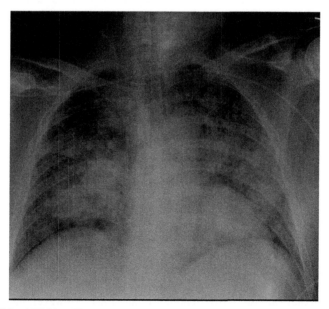

FIGURE 1.5 ARDS in a 22-year-old patient.

Lung ultrasound in the emergency department setting has moreover been shown to identify patients with a risk of ARDS in blunt chest trauma patients (Leblanc et al., 2014).

There is evidence, that the amount of contusioned lung volume detected on a CT chest can guide clinicians in the identification of patients at high risk for ARDS (Mahmood et al., 2017).

Differential diagnosis

The most important differential diagnosis of ARDS in trauma is cardiogenic pulmonary edema. In ARDS, the more even horizontal bilateral distribution in dependent areas of the pulmonary infiltrates is common, predominantly upper-lobe craniocaudal distribution and more central horizontal distribution is more likely associated with cardiogenic pulmonary edema (Komiya et al., 2013). Cardiogenic pulmonary edema is accompanied by abnormal findings in transthoracic echocardiographic imaging and elevation of brain natriuretic peptide which have to be interpreted within the clinical context. One however always has to take into consideration that volume overload or heart failure coexist in up to 33% of patients with ARDS (Thompson et al., 2017).

Therapy

Despite the high incidence of thoracic trauma and wide knowledge of high mortality (Rubenfeld et al., 2005), there remains the lack of a standardized treatment (including pharmacological treatment) other than supportive ventilation and fluid management for ARDS due to the variety of etiologies and lack of strong treatment consensus (Boyle et al., 2013; Huang et al., 2022). Imminent treatment and early identification of high-risk patients for developing respiratory failure, however, is necessary facing high mortality rates of up to a mean of cumulative 45% mortality between the three groups of mild-moderate-severe ARDS (Bellani et al., 2016; Huang et al., 2020; Martin et al., 2021). Ventilatory support is still considered to be the most important therapeutic option (Fan et al., 2005).

Volume- and pressure-limited lung-protective ventilation is beneficial in ARDS (Ma et al., 2021). Mechanical ventilation with tidal volumes of 6–8 mL/kg of ideal patient body weight, a frequency of 18–20 breaths/min, positive end-expiratory pressure (PEEP) between 5 and 15 cmH$_2$O and inspiration: expiration ratio of 1:1.0–2.0 has been used with a reduction of diffuse alveolar damage in post-mortem examination over the last years (Ma et al., 2021; Thille et al., 2013).

Carefully applied lung recruitment maneuvers and PEEP ventilation to open atelectatic lung parenchyma have proven to increase arterial oxygenation (Cohn and Dubose, 2010; Schreiter et al., 2004). However, caution has to be applied as mechanical ventilation potentially results in barotrauma or volutrauma as well as can lead to adverse systemic effects via cytokine release ultimately leading to multi-organ failure (Fan et al., 2005). The optimal pressure, duration, and frequency of recruitment maneuvers have not been defined. High-frequency ventilation was reported to be specifically beneficial as it allows for higher mean airway pressures with reduced tidal volumes (Imai & Slutsky, 2005).

Adjunct therapies

Contrary to earlier studies, where there was no significant benefit seen in the prone position (Effect of Prone Positioning on the Survival of Patients with Acute Respiratory Failure and NEJM Internet, 2023; Guerin et al., 2004), proning in mechanically ventilated ARDS-patients with lung protective ventilation modes has overall shown to improve survival in a large meta-analysis of the currently available data (Sud et al., 2014). The benefit of the prone position results from the recruitment of the nondependent atelectatic lung areas with all the consecutive improvements in respiratory mechanics and resulting physiological changes in oxygenation. However, the prone position might be contraindicated in polytraumatized patients due to concomitant injuries.

Inhaled nitric oxide as an adjunct provides selective vasodilation in ventilated lung parenchyma with improvement of ventilation-perfusion mismatch, hypoxemia, and pulmonary hypertension (Fan et al., 2005). Even though It has been applied and studied in ARDS treatment for many years, the effects of increased oxygenation remain transient and there was no benefit in terms of mortality (Gebistorf et al., 2016).

Extracorporeal membrane oxygenation has been used to stabilize patients in small case studies for gas exchange management with so far mainly positive outcomes in trauma patients with extensive lung injury as salvage therapy when ventilation was deemed to improve oxygenation due to excessive lung tissue damage (Cordell-Smith et al., 2006; Madershahian et al., 2007).

Recovery and long-term results

Few research has been conducted on long-term outcomes following blunt thoracic trauma and the recovery of survivors. Therefore, minimal is known about potential sequelae and lung functions as well as the functional status of the patients who recover from blunt chest injuries. A clear separation has to be made between patients suffering from blunt chest trauma with flail chest, ARDS, and respiratory failure and patients with pulmonary contusion.

More than 50% of the patients with flail chest injuries complained of chest tightness or pain in a 5-year follow-up and were reported to have persistently impaired spirometry (Landercasper et al., 1984). Some studies moreover showed that patients surviving ARDS had reduced quality of life with significant emotional, physical, and neurocognitive morbidity (Hopkins et al., 2005).

Research/future therapy options

Prospective randomized research in the investigation of blunt chest trauma is difficult, whereas animal models have been shown to deliver relevant information on potential treatment options. The agent Levosimendan was shown to reduce the local and systemic inflammatory response in a rat model after inducing blunt chest trauma (Ateş et al., 2017). The Leukotriene-receptor antagonist Montelukast, an anti-inflammatory drug regularly administered in airway inflammatory disease, did not provide any therapeutic effect in pulmonary contusion or respiratory failure (Heydari et al., 2023). The identification of potential biomarkers for future targeted therapies is the subject of current ongoing research (Qi et al., 2023).

Conclusion

Blunt thoracic trauma occurs in approximately 60% of polytraumatized patients with a mortality of up to 20%−25% (Dogrul et al., 2020). Pulmonary contusion is often accompanied by serial rib fractures and flail chest, being highly predictive for the development of ARDS (Bellani et al., 2016). Early detection and aggressive immediate treatment are inevitable to prevent adverse events like pneumonia, ARDS, and respiratory failure to reduce morbidity and mortality (Martin et al., 2021).

While sufficient pain management and lung-protective ventilatory support are the standard treatment, surgical management of flail chest is increasingly popular and has been shown to reduce ICU stay and mechanical ventilation, overall LOS, and mortality in highly selected patients (Coughlin et al., 2016).

References

Abbasi, S., Shaker, H., Zareiee, F., Farsi, D., Hafezimoghadam, P., Rezai, M., et al. (2018). Screening performance of ultrasonographic B-lines in detection of lung contusion following blunt trauma; A diagnostic accuracy study. *Emergency*, 6(1), e55.

Ageta, K., Hirayama, T., Aokage, T., Seya, M., Meng, Y., Nojima, T., et al. (2023). Hydrogen inhalation attenuates lung contusion after blunt chest trauma in mice. *Surgery, 174*(2), 343–349.

A new method to analyze lung compliance when pressure-volume relationship is nonlinear. Retrieved September 14, 2023, from https://pubmed.ncbi.nlm.nih.gov/9769260/.

ARDS Definition Task Force., Ranieri, V. M., Rubenfeld, G. D., Thompson, B. T., Ferguson, N. D., Caldwell, E., et al. (2012). Acute respiratory distress syndrome: The Berlin definition. *The Journal of the American Medical Association, 307*(23), 2526–2533.

Ateş, G., Yaman, F., Bakar, B., Kısa, Ü., Atasoy, P., & Büyükkoçak, Ü. (2017). Evaluation of the systemic antiinflammatory effects of levosimendan in an experimental blunt thoracic trauma model. *Turkish Journal of Trauma & Emergency Surgery, 23*(5), 368–376.

Baker, S. P., O'Neill, B., Haddon, W., & Long, W. B. (1974). The injury severity score: A method for describing patients with multiple injuries and evaluating emergency care. *The Journal of Trauma, 14*(3), 187–196.

Bakowitz, M., Bruns, B., & McCunn, M. (2012). Acute lung injury and the acute respiratory distress syndrome in the injured patient. *Scandinavian Journal of Trauma, Resuscitation and Emergency Medicine, 20*, 54, Aug 10.

Bastos, R., Calhoon, J. H., & Baisden, C. E. (2008). Flail chest and pulmonary contusion. *Seminars in Thoracic and Cardiovascular Surgery, 20*(1), 39–45.

Battle, C. E., Hutchings, H., & Evans, P. A. (2012). Risk factors that predict mortality in patients with blunt ches wall trauma: A systematic review and meta-analysis. *Injury, 43*(1), 8–17.

Bellani, G., Laffey, J. G., Pham, T., Fan, E., Brochard, L., Esteban, A., et al. (2016). Epidemiology, patterns of care, and mortality for patients with acute respiratory distress syndrome in intensive care units in 50 countries. *The Journal of the American Medical Association, 315*(8), 788–800.

Bernard, G. R., Artigas, A., Brigham, K. L., Carlet, J., Falke, K., Hudson, L., et al. (1994). Report of the American-European Consensus conference on acute respiratory distress syndrome: definitions, mechanisms, relevant outcomes, and clinical trial coordination. Consensus Committee. *Journal of Critical Care, 9*(1), 72–81.

Boyd, C. R., Tolson, M. A., & Copes, W. S. (1987). Evaluating trauma care: The TRISS method. Trauma Score and the Injury Severity Score. *Journal of Trauma, 27*(4), 370–378.

Boyle, A. J., Sweeney, R. M., & McAuley, D. F. (2013). Pharmacological treatments in ARDS; A state-of-the-art update. *BMC Medicine, 11*, 166, Aug 20.

Bulger, E. M., Edwards, T., Klotz, P., & Jurkovich, G. J. (2004). Epidural analgesia improves outcome after multiple rib fractures. *Surgery, 136*(2), 426–430.

Carboni Bisso, I., Gemelli, N. A., Barrios, C., & Las Heras, M. (2021). Pulmonary laceration. *Trauma Case Reports, 32*, 100449.

Chapman, B. C., Herbert, B., Rodil, M., Salotto, J., Stovall, R. T., Biffl, W., et al. (2016). RibScore: A novel radiographic score based on fracture pattern that predicts pneumonia, respiratory failure, and tracheostomy. *Journal of Trauma and Acute Care Surgery, 80*(1), 95–101.

Cohn, S. M., & Dubose, J. J. (2010). Pulmonary contusion: an update on recent advances in clinical management. *World Journal of Surgery, 34*(8), 1959–1970.

Cordell-Smith, J. A., Roberts, N., Peek, G. J., & Firmin, R. K. (2006). Traumatic lung injury treated by extracorporeal membrane oxygenation (ECMO). *Injury, 37*(1), 29–32.

Coughlin, T. A., Ng, J. W. G., Rollins, K. E., Forward, D. P., & Ollivere, B. J. (2016). Management of rib fractures in traumatic flail chest: A meta-analysis of randomised controlled trials. *The Bone & Joint Journal, 98-B*(8), 1119–1125.

Dehghan, N., de Mestral, C., McKee, M. D., Schemitsch, E. H., & Nathens, A. (2014). Flail chest injuries: A review of outcomes and treatment practices from the National Trauma Data Bank. *Journal of Trauma and Acute Care Surgery, 76*(2), 462–468.

Deunk, J., Poels, T. C., Brink, M., Dekker, H. M., Kool, D. R., Blickman, J. G., et al. (2010). The clinical outcome of occult pulmonary contusion on multidetector-row computed tomography in blunt trauma patients. *The Journal of Trauma, 68*(2), 387–394.

Diffuse alveolar damage—the role of oxygen, shock, and related factors. A review. Retrieved September 11, 2023, from https://www.ncbi.nlm.nih.gov/pmc/articles/PMC2032554/.

Dogrul, B. N., Kiliccalan, I., Asci, E. S., & Peker, S. C. (2020). Blunt trauma related chest wall and pulmonary injuries: An overview. *Chinese Journal of Traumatology, 23*(3), 125–138.

Effect of prone positioning on the survival of patients with acute respiratory failure. The New England Journal of MedicineRetrieved September 15, 2023, https://www.nejm.org/doi/full/10.1056/NEJMoa010043.

Elmali, M., Baydin, A., Nural, M. S., Arslan, B., Ceyhan, M., & Gürmen, N. (2007). Lung parenchymal injury and its frequency in blunt thoracic trauma: The diagnostic value of chest radiography and thoracic CT. *Diagnostic and Interventional Radiology, 13*(4), 179–182.

Fan, E., Needham, D. M., & Stewart, T. E. (2005). Ventilatory management of acute lung injury and acute respiratory distress syndrome. *The Journal of the American Medical Association, 294*(22), 2889–2896.

Flagel, B. T., Luchette, F. A., Reed, R. L., Esposito, T. J., Davis, K. A., Santaniello, J. M., et al. (2005). Half-a-dozen ribs: The breakpoint for mortality. *Surgery, 138*(4), 717–723, discussion 723-725.

Fokin, A. A., Hus, N., Wycech, J., Rodriguez, E., & Puente, I. (2020). Surgical stabilization of rib fractures. *JBJS Essential Surgical Techniques, 10*(2), e0032.

Gayzik, F. S., Hoth, J. J., Daly, M., Meredith, J. W., & Stitzel, J. D. (2007). *A finite element-based injury metric for pulmonary contusion: Investigation of candidate metrics through correlation with computed tomography*. SAE Technical Paper, Report No.: 2007-22-0009. Retrieved September 13, 2023, from https://www.sae.org/publications/technical-papers/content/2007-22-0009/.

Gebistorf, F., Karam, O., Wetterslev, J., & Afshari, A. (2016). Inhaled nitric oxide for acute respiratory distress syndrome (ARDS) in children and adults. *Cochrane Database of Systematic Reviews* (6), Retrieved September 15, 2023, from https://www.cochranelibrary.com/cdsr/doi/10.1002/14651858.CD002787.pub3/full.

Guerin, C., Gaillard, S., Lemasson, S., Ayzac, L., Girard, R., Beuret, P., et al. (2004). Effects of systematic prone positioning in hypoxemic acute respiratory failure: A randomized controlled trial. *The Journal of the American Medical Association, 292*(19), 2379–2387.

Gupta, A. K., Ansari, A., Gupta, N., Agrawal, H., B, M., Bansal, L. K., et al. (2021). Evaluation of risk factors for prognosticating blunt trauma chest. *Polish Journal of Surgery, 94*(1), 12–19.

Heydari, S., Khoshmohabat, H., Akerdi, A. T., Ahmadpour, F., & Paydar, S. (2023). Evaluating the effect of montelukast tablets on respiratory complications in patients following blunt chest wall trauma: A double-blind, randomized clinical trial. *Chinese Journal of Traumatology, 26*(2), 116–120.

Hopkins, R. O., Weaver, L. K., Collingridge, D., Parkinson, R. B., Chan, K. J., & Orme, J. F. (2005). Two-year cognitive, emotional, and quality-of-life outcomes in acute respiratory distress syndrome. *American Journal of Respiratory and Critical Care Medicine, 171*(4), 340–347.

Hosseini, M., Ghelichkhani, P., Baikpour, M., Tafakhori, A., Asady, H., Haji Ghanbari, M. J., et al. (2015). Diagnostic accuracy of ultrasonography and radiography in detection of pulmonary contusion; A systematic review and meta-analysis. *Emergency, 3*(4), 127–136.

Huang, S., Wang, Y. C., & Ju, S. (2022). Advances in medical imaging to evaluate acute respiratory distress syndrome. *Chinese Journal of Academic Radiology, 5*(1), 1–9.

Huang, X., Zhang, R., Fan, G., Wu, D., Lu, H., Wang, D., et al. (2020). Incidence and outcomes of acute respiratory distress syndrome in intensive care units of mainland China: A multicentre prospective longitudinal study. *Critical Care, 24*(1), 515.

Imai, Y., & Slutsky, A. S. (2005). High-frequency oscillatory ventilation and ventilator-induced lung injury. *Critical Care Medicine, 33*(3), S129.

Ingoe, H. M., Coleman, E., Eardley, W., Rangan, A., Hewitt, C., & McDaid, C. (2019). Systematic review of systematic reviews for effectiveness of internal fixation for flail chest and rib fractures in adults. *BMJ Open, 9*(4), e023444.

Kollmorgen, D. R., Murray, K. A., Sullivan, J. J., Mone, M. C., & Barton, R. G. (1994). Predictors of mortality in pulmonary contusion. *American Journal of Surgery, 168*(6), 659–663, discussion 663-664.

Komiya, K., Ishii, H., Murakami, J., Yamamoto, H., Okada, F., Satoh, K., et al. (2013). Comparison of chest computed tomography features in the acute phase of cardiogenic pulmonary edema and acute respiratory distress syndrome on arrival at the emergency department. *Journal of Thoracic Imaging, 28*(5), 322–328.

Landercasper, J., Cogbill, T. H., & Lindesmith, L. A. (1984). Long-term disability after flail chest injury. *The Journal of Trauma, 24*(5), 410–414.

Leblanc, D., Bouvet, C., Degiovanni, F., Nedelcu, C., Bouhours, G., Rineau, E., et al. (2014). Early lung ultrasonography predicts the occurrence of acute respiratory distress syndrome in blunt trauma patients. *Intensive Care Medicine, 40*(10), 1468–1474.

Ludwig, C., & Koryllos, A. (2017). Management of chest trauma. *Journal of Thoracic Disease, 9*(Suppl 3), S172–S177.

Ma, X., Dong, Z., Wang, Y., Gu, P., Fang, J., & Gao, S. (2021). Risk factors analysis of thoracic trauma complicated with acute respiratory distress syndrome and observation of curative effect of lung-protective ventilation. *Frontiers in Surgery, 8*, 826682.

Madershahian, N., Wittwer, T., Strauch, J., Franke, U. F. W., Wippermann, J., Kaluza, M., et al. (2007). Application of ECMO in multitrauma patients with ARDS as rescue therapy. *Journal of Cardiac Surgery, 22*(3), 180–184.

Mahmood, I., El-Menyar, A., Younis, B., Ahmed, K., Nabir, S., Ahmed, M. N., et al. (2017). Clinical significance and prognostic implications of quantifying pulmonary contusion volume in patients with blunt chest trauma. *Medical Science Monitor, 23*, 3641–3648.

Bozzay, Joseph D., & Bradley, Matthew J. (Eds.), (2019). *Management of post-traumatic retained hemothorax*. Retrieved September 10, 2023, from [cited 2023 Sep 10]. Available from: https://journals.sagepub.com/doi/full/10.1177/1460408617752985.

Martin, C. S., Lu, N., Inouye, D. S., Nakagawa, K., Ng, K., Yu, M., et al. (2021). Delayed respiratory failure after blunt chest trauma. *The American Surgeon, 87*(9), 1468–1473.

Mason, R. J., Broaddus, V. C., Martin, T. R., King, T. E., Schraufnagel, D., Murray, J. F., et al. (2010). *Murray and Nadel's textbook of respiratory medicine e-book: 2-Volume set* (p. 2457) Elsevier Health Sciences.

Navarrete-Navarro, P., Rivera-Fernández, R., Rincón-Ferrari, M. D., García-Delgado, M., Muñoz, A., Jiménez, J. M., et al. (2006). Early markers of acute respiratory distress syndrome development in severe trauma patients. *Journal of Critical Care, 21*(3), 253–258.

Ono, H., Nishijima, Y., Ohta, S., Sakamoto, M., Kinone, K., Horikosi, T., et al. (2017). Hydrogen gas inhalation treatment in acute cerebral infarction: A randomized controlled clinical study on safety and neuroprotection. *Journal of Stroke & Cerebrovascular Diseases, 26*(11), 2587–2594.

Pain management for blunt thoracic trauma: A joint practice management guideline from the Eastern Association for the Surgery of Trauma and Trauma Anesthesiology Society. *Journal of Trauma and Acute Care Surgery*. Retrieved September 13, 2023, from https://journals.lww.com/jtrauma/Fulltext/2016/11000/Pain_management_for_blunt_thoracic_trauma_A.20.aspxCrandall/Firearm.

Palas, J., Matos, A. P., Mascarenhas, V., Herédia, V., & Ramalho, M. (2014). Multidetector computer tomography: Evaluation of blunt chest trauma in adults. *Radiology Research and Practice, 2014*, 864369.

Pape, H. C., Remmers, D., Rice, J., Ebisch, M., Krettek, C., & Tscherne, H. (2000). Appraisal of early evaluation of blunt chest trauma: Development of a standardized scoring system for initial clinical decision making. *The Journal of Trauma, 49*(3), 496−504.

Peek, J., Kremo, V., Beks, R., van Veelen, N., Leiser, A., Link, B. C., et al. (2022). Long-term quality of life and functional outcome after rib fracture fixation. *European Journal of Trauma and Emergency Surgery, 48*(1), 255−264.

Peters, S., Nicolas, V., & Heyer, C. M. (2010). Multidetector computed tomography-spectrum of blunt chest wall and lung injuries in polytraumatized patients. *Clinical Radiology, 65*(4), 333−338.

Pharaon, K. S., Marasco, S., & Mayberry, J. (2015). Rib fractures, flail chest, and pulmonary contusion. *Current Trauma Reports, 1*(4), 237−242, 1.

Prunet, B., Bourenne, J., David, J. S., Bouzat, P., Boutonnet, M., Cordier, P. Y., et al. (2019). Patterns of invasive mechanical ventilation in patients with severe blunt chest trauma and lung contusion: A French multicentric evaluation of practices. *Journal of the Intensive Care Society, 20*(1), 46−52.

Qi, P., Huang, M., & Li, T. (2023). Identification of potential biomarkers and therapeutic targets for posttraumatic acute respiratory distress syndrome. *BMC Medical Genomics, 16*(1), 54.

Rendeki, S., & Molnár, T. F. (2019). Pulmonary contusion. *Journal of Thoracic Disease, 11* (Suppl 2), S141−S151.

Rubenfeld, G. D., Caldwell, E., Peabody, E., Weaver, J., Martin, D. P., Neff, M., et al. (2005). Incidence and outcomes of acute lung injury. *The New England Journal of Medicine, 353* (16), 1685−1693.

Sawyer, E., Wullschleger, M., Muller, N., & Muller, M. (2022). Surgical rib fixation of multiple rib fractures and flail chest: A systematic review and meta-analysis. *The Journal of Surgical Research, 276*, 221−234.

Sayed, M. S., Elmeslmany, K. A., Elsawy, A. S., & Mohamed, N. A. (2022). The validity of quantifying pulmonary contusion extent by Lung Ultrasound Score for predicting ARDS in blunt thoracic trauma. *Critical Care Research and Practice, 2022*, 3124966.

Schnyder, P., & Wintermark, M. (2000). *Radiology of blunt trauma of the chest* (p. 170) Springer Science & Business Media.

Schreiter, D., Reske, A., Stichert, B., Seiwerts, M., Bohm, S. H., Kloeppel, R., et al. (2004). Alveolar recruitment in combination with sufficient positive end-expiratory pressure increases oxygenation and lung aeration in patients with severe chest trauma. *Critical Care Medicine, 32*(4), 968−975.

Seitz, D. H., Niesler, U., Palmer, A., Sulger, M., Braumüller, S. T., Perl, M., et al. (2010). Blunt chest trauma induces mediator-dependent monocyte migration to the lung. *Critical Care Medicine, 38*(9), 1852−1859.

Simmonds, A., Smolen, J., Ciurash, M., Alexander, K., Alwatari, Y., Wolfe, L., et al. (2023). Early surgical stabilization of rib fractures for flail chest is associated with improved patient outcomes: An ACS-TQIP review. *Journal of Trauma and Acute Care Surgery, 94*(4), 532.

Simon, B., Ebert, J., Bokhari, F., Capella, J., Emhoff, T., Hayward, T., et al. (2012). Management of pulmonary contusion and flail chest: An Eastern Association for the Surgery of Trauma practice management guideline. *Journal of Trauma and Acute Care Surgery, 73*(5), S351−S361.

Staub, L. J., Biscaro, R. R. M., Kaszubowski, E., & Maurici, R. (2018). Chest ultrasonography for the emergency diagnosis of traumatic pneumothorax and haemothorax: A systematic review and meta-analysis. *Injury, 49*(3), 457–466.

Sud, S., Friedrich, J. O., Adhikari, N. K. J., Taccone, P., Mancebo, J., Polli, F., et al. (2014). Effect of prone positioning during mechanical ventilation on mortality among patients with acute respiratory distress syndrome: A systematic review and meta-analysis. *Canadian Medical Association Journal, 186*(10), E381–E390.

Thille, A. W., Esteban, A., Fernández-Segoviano, P., Rodriguez, J. M., Aramburu, J. A., Peñuelas, O., et al. (2013). Comparison of the Berlin definition for acute respiratory distress syndrome with autopsy. *American Journal of Respiratory and Critical Care Medicine, 187*(7), 761–767.

Thompson, B. T., Chambers, R. C., & Liu, K. D. (2017). Acute respiratory distress syndrome. *The New England Journal of Medicine, 377*(6), 562–572, Drazen, J.M. (ed.).

Tran, A., Fernando, S. M., Brochard, L. J., Fan, E., Inaba, K., Ferguson, N. D., et al. (2022). Prognostic factors for development of acute respiratory distress syndrome following traumatic injury: A systematic review and meta-analysis. *The European Respiratory Journal, 59*(4), 2100857.

Tremblay, L. N., & Slutsky, A. S. (1998). Ventilator-induced injury: From barotrauma to biotrauma. *Proceedings of the Association of American Physicians, 110*(6), 482–488.

Wagner, R. B., Crawford, W. O., & Schimpf, P. P. (1988). Classification of parenchymal injuries of the lung. *Radiology, 167*(1), 77–82.

Chapter 2

Anesthesiological aspects in the chest trauma patient

Emiliano Gamberini[1], Maria Maddalena Bitondo[1], Emanuele Russo[2], Domenico Pietro Santonastaso[2] and Vanni Agnoletti[2]

[1]*Anesthesia and Intensive Care Unit, "Infermi" Hospital, Local Health Authority of Romagna, Rimini (RN), Italy,* [2]*Anesthesia and Intensive Care Unit, "M. Bufalini" Hospital, Local Health Authority of Romagna, Cesena (FC), Italy*

Initial management of chest blunt trauma

The general principles for managing blunt trauma remain similar to the original approach set out in Advanced Trauma Life Support (Stewart, 2018).

The main causes of death in a trauma patient are airway obstruction, respiratory failure, massive bleeding, and brain injuries. Therefore, these are the areas targeted during the primary survey. A primary survey based on the C-ABCDE approach identifies most life-threatening injuries.

Catastrophic hemorrhage
Airway (with c-spine protection)
Breathing
Circulation
Disability
Exposure

Each stage of the C-ABCDE approach involves clinical assessment, investigations, and interventions; as a problem is identified, it should be addressed and treated. The patient is reassessed regularly to monitor the response to treatment. Reviewing step by step the algorithm is beyond the scope of this chapter; focusing on chest blunt trauma, the *immediate* life-threatening injuries that must be ruled out and, if identified, quickly treated are:

Airway injury
Tracheobronchial injury
Tension pneumothorax
Open pneumothorax

Massive haemothorax
Cardiac tamponade

Afterwards, other potentially life-threatening injuries that should be excluded are (Cubasch & Degiannis, 2004):

pulmonary contusion
tracheobronchial injuries
diaphragmatic injuries
myocardial injury
thoracic aortic disruption
esophageal injury.

Of note, around 10% of blunt chest trauma patients require surgical treatment, the remaining can be treated conservatively, with appropriate airway assessment and respiratory support, early chest tube insertion if necessary, aggressive pain control, and early respiratory physiotherapy (Chrysou et al., 2017).

Head-to-toe clinical examination is always useful in trauma patients, but it lacks sensitivity for identifying many life-threatening thoracic injuries (Dogrul et al., 2020). In the primary survey, extended focused assessment with sonography for trauma (E-FAST) and a bedside radiological assessment with chest X-ray are essential investigations. Once the patient's condition is stable, more definitive information can be obtained with computed tomography (CT) scan imaging. If the patient is rapidly deteriorating with loss of pulses in the emergency room (ER) and no obvious unsurvivable injury, the first thing to do is decompression of the chest with a mini-thoracostomy. As a general rule, a thoracic trauma patient should not die without opening the chest.

Tracheobronchial injury and airway management

Severe traumatic tracheobronchial injuries (TIs), if not lethal on scene, represent a challenge for the anesthesiologist. While penetrating injuries are quickly identified, in blunt trauma a high index of suspicion is required to recognize life-threatening TIs (Shemmeri & Vallières, 2018). They are generally caused by acceleration/deceleration mechanisms or by compression of the airway with the glottis closed. TIs are mostly located within 2 cm of the carina, and on the right bronchus for anatomical reasons. Presentation may vary from little pneumomediastinum, to variable grade of subcutaneous emphysema, to severe respiratory distress with or without hemoptysis. TIs are usually worsened by positive pressure ventilation. Continuous leakage of air after the correct placement of a chest tube is a sign that must raise suspicion for TIs (Owston et al., 2020).

The first step for the anesthesiologist is to determine if the airways need to be secured immediately or if there is time to organize a diagnostic and workup (time-critical vs no time-critical scenario). Uncooperative patients could undergo a trial of facilitated cooperation with spontaneous breathing sedation, to achieve if possible awake intubation (Mercer et al., 2016). The drug of choice is ketamine, for its characteristic of ensuring comfort and cooperation while maintaining airway reflexes and spontaneous breathing. Ketamine has no sympatholytic effect and permits the maintenance of mean arterial blood pressure in hemodynamically unstable patients. Concerns about ketamine's potential to increase intracranial pressure have been questioned and refuted by literature, as preserving mean arterial pressure is furthermore important for ensuring adequate cerebral perfusion.

A supraglottic device is not a choice in this case, hence it is of paramount importance to be ready for a different plan B in case of intubation failure, consisting of cricothyroid or tracheal rapid access (frontal neck airway — FONA), depending on the lesion level. The 2015 Difficult Airway Society guidelines recommend that all clinicians responsible for airway management be able to perform a FONA, and suggest that the scalpel-finger-bougie technique is of choice, because of the standardized few steps with well-known devices (Frerk et al., 2015). The option of awakening a difficult airway trauma patient in case of intubation failure is simply not feasible. Adequate and continuous oxygenation delaying positive pressure ventilation must be ensured throughout the workout, first via a reservoir mask plus nasal cannulas, then applying continuous oxygen supplement to the device used for intubation. The endotracheal tube should be advanced on a video/fiberoptic guide and fixed beyond the margin of the lesion. If the lesion is at a bronchus level, a double-lumen tube or a bronchial blocker will exclude the damaged lung; if not possible the endotracheal tube should be advanced in the opposite bronchus to perform a one-lung ventilation.

An algorithm for difficult airway management in tracheobronchial injury is provided in Fig. 2.1 (Mercer et al., 2016).

Besides the specific issues of tracheobronchial injuries, trauma airway management remains challenging for the anesthesiologist because of the pathology itself and suboptimal intubating conditions (Kovacs & Sowers, 2018). Patients with chest blunt trauma can present agitated, uncooperative, hypoxemic, with abnormal respiratory mechanics, hemodynamically unstable, and at risk for aspiration and airway obstruction. Signs of airway obstruction are stridor and dyspnea, drooling, trismus, and painful swallowing. Clinical examination can show expanding hematoma, subcutaneous emphysema, tracheal deviation, or other evident anatomical abnormalities of the neck.

As described above, anesthesiologists should distinguish a time-critical scenario from a no-time-critical scenario. Uncooperative patients in a no-time-critical scenario could undergo a trial of facilitated cooperation using ketamine, in order to achieve if possible awake intubation. A supraglottic device can be considered a plan B when direct airway trauma and airway obstruction are excluded.

To minimize the risk of unpreventable complications during intubation, all the key drugs and airway management tools useful in these conditions should be available and regularly checked, and a local algorithm for difficult-to-intubate

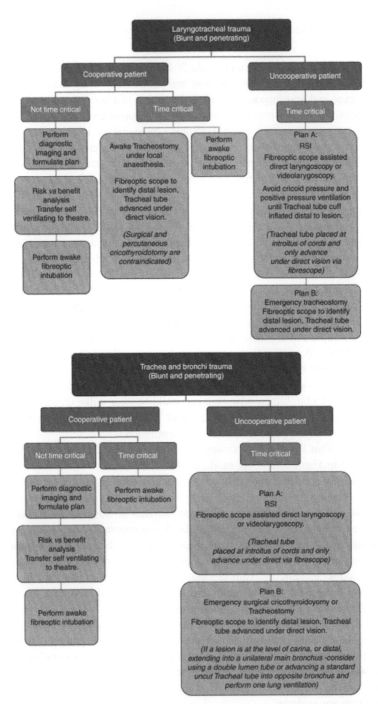

FIGURE 2.1 A shared mental model map for laryngotracheal trauma (blunt and penetrating). *RSI*, Rapid sequence induction. *From Mercer, S. J., Jones, C. P., Bridge, M., Clitheroe, E., Morton, B., & Groom, P. (2016). Systematic review of the anaesthetic management of non-iatrogenic acute adult airway trauma.* British Journal of Anaesthesia, 117, i49–i59.

TABLE 2.1 Essential emergency airway equipment.

Oxygen source and administration devices (venturi mask, reservoir mask, nasal cannulas)

Bag mask ventilation device with a positive pressure valve

Suction with tonsil tip and suction tubing devices

Monitor including capnography/capnometry and pulse oximetry

Labeled syringes containing induction and resuscitation drugs

Rhino/oro- pharyngeal cannulas, supraglottic devices

Intubation equipment: laryngoscope with multiple size blades, stylet and pretested multiple size endotracheal tubes

Difficult airway devices according to guidelines and local policy: videolaryngoscope, flexible video/fiberoptic bronchoscope, video stylet, bougies, frova intubating introducer, percutaneous cricothyroidotomy set

---

scenarios should be developed (Brown, n.d.). High-fidelity training can provide the clinician with adequate skill and confidence with equipment and rescue procedures, especially if infrequently performed, like FONA.

A list of essential emergency airway equipment is provided in Table 2.1.

Tension pneumothorax

A tension pneumothorax is the rapid accumulation of extrapleural air that compresses lung and intrathoracic vessels and obstructs venous returns to the heart (Kim & Moore, 2020). Treatment is mandatory.

Diagnosis is usually based on clinical signs, such as severe respiratory distress, hemodynamic instability, decreased or absent breath sounds on one side of the chest, and tracheal deviation away from the affected side. During the primary survey both ultrasound examination of the chest (CUS) and chest X-rays (CXR) can provide confirmation, but chest decompression should be immediate if high clinical suspicion is associated with rapid hemodynamic or respiratory worsening. In the last decades, CUS use has become widespread due to its high sensitivity and specificity in diagnosing pneumothorax. A recent review of the literature showed that the diagnostic accuracy of CUS performed by emergency non-radiologist physicians is superior to supine CXR, regardless of the type of trauma, type of CUS operator, or type of CUS probe used (Chan et al., 2020).

Chest decompression in the ER setting is typically achieved by performing a mini-thoracostomy to allow the trapped air to escape. A chest tube is then inserted after completing the primary survey, to reexpand the lung and manage the pneumothorax.

Open pneumothorax

Although relatively rare, open pneumothorax can be associated with chest blunt trauma, in this case when the subject inhales air enters the chest and accumulates between the chest wall and the lung, leading to lung collapse. Emergency treatment consists of applying a three-way dressing to the wound and taping the dressing on three of the edges. This prevents the patient from inhaling air through the opening in the chest wall while allowing them to exhale. As soon as possible a chest tube should be inserted. Definitive treatment of vast lesions requires surgical repair of the open wound or defect in the chest.

Hemothorax

Hemothorax and hemopneumothorax occur in almost one in three trauma patients with thoracic injuries. In blunt trauma, the hemothorax is usually caused by injury to the intercostal vessels or intraparenchymal pulmonary vessels due to rib fractures. Suffering three or more rib fractures is a risk factor for developing hemothorax. Rarely it can be associated with vertebral fractures. Treatment in the Emergency Department (ED) consists of inserting a chest tube, this will provide reexpansion of the lungs; contact of the lung with the parietal pleura can compress the bleeding source. The clinician needs to measure the amount of blood loss.

If the patient has massive hemothorax, defined as >1500 mL of blood or 200 mL of bleeding per hour for at least 4 h from the chest tube, thoracotomy is evaluated as an acute intervention. Fortunately, most of the time bleeding is self-limited. In case of persistent but not massive bleeding, after stabilizing the patient's conditions, interventional radiology can be evaluated to perform endovascular treatment of intercostal arterial bleeding. Blood in the pleural cavity should always be drained as much as possible, to reduce the risk of complications of retained hemothorax, such as pneumonia, empyema, fibrothorax, and subsequent acute and chronic respiratory distress. With referral to the European Guidelines on the management of major bleeding and coagulopathy following trauma, the entire equipment should be aware and ready to apply a massive bleeding protocol (Spahn et al., 2019). Large-bore peripheral veins cannulas, central venous catheters, or intraosseous catheters should be inserted, keeping in mind that large-bore peripheral lines may have higher flow rates than central catheters. High-speed transfusion devices and blood warming devices should ideally be quickly available, such as bedside tools for goal-directed coagulation management (thromboelastography or thromboelastometry).

Cardiac tamponade

Direct injury of the heart is instantly fatal because of acute tamponade or massive bleeding in most of the cases, hence the incidence of this kind of lesion is very rare at the arrival to the ED. Patients that survive present frequently with a tear in a low-pressure camera, such as an atrium. A cardiac cause of shock must be suspected in any patient with severe chest trauma not responding to fluid resuscitation or with a shock status not justified by the amount of blood loss (Offenbacher et al., 2021). The widespread use of echo FAST assessment has improved the workout and treatment of these patients, who must immediately undergo a thoracotomy. If a delay to the procedure is foreseen, a subxiphoid drainage under ultrasound guidance can be placed in the ED to gain time. The trauma leader must follow those patients in the operating room because of the rapid evolutive situation. The anesthesiologist has to secure the airway adding minimal hemodynamic impairment, as cardiac arrest after induction of general anesthesia is common in cardiac tamponade due to general anesthesia sympatholysis. If possible intubation should be delayed until the surgeon is ready to open the chest. Induction drug dose should be very conservative and the drugs of choice poorly sympatholytic, for example, ketamine in association with a neuromuscular blocker. Hemodynamic support must be initiated before induction, and the anesthesiologist has to be ready to restore blood loss and treat coagulation impairment. Transoesophageal echocardiography can guide intraoperative management and help the surgeon.

For patients with severe hypotension or in cardiac arrest, the treatment of choice is ED resuscitative thoracotomy with pericardiotomy, because closed cardiopulmonary resuscitation is ineffective. Resuscitative thoracotomy permits evacuation of pericardial tamponade, direct control of massive intrathoracic hemorrhage, cross-clamping of the descending aorta, and open cardiac massage. It ensures rapid access to the heart and major thoracic vessels through an anterolateral chest incision. Survival of those patients remains extremely poor (Fitzgerald et al., 2005; Pendleton & Leichtle, 2022).

Extracorporeal membrane oxygenation

Since the first application in 1972 on a surviving adult patient, extracorporeal membrane oxygenation (ECMO) has been increasingly used. Mainly two configurations of the circuit can be employed, differing in the site of oxygenated blood reinfusion:

Veno-arterial ECMO: for cardiac or cardiopulmonary failure
Veno-venous ECMO: for respiratory failure

In chest blunt trauma both cardiac and pulmonary failure can be severe, rapidly deteriorating, and fatal, hence the attempt to implement ECMO as a supportive therapy.

Systemic anticoagulation is used during ECMO, to prevent clotting of the membrane lung and the circuit. This may seem a contraindication to its use in a potentially bleeding condition, such as trauma, but technological advancements and high performance of most recent ECMO cannulas and circuits, even with reduced or without any level of anticoagulation, have opened the way for more extended use. A recent retrospective study on the Extracorporeal Life Support Organization database showed that the use of ECMO in blunt thoracic trauma is feasible and resulted in favorable outcomes, and the authors advocate for further investigations. The selection of patients and timing for starting the treatment are under debate, and at the moment the decision to start the treatment must be evaluated on a single case risk-benefit balance (Della Torre et al., 2019; Jacobs et al., 2015).

Anesthesia for surgical treatment

Once the primary and secondary surveys are completed, patients in need of surgical intervention are immediately transported to the operating room. Surgery strategies are discussed in detail elsewhere in this book. If the patient is not intubated, at this point a rapid sequence induction following the general airway management guidelines is performed (Apfelbaum et al., 2022). Rapid desaturation and hemodynamic instability soon after intubation and mechanical ventilation in chest trauma patients should lead the clinician to reassess for tension pneumothorax and eventually treat it. At least an arterial line should be placed, in aortic dissection placing a second arterial line on the left arm is suggested. While obtaining large and well-functioning peripheral venous access is mandatory, placing a central venous line, even if useful for fluids and drug infusions, should not delay damage control surgery. In case of hemodynamically significant arrhythmias due to myocardial injury, defibrillator pads should be available and ideally placed before the induction. Hypovolemia should be aggressively corrected, but once the patient is stabilized accurate volume status and fluid responsiveness assessment is mandatory, due to the risk of fluid overload that will worsen the respiratory distress and the risk for hemodilution that contributes to trauma coagulopathy. Depending on local expertise, goal-directed fluid therapy can be guided with different techniques (transesophageal echocardiography, transpulmonary thermodilution, etc.), and dynamic parameters and trend analysis should be preferred over static ones. Frequent blood gas analysis and laboratory exams are required during the acute phase. Ideally, coagulation management should be guided by bedside point-of-care tools, such as thromboelastography or thromboelastometry. From the intraoperative period, a protective ventilation strategy is recommended: low tidal volume (6 ml per kg ideal body weight), adequate positive end-expiratory pressure (PEEP), minimize lung stress and mechanical power reducing respiratory rate and keeping low respiratory pressures (plateau pressure <28 cmH$_2$O and driving pressure $<13-15$ cmH$_2$O). Monitoring transpulmonary pressure with an esophageal balloon can help to personalize PEEP level and ventilation strategy, avoiding overdistension of the lung and subsequent secondary ventilator-induced lung injury. Keeping adequate intrathoracic pressures can minimize the impact on the right ventricle, especially in case of myocardial-associated contusion or injury. In severe cases of lung contusion patients will require a postoperative Intensive Care Unit (ICU) stay for prolonged mechanical ventilation and protected weaning. Non-invasive ventilation and high-flow oxygen after extubation, along with optimized analgesia and early respiratory physiotherapy, reduce the incidence of weaning failure and reintubation rate (Chiumello et al., 2013; Hernandez et al., 2010). Pneumonia is a frequent complication in chest trauma patients. In a recent retrospective study on a large number of traumatic patients, the overall incidence of pneumonia was 27.5% (Wutzler et al., 2019). Compared to patients without pneumonia, patients with pneumonia had sustained more severe injuries, were older, and spent longer periods under mechanical ventilation. Thus other indications for postoperative monitoring and treatment in the ICU are multiple rib fractures and flail chest, especially in elderly patients, even if surgery is not performed.

Analgesia in blunt chest trauma

The clinical course of patients experiencing chest trauma is frequently characterized by several complications, including atelectasis, infection, hypoventilation resulting in hypoxia and hypercapnia, along with the potential development of respiratory failure. One of the major drivers of the above complications is chest pain, as it hampers normal chest wall expansion and effective coughing. Setbacks can arise in patients who do not have primary traumatic lung injury but rather experience pain from rib fractures. Indeed, the probability of developing complications is related to the number of fractured ribs (Hakim et al., 2012; Unsworth et al., 2015).

Failure to manage pain adequately can result in another significant complication: delirium. This is particularly true in the case of the elderly population (O'Connell et al., 2021). Moreover, inadequate pain control may lead to sleep disorders, undernutrition, and psychological stress (Dogrul et al., 2020).

A modern approach to managing patients with closed chest trauma increasingly involves non-invasive ventilation or high-flow nasal oxygen therapy, aiming to avoid the risks (both infectious and non-infectious) associated with orotracheal intubation. Analgesia continues to hold a key role in this context; it is an essential tool for facilitating patient adaptation and preventing the failure of non-invasive ventilation, leading to invasive ventilation instead (Carrié et al., 2023).

Pain control is necessary also to start early physiotherapy, which is another essential element of trauma patient care.

Analgesia thus becomes a pivotal component in the management strategies for chest trauma, alongside the optimization of gas exchange, ventilation, physiotherapy, and potential rib fracture repairs.

Early initiation of analgesic therapy is another key point; treatments should commence already in the prehospital setting and EDs. This is highlighted by an interesting document from Société française d'anesthésie et de réanimation — SFAR suggests this recommendation: "In case of chest trauma, relieving pain is an emergency" (Bouzat et al., 2017).

Readers interested in the history of medicine can retrieve the papers by Trinkle and Dittmann dating back to 1975, which introduce modern concepts of thoracic trauma management (Dittmann et al., 1975; Trinkle et al., 1975).

Pain control techniques can be pharmacological and non-pharmacological, and in particular, pharmacological approaches are divided into two macro-areas, namely systemically administered drugs and locoregional anesthesia techniques.

Historically used systemic drugs belong to the classes of Nonsteroidal anti-inflammatory drugs (NSAIDs) and opioids.

Intravenous opiates for decades represented the workhorse. The benefits assured are effectiveness, easiness of administration compared to invasive locoregional anesthesia techniques, and the feasibility of titrating dosages using patient-controlled analgesia (Simon et al., 2005).

Opiates may also be administered enterally and transdermally.

However, opioids are associated with well-known side effects, including oversedation, respiratory depression, nausea, constipation, itching, the need for increasing dosages due to tolerance, and the possibility of generating dependency phenomena. Additionally, in recent years, there has been a growing interest in their potential immunomodulatory role (Russo et al., 2023).

The classical paradigm consisted of a step-by-step sequence with the first step represented by systemic pharmacological therapy and the subsequent recourse to locoregional interventional techniques.

An up-to-date approach is based on very early pain control and precocious use of locoregional techniques. In particular, such strategies can have a major impact on the outcome of elderly, more frail patients in whom complications could have irreparable consequences (Proaño-Zamudio et al., 2023).

Non-steroidal anti-inflammatory drugs have been widely used, but side effects limit their use as long-term administration is often required (Birse et al., 2020).

A very interesting molecule used when a sedative effect is required in addition to the analgesic action is ketamine, which is employed especially for procedural sedation.

However, up to now, no definitive data favors one approach over another, and adopting multimodal strategies is a highly commendable option (Galvagno et al., 2016).

In recent years, there has been a growing trend in surgical fixation experiences for rib fractures aimed at reducing mechanical ventilation times and associated complications; this is particularly evident in the case of flail chest.

The purpose of surgery is clearly to solve the mechanical problem, however, limiting preternatural motility also benefits in terms of analgesia.

While the available literature is experiencing rapid expansion, specific indications and quantification of benefits remain to be fully elucidated (Beks et al., 2019).

Regional anesthesia

Regional anesthesia offers great benefits to patients with blunt chest trauma, especially in the management of pain from rib fractures. Pain relief is usually immediate and very satisfying with minimal adverse effects.

There are many options in choosing regional anesthesia in patients with blunt chest trauma.

Thoracic epidural analgesia (TEA) has traditionally been considered the gold standard regional technique for treating patients with multiple rib fractures but in recent years the advent of fascial blocks has provided many alternatives in choosing which regional anesthesia technique to practice.

Thoracic epidural analgesia

Thoracic epidural analgesia placed at a level that corresponds to approximately the middle of multiple fractures, using local anesthetic with or without opioids, provides multilevel and bilateral coverage with excellent pain relief in patients with rib fractures.

The use of TEA can spare the use of opioids and NSAIDs and minimize their associated side effects of respiratory depression, constipation, gastritis, nephrotoxicity, etc. Furthermore, provides benefits to the patient not only in terms of pain control but also in the improvement of pulmonary compliance, functional residual capacity, PaO2, and vital capacity (Mackersie et al., 1991; Moon et al., 1999; Simon et al., 2005).

Potential complications of TEA include hypotension, dural puncture, headache, bradycardia, systemic local anesthetic toxicity, meningitis, epidural hematoma, and direct spinal cord injury.

Furthermore, it is important to consider the risks of motor blockade, pruritis, and urinary retention.

A limitation of TEA is that it is contraindicated in patients with hypotension, hypovolemia, coagulopathy, significant spinal or traumatic brain injuries, and systemic infection, all of which are common in multi-trauma patients.

Thoracic paravertebral block

Thoracic paravertebral (TPV) block may be an alternative to TEA in patients who refuse TEA or in whom TEA is contraindicated. The TPV block consists of the the injection of local anesthetic alongside the thoracic vertebral body, in the TPV space, close to where the spinal nerves emerge from the intervertebral foramen with unilateral (ipsilateral), somatic, segmental, and sympathetic nerve block of multiple contiguous thoracic dermatomes.

Analgesia via TPV Block can be obtained by single shot bolus, preferably with the addition of local anesthetic adjuvants, such as dexamethasone, to prolong its effect, or by placing a catheter in the paravertebral space and using continuous administration of local anesthetic (Gilbert & Hultman, 1989; Karmakar & Ho, 2003; McKnight & Marshall, 1984; Williamson & Kumar, 1997).

TPV Block can be used to manage patients with multiple rib fractures because of its relative technical ease, above all thanks to the ultrasound-guided technique, liberal anticoagulation guidelines, lower incidence of sympathectomy, and absent risk of spinal cord injury.

The only absolute contraindications to the execution of the TPV Block are: patient refusal, allergy to local anesthetic, and local infection or sepsis.

Complications of this technique include infection, pneumothorax, local anesthetic toxicity, and neuraxial spread of the local anesthetic resulting in epidural or intrathecal block; the execution of the thoracic paravertebral block with ultrasound-guided technique improved the safety profile of the technique (Karmakar, 2001).

Serratus anterior plane block

Serratus anterior plane (SAP) block aims to provide anesthesia for the hemithorax and it has been used in patients with rib fractures as an excellent alternative to thoracic paravertebral blocks and thoracic epidurals (Millerchip et al., 2014; Nair & Diwan, 2022).

The serratus anterior muscle originates on the anterior surface of ribs 1−8 and inserts on the medial border of the scapula (Blanco et al., 2013).

A potential space exists both superficial and deep to the serratus anterior muscle and the SAP block can be performed, under ultrasound guide, injecting into the fascial plane either superficial to the serratus anterior muscle, between the latissimus dorsi and serratus anterior muscles, or deep to the serratus anterior muscle, between the serratus anterior muscle and intercostal muscles and ribs.

This provides analgesia to the lateral branches of the intercostal nerves T2−T9 (Nair & Diwan, 2022; Xie et al., 2021).

There are very few contraindications to performing a SAP block: patient refusal, allergy to local anesthetics, and local infection are the only standard absolute reasons.

Complications of this technique include pneumothorax, vascular puncture, nerve damage, failure/inadequate block, local anesthetic toxicity, and infection.

Erector spinae plane block

The erector spinae plane (ESP) block is a simple and increasingly popular new block. Local anesthetic is deposited between the erector spinae muscle and the adjacent transverse process blocking the dorsal and ventral rami of the thoracic and abdominal spinal nerves (Forero et al., 2016).

This blockage of the dorsal and ventral rami of the spinal nerves helps to achieve a multi-dermatomal sensory block of the anterior, posterior, and lateral thoracic and abdominal walls.

The clinician simply advances the block needle, with ultrasound guidance, until its tip hits the transverse process, whereupon local anesthetic is deposited followed by catheterization.

Current literature surrounding ESP Block in rib fracture management provides a positive qualitative evaluation of efficacy and safety. Improvements in pain and respiratory parameters were almost universal (Jiang et al., 2023).

Infection at the site of injection in the paraspinal region or patient refusal, are absolute contraindications for performing an ESP block.

Complications are very rare because the site of injection is far from the pleura, major blood vessels, and the spinal cord. Infection at the needle insertion site, local anesthetic toxicity/allergy, vascular puncture, pleural puncture, pneumothorax, and failed block are the primary complications (Krishnan & Cascella, 2023).

Intercostal nerve block

The intercostal nerves (ICNs) innervate the major parts of the skin and musculature of the chest and abdominal wall and ICN lock provides excellent analgesia in patients with rib fractures.

ICN Block is a technique very simple to perform but injection both above and below the affected rib(s) is required to produce adequate coverage, which necessitates multiple injections, increasing the risk for complications such as pneumothorax and vascular puncture.

Infection at the site of injection in the paraspinal region or patient refusal, are absolute contraindications for performing an ICN block (Kim & Moore, 2020).

References

Apfelbaum, J. L., Hagberg, C. A., Connis, R. T., Abdelmalak, B. B., Agarkar, M., Dutton, R. P., Fiadjoe, J. E., Greif, R., Klock, P. A., Mercier, D., Myatra, S. N., O'Sullivan, E. P., Rosenblatt, W. H., Sorbello, M., & Tung, A. (2022). 2022 American Society of Anesthesiologists practice guidelines for management of the difficult airway. *Anesthesiology*, *136*(1), 31–81. Available from https://doi.org/10.1097/ALN.0000000000004002.

Beks, R. B., Peek, J., de Jong, M. B., Wessem, K. J. P., Öner, C. F., Hietbrink, F., Leenen, L. P. H., Groenwold, R. H. H., & Houwert, R. M. (2019). Fixation of flail chest or multiple rib fractures: Current evidence and how to proceed. A systematic review and meta-analysis. *European Journal of Trauma and Emergency Surgery*, *45*(4), 631–644. Available from https://doi.org/10.1007/s00068-018-1020-x.

Birse, F., Williams, H., Shipway, D., & Carlton, E. (2020). Blunt chest trauma in the elderly: An expert practice review. *Emergency Medicine Journal*, *37*(2), 73–78. Available from https://doi.org/10.1136/emermed-2019-209143.

Blanco, R., Parras, T., McDonnell, J. G., & Prats-Galino, A. (2013). Serratus plane block: A novel ultrasound-guided thoracic wall nerve block. *Anaesthesia*, *68*(11), 1107−1113. Available from https://doi.org/10.1111/anae.12344.

Bouzat, P., Raux, M., David, J. S., Tazarourte, K., Galinski, M., Desmettre, T., Garrigue, D., Ducros, L., Michelet, P., Freysz, M., Savary, D., Rayeh-Pelardy, F., Laplace, C., Duponq, R., Monnin Bares, V., D'Journo, X. B., Boddaert, G., Boutonnet, M., Pierre, S., ... Vardon, F. (2017). Chest trauma: First 48 hours management. *Anaesthesia Critical Care and Pain Medicine*, *36*(2), 135−145. Available from https://doi.org/10.1016/j.accpm.2017.01.003.

Brown (n.d.). The walls manual of emergency airway management. In: *Wolters Kluwer, 2022 and the difficult airway course.*

Carrié, C., Rieu, B., Benard, A., Trin, K., Petit, L., Massri, A., Jurcison, I., Rousseau, G., Tran Van, D., Reynaud Salard, M., Bourenne, J., Levrat, A., Muller, L., Marie, D., Dahyot-Fizelier, C., Pottecher, J., David, J. S., Godet, T., & Biais, M. (2023). Early non-invasive ventilation and high-flow nasal oxygen therapy for preventing endotracheal intubation in hypoxemic blunt chest trauma patients: The OptiTHO randomized trial. *Critical Care*, *27*(1). Available from https://doi.org/10.1186/s13054-023-04429-2.

Chan, K. K., Joo, D. A., McRae, A. D., Takwoingi, Y., Premji, Z. A., Lang, E., & Wakai, A. (2020). Chest ultrasonography versus supine chest radiography for diagnosis of pneumothorax in trauma patients in the emergency department. *Cochrane Database of Systematic Reviews*, *2020*(8). Available from https://doi.org/10.1002/14651858.cd013031.pub2.

Chiumello, D., Coppola, S., Froio, S., Gregoretti, C., & Consonni, D. (2013). Noninvasive ventilation in chest trauma: Systematic review and meta-analysis. *Intensive Care Medicine*, *39*(7), 1171−1180. Available from https://doi.org/10.1007/s00134-013-2901-4.

Chrysou, K., Halat, G., Hoksch, B., Schmid, R. A., & Kocher, G. J. (2017). Lessons from a large trauma center: Impact of blunt chest trauma in polytrauma patients-still a relevant problem? *Scandinavian Journal of Trauma, Resuscitation and Emergency Medicine*, *25*(1). Available from https://doi.org/10.1186/s13049-017-0384-y.

Cubasch, H., & Degiannis, E. (2004). The deadly dozen of chest trauma. *Continuing Medical Education*, 1−8.

Della Torre, V., Robba, C., Pelosi, P., & Bilotta, F. (2019). Extra corporeal membrane oxygenation in the critical trauma patient. *Current Opinion in Anaesthesiology*, *32*(2), 234−241. Available from https://doi.org/10.1097/ACO.0000000000000698.

Dittmann, M., Ferstl, A., & Wolff, G. (1975). Epidural analgesia for the treatment of multiple ribfractures. *European Journal of Intensive Care Medicine*, *1*(2), 71−75. Available from https://doi.org/10.1007/BF00626429.

Dogrul, B. N., Kiliccalan, I., Asci, E. S., & Peker, S. C. (2020). Blunt trauma related chest wall and pulmonary injuries: An overview. *Chinese Journal of Traumatology*, *23*(3), 125−138. Available from https://doi.org/10.1016/j.cjtee.2020.04.003.

Fitzgerald, M., Spencer, J., Johnson, F., Marasco, S., Atkin, C., & Kossmann, T. (2005). Definitive management of acute cardiac tamponade secondary to blunt trauma. *Emergency Medicine Australasia*, *17*(5−6), 494−499. Available from https://doi.org/10.1111/j.1742-6723.2005.00782.x.

Forero, M., Adhikary, S. D., Lopez, H., Tsui, C., & Chin, K. J. (2016). The erector spinae plane block a novel analgesic technique in thoracic neuropathic pain. *Regional Anesthesia and Pain Medicine*, *41*(5), 621−627. Available from https://doi.org/10.1097/AAP.0000000000000451.

Frerk, C., Mitchell, V. S., McNarry, A. F., Mendonca, C., Bhagrath, R., Patel, A., O'Sullivan, E. P., Woodall, N. M., & Ahmad, I. (2015). Difficult Airway Society 2015 guidelines for management of unanticipated difficult intubation in adults. *British Journal of Anaesthesia*, *115*(6), 827−848. Available from https://doi.org/10.1093/bja/aev371.

Galvagno, S. M., Smith, C. E., Varon, A. J., Hasenboehler, E. A., Sultan, S., Shaefer, G., To, K. B., Fox, A. D., Alley, D. E. R., Ditillo, M., Joseph, B. A., Robinson, B. R. H., & Haut, E. R. (2016). Pain management for blunt thoracic trauma: A joint practice management guideline from the Eastern Association for the Surgery of Trauma and Trauma Anesthesiology Society. *Journal of Trauma and Acute Care Surgery*, *81*(5), 936−951. Available from https://doi.org/10.1097/TA.0000000000001209.

Gilbert, J., & Hultman, J. (1989). Thoracic paravertebral block: A method of pain control. *Acta Anaesthesiologica Scandinavica*, *33*(2), 142−145. Available from https://doi.org/10.1111/j.1399-6576.1989.tb02877.x.

Hakim, S. M., Latif, F. S., & Anis, S. G. (2012). Comparison between lumbar and thoracic epidural morphine for severe isolated blunt chest wall trauma: A randomized open-label trial. *Journal of Anesthesia*, *26*(6), 836−844. Available from https://doi.org/10.1007/s00540-012-1424-4.

Hernandez, G., Fernandez, R., Lopez-Reina, P., Cuena, R., Pedrosa, A., Ortiz, R., & Hiradier, P. (2010). Noninvasive ventilation reduces intubation in chest trauma-related hypoxemia: A randomized clinical trial. *Chest*, *137*(1), 74−80. Available from https://doi.org/10.1378/chest.09-1114.

Jacobs, J. V., Hooft, N. M., Robinson, B. R., Todd, E., Bremner, R. M., Petersen, S. R., & Smith, M. A. (2015). The use of extracorporeal membrane oxygenation in blunt thoracic trauma: A study of the Extracorporeal Life Support Organization database. *Journal of Trauma and Acute Care Surgery*, *79*(6), 1049−1053. Available from https://doi.org/10.1097/TA.0000000000000790.

Jiang, M., Peri, V., Yang, B. O., Chang, J., & Hacking, D. (2023). Erector spinae plane block as an analgesic intervention in acute rib fractures: A scoping review. *Local and Regional Anesthesia*, *16*, 81−90. Available from https://doi.org/10.2147/LRA.S414056.

Karmakar, M. K. (2001). Thoracic paravertebral block. *Anesthesiology*, *95*(3), 771−780. Available from https://doi.org/10.1097/00000542-200109000-00033.

Karmakar, M. K., & Ho, A. M. H. (2003). Acute pain management of patients with multiple fractured ribs. *Journal of Trauma*, *54*(3), 615−625. Available from https://doi.org/10.1097/01.TA.0000053197.40145.62.

Kim, M., & Moore, J. E. (2020). Chest trauma: Current recommendations for rib fractures, pneumothorax, and other injuries. *Current Anesthesiology Reports*, *10*(1), 61−68. Available from https://doi.org/10.1007/s40140-020-00374-w.

Kovacs, G., & Sowers, N. (2018). Airway management in trauma. *Emergency Medicine Clinics of North America*, *36*(1), 61−84. Available from https://doi.org/10.1016/j.emc.2017.08.006.

Krishnan, S., & Cascella. (2023). *Erector spinae plane block*. StatPearls Publishing, In StatPearls Internet.

Mackersie, R. C., Karagianes, H. G., Hoyt, D. B., & Davis, J. W. (1991). Prospective evaluation of epidural and intravenous administration of fentanyl for pain control and restoration of ventilatory function following multiple rib fractures. *Journal of Trauma - Injury, Infection and Critical Care*, *31*(4), 443−451. Available from https://doi.org/10.1097/00005373-199104000-00002.

McKnight, C. K., & Marshall, M. (1984). Monoplatythela and paravertebral block. *Anaesthesia*, *39*(11), 1147. Available from https://doi.org/10.1111/j.1365-2044.1984.tb08953.x.

Mercer, S. J., Jones, C. P., Bridge, M., Clitheroe, E., Morton, B., & Groom, P. (2016). Systematic review of the anaesthetic management of non-iatrogenic acute adult airway trauma. *British Journal of Anaesthesia*, *117*, i49−i59. Available from https://doi.org/10.1093/bja/aew193.

Millerchip, S., May, L., & Hillermann, C. (2014). Continuous serratus anterior block: An alternative regional analgesia technique in patients with multiple rib fractures. *Regional Anesthesia and Pain Medicine, 39*.

Moon, M. R., Luchette, F. A., Gibson, S. W., Crews, J., Sudarshan, G., Hurst, J. M., Davis, K., Johannigman, J. A., Frame, S. B., & Fischer, J. E. (1999). Prospective, randomized comparison of epidural versus parenteral opioid analgesia in thoracic trauma. *Annals of Surgery, 229* (Issue 5), 684−692. Available from https://doi.org/10.1097/00000658-199905000-00011.

Nair, A., & Diwan, S. (2022). Efficacy of ultrasound-guided serratus anterior plane block for managing pain due to multiple rib fractures: A scoping review. *Cureus*. Available from https://doi.org/10.7759/cureus.21322.

O'Connell, K. M., Patel, K. V., Powelson, E., Robinson, B. R. H., Boyle, K., Peschman, J., Blocher-Smith, E. C., Jacobson, L., Leavitt, J., McCrum, M. L., Ballou, J., Brasel, K. J., Judge, J., Greenberg, S., Mukherjee, K., Qiu, Q., Vavilala, M. S., Rivara, F., & Arbabi, S. (2021). Use of regional analgesia and risk of delirium in older adults with multiple rib fractures: An Eastern Association for the Surgery of Trauma multicenter study. *Journal of Trauma and Acute Care Surgery, 91*(Issue 2), 265−271. Available from https://doi.org/10.1097/TA.0000000000003258, Lippincott Williams and Wilkins.

Offenbacher, J., Liu, R., Venitelli, Z., Martin, D., Fogel, K., Nguyen, V., & Kim, P. K. (2021). Hemopericardium and cardiac tamponade after blunt thoracic trauma: A case series and the essential role of cardiac ultrasound. *Journal of Emergency Medicine, 61*(3), e40−e45. Available from https://doi.org/10.1016/j.jemermed.2021.05.013.

Owston, H., Jones, C., Groom, P., & Mercer, S. J. (2020). The anaesthetic management of the airway after blunt and penetrating neck injury. *Trauma, 22*(1), 3−6. Available from https://doi.org/10.1177/1460408619886216.

Pendleton, A. C., & Leichtle, S. W. (2022). Cardiac tamponade from blunt trauma. *American Surgeon, 88*(6), 1319−1321. Available from https://doi.org/10.1177/0003134820942170.

Proaño-Zamudio, J. A., Argandykov, D., Renne, A., Gebran, A., Ouwerkerk, J. J. J., Dorken-Gallastegi, A., de Roulet, A., Velmahos, G. C., Kaafarani, H. M. A., & Hwabejire, J. O. (2023). Timing of regional analgesia in elderly patients with blunt chest-wall injury. *Surgery (United States), 174*(Issue 4), 901−906. Available from https://doi.org/10.1016/j.surg.2023.07.006, Elsevier Inc.

Russo, E., Latta, M., Santonastaso, D. P., Bellantonio, D., Cittadini, C., Pietrantozzi, D., Circelli, A., Gamberini, E., Martino, C., Spiga, M., Agnoletti, V., Avolio, F., Benini, B., Benni, M., Bergamini, C., Bini, G., Bissoni, L., Bolondi, G., Campagna, D., ... Zecchini, M. C. (2023). Regional anesthesia in the intensive care unit: A single center's experience and a narrative literature review. *Discover Health Systems, 2*(1). Available from https://doi.org/10.1007/s44250-023-00018-w.

Shemmeri, E., & Vallières, E. (2018). Blunt tracheobronchial trauma. *Thoracic Surgery Clinics, 28*(3), 429−434. Available from https://doi.org/10.1016/j.thorsurg.2018.04.008.

Simon, B. J., Cushman, J., Barraco, R., Lane, V., Luchette, F. A., Miglietta, M., Roccaforte, D. J., & Spector, R. (2005). Pain management guidelines for blunt thoracic trauma. *Journal of Trauma − Injury, Infection and Critical Care, 59*(5), 1256−1267. Available from https://doi.org/10.1097/01.ta.0000178063.77946.f5.

Spahn, D. R., Bouillon, B., Cerny, V., Duranteau, J., Filipescu, D., Hunt, B. J., Komadina, R., Maegele, M., Nardi, G., Riddez, L., Samama, C. M., Vincent, J. L., & Rossaint, R. (2019). The European guideline on management of major bleeding and coagulopathy following trauma: Fifth edition. *Critical Care (London, England), 23*(1), 98. Available from https://doi.org/10.1186/s13054-019-2347-3.

Stewart, R. (2018). *Advanced trauma life support: Student course manual*. American College of Surgeons. Committee On Trauma.

Trinkle, J. K., Richardson, J. D., Franz, J. L., Grover, F. L., Arom, K. V., & Holmstrom, F. M. G. (1975). Management of flail chest without mechanical ventilation. *Annals of Thoracic Surgery, 19*(4), 355–363. Available from https://doi.org/10.1016/S0003-4975(10)64034-9.

Unsworth, A., Curtis, K., & Asha, E. E. (2015). Treatments for blunt chest trauma and their impact on patient outcomes and health service delivery. *Scandinavian Journal of Trauma, Resuscitation and Emergency Medicine, 23*(1). Available from https://doi.org/10.1186/s13049-015-0091-5.

Williamson, S., & Kumar, C. M. (1997). Paravertebral block in head injured patient with trauma [9]. *Anaesthesia, 52*(3), 284–285.

Wutzler, S., Bläsius, F. M., Störmann, P., Lustenberger, T., Frink, M., Maegele, M., Weuster, M., Bayer, J., Caspers, M., Seekamp, A., Marzi, I., Andruszkow, H., & Hildebrand, F. (2019). Pneumonia in severely injured patients with thoracic trauma: Results of a retrospective observational multi-centre study. *Scandinavian Journal of Trauma, Resuscitation and Emergency Medicine, 27*(1). Available from https://doi.org/10.1186/s13049-019-0608-4.

Xie, C., Ran, G., Chen, D., & Lu, Y. (2021). A narrative review of ultrasound-guided serratus anterior plane block. *Annals of Palliative Medicine, 10*(1), 700–706. Available from https://doi.org/10.21037/apm-20-1542.

Chapter 3

Scoring systems of blunt thoracic trauma and rib fractures

Jennifer E. Baker, Kevin N. Harrell and Fredric M. Pieracci
Department of Surgery, Division of Trauma, Denver Health Medical Center, Denver, CO, United States

Introduction

Clinical scoring systems for injury patterns provide efficacious and algorithmic approaches to the diagnosis and treatment of common medical conditions. They also allow for a common language in which providers can communicate, prognosticate, educate, and conduct research. The purpose of a scoring system is to organize the variability of injury patterns in a graded fashion to stratify patients by risk of developing complications and overall outcomes. Once a specific scoring system has been developed, novel therapies can then be studied to identify the ideal target population. Scoring systems must be predictable, easily utilized, and perform reliably. They must be validated rigorously and hold up to performance characteristics in multiple populations to be useful in practice (Challener et al., 2019).

Challenges remain in the development of a chest wall injury scoring system. Some of these include the lack of an adopted universal nomenclature to describe these injuries, overreliance on radiographic parameters, exclusion of long-term outcomes, and lack of validation in large, diverse patient populations. However, several scoring systems are available and have been validated to aid in risk stratification and resource allocation in the treatment of chest injuries (Chapman et al., 2016; Easter, 2001; Hardin et al., 2019; Manay et al., 2017; Moore et al., 1992; Pape et al., 2000; Pressley et al., 2012).

General properties of scoring systems

The ideal diagnostic test predicts an outcome perfectly; if a test is positive, then the outcome must be present, if a test is negative, the patient does not have the outcome. The use of positive predictive value (PPV), negative predictive value (NPV), specificity, and sensitivity are important basic test properties (Table 3.1). These parameters are uniquely related to help aid in the development of scoring systems and the development of clinical tests.

Sensitivity is the proportion of patients with the outcome that test positive (true positive rate). Specificity is defined as the proportion of patients without the outcome that test negative (true negative rate). Highly sensitive tests minimize false negatives and are thus useful to rule in a condition, whereas highly specific tests, minimize false positives and are useful to rule out conditions. Values greater than 80% are considered favorable tests. Neither specificity nor sensitivity rely on the prevalence of the outcome (proportion of individuals with the outcome in the population at a time point). However, these tests rely upon knowing the outcome status of the patient to apply the test clinically.

PPV is the proportion of patients who test positive and have the outcome. NPV is the inverse; the proportion of patients who test negative and do not have the outcome. These two tests are dependent upon the prevalence of the outcome. Overall, PPV and NPV are oftentimes considered more practical for prognostic purposes than sensitivity and specificity.

Let's consider a hypothetical scenario in which there is a test/score that determines the relationship between forced vital capacity (FVC) <1500 mL and pneumonia in rib fracture patients. If this test were 80% sensitive, it would mean that 80% of patients who developed pneumonia had an FVC <1500 mL. This test would not be clinically relevant because on admission we would not know who ultimately would develop pneumonia. However, if the PPV of the test was 80%, this would mean that 80% of the patients with an FVC <1500 mL would develop pneumonia. This information would be incredibly useful to help prognosticate and allow clinicians to intervene in the patient group with an FVC <1500 mL before the outcome is observed.

Prediction models may have either categorical or continuous variables. Categorical models are further divided into binary or multilevel variables. Using the aforementioned example, a FVC <1500 mL is binary with two

TABLE 3.1 Sensitivity, specificity, positive predictive value, negative predictive value.

	Outcome is positive	Outcome is negative
Test is positive	True positive (TP)	False positive (FP)
Test is negative	False negative (FN)	True negative (TN)

2 × 2 table illustrating the four basic test parameters. Sensitivity = TP/(TP + FN); Specificity = TN/(TN + FP); Positive predictive value = TP/(TP + FP); Negative predictive value = TN/(TN + FP).

outcomes. However, FVC could be a continuous variable in which the raw number is utilized for each patient. FVC can also be operationalized as poor, adequate, or good, this would represent a multi-level, ordinal categorical variable. Most predictive models are continuous or ordinal categorical, thus consisting of multiple thresholds for predicting the outcome. As an example, consider using age as a continuous variable to predict the need for tracheostomy and ventilator weaning among rib fracture patients. There are an infinite number of age thresholds (i.e., age >20 years old, age >21 years old, etc.) for which patients may be grouped. By calculating sensitivity (true positive) and 1 − specificity (false positive) for the outcome at each of these thresholds, a curved line is generated when plotted on a graph Fig. 3.1. This graphical representation is known as a receiver operating characteristic (ROC) curve representing a test's predictive ability. A likelihood ratio (LR) is obtained by dividing sensitivity by 1 − specificity; an LR ratio close to 1 indicates that the test adds little information to the clinical picture.

The area under curve (AUC) for the ROC is the standard statistical expression of a test's predictive ability. The AUC ranges from 0.5 to 1.0; a value of 0.5 indicates that the test is no better than chance while a value of 1.0 is a perfect prediction. In general, a ROC AUC of moderate predictive ability correlates to a value of 0.7−0.8, a value of 0.8−0.9 is good predictive ability, and an excellent predictive value correlates to a value of 0.9-0.99.

Lastly, the point of maximal distance between the ROC curve and the change line is known as Youden's J; this summarizes the overall performance

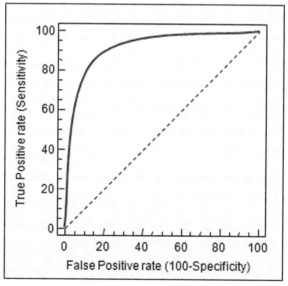

FIGURE 3.1 **Receiver operator characteristic (ROC) curve.** The dotted line represents a test with no predictive ability (area under the curve = 0.50).

of the test. Youden's J is calculated as sensitivity + specificity -1 and the value ranges from -1 to 1. A value of 0 indicates that there is the same proportion of positive tests in both the population with the outcome and without, thus rendering the test worthless. A value of 1 indicates a perfect test without false negatives or false positives (Youden, 1950).

Ideal characteristics of a chest wall scoring systems

To design a high-fidelity chest wall injury scoring system, many challenges exist and multiple considerations must be made. Specifically, looking at predictor and outcome variables remains a challenge. There also are challenges with how to weigh each characteristic and whether a linear or nonlinear relationship between injury and score exists. Furthermore, a standardized grading system for chest wall injuries has not yet been universally adopted and utilized in a scoring system.

Predictor variables selected for inclusion into models must be evaluated for association with outcome, overall prevalence, ease of abstraction, objectivity, and interobserver reliability. For example, utilizing the variable "degree of fracture displacement" as a marker of injury leaves too much subjective interpretation of the injury pattern. While many agree that it is an important variable in predicting outcomes, how exactly would this be measured, who would measure it, and when would it be measured? Does the location of the fracture matter? Would different locations of fractures be weighted differently? A more interpretable variable may instead be "four or more fractures with biocortical displacement." Additionally, the factors chosen must occur commonly enough to be of value; utilizing the presence of an open fracture would not be meaningful, even though it carries high morbidity, given its low incidence. Utilizing "total number of ribs fractures" provides low interobserver variability, is easily determined, and is commonly obtained. However, it is simplistic and does not account for the differences in outcome based on fracture location, displacement, and presence/absence of a flail segment. Nor does it account for the phenotype of the patient, which is variable among patients and may also be more clinically relevant.

The selection of outcome variables also requires careful consideration. Mortality, while a grave complication, is relatively rare in chest wall-injured patients. In a recent multicenter analysis of the timing of surgical stabilization of rib fractures, morality was noted to be only 1.2% (Pieracci et al., 2018). Utilizing an outcome like mortality also does not take into account other patient outcomes. Many patients who survive their initial chest wall injury remain disabled, on a ventilator, and unable to work or enjoy other activities (Fabricant et al., 2013). Not incorporating these outcomes into a scoring system and only utilizing mortality would not capture the true impact of chest wall injuries.

Lastly, there remain statistical challenges in developing a scoring system for chest wall injuries. Many trauma scoring systems are linear, giving equal weight to each predictive value and also to each increase in injury severity. For example, the blunt Pulmonary Contusion Score has incremental increases in point assignment for mild contusion (1 point), moderate contusion (2 points), and severe contusions (3 points) (Tyburski et al., 1999). Realistically, these differences are likely not linear and a moderate contusion has more than twice the impact of a simple one. Partnerships with biostatisticians to aid in the creation of a scoring system should be utilized to correctly model and determine the weight of each variable.

Thoracic chest wall taxonomy

One of the largest limitations in the development of standardized chest wall scoring systems remains the lack of a universal nomenclature for chest wall injuries before 2018, there remained limited standardization of the nomenclature surrounding rib fractures and bony thoracic trauma. Significant provider discrepancies existed between the interpretation of flail chest, fracture displacement, and lateral rib fractures. This made utilizing fracture pattern differences in chest wall injury scoring systems challenging.

In 2018 the AO Association in collaboration with the Orthopedic Trauma Association (OTA) released an international comprehensive classification of fractures and dislocations of the thorax. This classification was developed to allow for a standardization of describing fractures and dislocations (Meinberg, Agel, Roberts, Karam, et al., 2018). Within this rating, the anatomy of the rib was further defined into the posterior end segment, shaft, and anterior end segment. Additionally, the classification categories of the different types of fractures were further illustrated and delineated (Meinberg, Agel, Roberts, et al., 2018). The Ohio State Bioinformatics Research Center undertook dynamic anterior-posterior blunt force loading to cadaveric ribs to further study this classification system. Overall, the group was able to classify 99.2% of fractures with this model allowing for improved detail and clarity for the features of the fracture patterns. Within this basic study, the classification system appears to be valid (Harden et al., 2019). However, additional studies should be undertaken to further validate this system in clinical settings.c

More recently, the Chest Wall Injury Society (CWIS) released an international Delphi consensus statement to create a standardized taxonomy for rib fractures. A consensus was reached to define displacement, anatomic sectors, and flail chest/segment of rib fractures. A flail segment was delineated as a radiographic finding while a flail chest remained in paradoxical motion on a clinical exam. However, consensus on the borders of the three sectors of the rib (anterior/lateral/posterior) was not reached (Edwards et al., 2020). In a survey looking at agreement between clinicians and this taxonomy, strong interobserver agreement based on location (anterior/lateral/posterior), moderate agreement for fracture type (simple/wedge/complex), and fair agreement when evaluating displacement (displaced/offset/displaced) was observed. Interestingly, despite no clear definition of how to define the fracture location, this was the strongest agreement despite observers utilizing different anatomical locations (Van Wijck et al., 2022).

A validation study on CWIS taxonomy in over 500 patients was performed to further determine the clinical significance against outcomes. It was noted that patients with complex rib fractures, displaced rib fractures, and flail segments were more likely to have pulmonary complications (pneumonia, acute respiratory failure, mechanical ventilation, and tracheostomy) and more adverse outcomes (intesnive care unit (ICU) length of stay >7 days, pulmonary embolism, or death). On multivariate analysis, the presence of a flail segment and the number of ribs fractured were noted to be predictors of pulmonary complications and adverse outcomes. Overall, the authors deemed that their findings supported the definitions as proposed by CWIS, however, there were limitations and the sample size was not adequate to test all of the proposed definitions. Some of the variables with three or more categories did not appear to have a linear relationship with complications when analyzed. The strongest independent prognostic factor remained the number of rib fractures (Clarke et al., 2019).

While progress has been made in defining a standardized nomenclature and classification system for rib fractures, continued work remains. Joint collaborations between the CWIS, the American Association for the Surgery of Trauma (AAST), and the AO/OTA groups continue to remain a top priority.

Current chest wall scoring systems

Many single parameters have been utilized to predict outcomes after chest wall injury without creating a formal scoring system. These parameters include the degree of pulmonary contusion, the number of rib fractures, the number of displaced rib fractures, vital capacity, and age (Bulger et al., 2000; Butts et al., 2017; Carver et al., 2015; Chien et al., 2017; Flagel et al., 2005; Holcomb et al., 2003; Tyburski et al., 1999). However, the following discussion will focus on validated scoring systems which include at least two variables. Table 3.2 depicts the parameters making up each score.

Chest wall Organ Injury Scale

The oldest and most commonly used classification system is the chest wall Organ Injury Scale (OIS) (Table 3.3); this system has been endorsed by the AAST (Moore et al., 1992). Similar to other organ systems OIS, it grades chest wall injuries on a scale of I–V and assigns a grade for each laceration, contusion, and fracture. Overall, it is one of the most inclusive scoring

TABLE 3.2 Comparison of different validated chest wall scoring systems.

	OIS chest	TTSS	RFS	CTS	RibScore	SCARF score
Rib fracture variables						
Number of ribs fractured	X	X	X	X	X	
Flail chest	X				X	
Bilateral fractures	X		X	X	X	
Degree of displacement					X	
Fracture location					X	
First rib fractured					X	
Non-rib fracture variables						
Age		X	X	X		
Pulmonary contusion	X	X		X		
Clavicle/scapula/sternal fracture	X					
Soft tissue Injury	X					
Pleural involvement		X				
PaO$_2$/FiO$_2$		X				
Incentive spirometry						X
Respiratory rate						X
Numeric pain score						X
Strength of cough						X

CTS, Chest trauma score; *OIS*, organ injury score; *RFS*, rib fracture score; *SCARF*, sequential clinical assessment of respiratory function; *TTSS*, thorax trauma severity score.

systems for chest wall injury as it takes into account all of the bony structures of the thorax (sternum, clavicle, and scapula). Additionally, rib fractures are also dichotomized based on number, presence of flail chest, and if the fractures occur bilaterally. The presence and degree of thoracic soft tissue injury are also included in the scoring system. Despite the inclusion of multiple variables, it lacks detailed information regarding the location of fractures and fracture patterns and their relationship to pulmonary physiology. Additionally, the inclusivity of multiple injuries into one grade may

TABLE 3.3 Chest wall Organ Injury Scale.

Grade	Injury type	Description of injury
I	Contusion	Any size
	Laceration	Skin and subcutaneous
	Fracture	<3 ribs, closed
		Nondisplaced clavicle, closed
II	Laceration	Skin, subcutaneous, and muscle
	Fracture	≥3 adjacent ribs, closed
		Displaced clavicle
		Open clavicle
		Nondisplaced sternum, closed
		Scapular body, open or closed
III	Laceration	Full thickness including pleura penetration
	Fracture	Unilateral flail segment (<3 ribs)
		Displaced sternum
		Open sternum
		Flail sternum
IV	Laceration	Avulsion of chest wall tissues with underlying fracture
	Fracture	Unilateral flail chest (≥3 ribs)
V	Fracture	Bilateral flail chest (≥3 ribs bilaterally)

complicate and devalue the total injury burden if a patient has multiple additive injuries in one grade of injury (i.e., ≥3 ribs fractures, a displaced clavicle fracture, and a closed scapula fracture would only be a grade II injury).

The chest wall OIS was developed through the expert opinion of the OIS Committee of the AAST and was initially published in 1992. This system is frequently referenced in the literature; however, it has not been externally validated extensively. When looking at a small sample (385 patients) from our trauma institution, the ROC AUC for predicting pneumonia, respiratory failure, and tracheostomy were 0.60, 0.61, and 0.66 respectively (Chapman et al., 2016). An additional larger retrospective study looking at 3033 patients demonstrated the ROC AUC for mortality, tracheostomy, pulmonary complications, and cardiac complications to be 0.68, 0.67, 0.64, and 0.66, respectively (Baker et al., 2020). Overall, it appears that OIS has poor predictive ability on common complications and mortality after chest wall injury. Grouping patients using this scale may ultimately be of limited use when investigating interventions for rib fracture patients.

Thoracic trauma severity score

The thoracic trauma severity score (TTSS) was developed in 2000 and focused on utilizing chest wall injury, pulmonary contusion, pleural involvement (hemothorax and/or pneumothorax), age, and a PaO_2/FiO_2 ratio. Initially, each parameter was given a score of 0–4 depending on the severity of the injury and then subsequently added to give a final TTSS score. A score of 0 represented normal function, a grade of 1 represented low-grade dysfunction with less than 10% of chest-related complications while a grade of 4 represented a complication rate of 60% or higher. When the predictive value of each parameter was ultimately calculated, it was noted that the difference between a grade 3 and 4 injury was significantly greater, and thus an additional point (to a grade of 5) was added for each grade 4 injury. Of note, the assessment was based on chest radiography (CXR) and not computerized tomography (CT). Only patients with a glasgow coma scale (GCS) >8, who required mechanical ventilation for at least 4 days, had no signs of infection, and survived for over 2 days were included in the original score development. Validation performed by the authors demonstrated an ROC AUC of 0.91–0.92 when looking at posttraumatic respiratory complications (Pape et al., 2000). Further studies have attempted to validate the TTSS for in-hospital mortality in this patient population but failed to show that TTSS could predict in-hospital mortality (ROC AUC: 0.472) (Moon et al., 2017).

When inclusion criteria were expanded to all patients with any blunt thoracic trauma and utilizing chest CT instead of CXR, the ROC AUC in predicting acute respiratory distress syndrome (ARDS) was 0.82 in the overall population and 0.78 in the non-hypoxemic subgroups (Daurat et al., 2016). Additionally, in isolated chest wall trauma patients (Abbreviated Injury Scale [AIS] <2 except in non-chest areas), TTSS performed better at predicting pulmonary complications as compared to many other scoring systems (Seok et al., 2019). In fact, in non-traumatic brain injury (TBI) patients, TTSS performed better for predicting acute respiratory distress syndrome, multisystem organ failure, and mortality after blunt trauma than other scoring systems as well (Mommsen et al., 2012). However, all of these associations are only moderately predictive and not as strong when the population is more generalized from the original study.

One of the greatest limitations of this system remains the need to collect a large amount of clinical data on arrival to calculate the score. Also, when more diverse populations are utilized for validation, predictive value diminishes.

Rib fracture score

The rib fracture score (RFS) was developed in 2001 and is the simplest to use as it only involves three variables. The total score is calculated as: (number of fractures x sides [unilateral = 1 and bilateral = 2]) + age factor (50–60 years = 1 point, 61–70 years = 2 points, 71–80 years = 3 points, >80 years = 4 points). This score was initially developed to triage older patients with rib fractures and assess the need for respiratory support, mobilization, and pain management (Easter, 2001). Ultimately, validation of this score revealed poor predictive value for mortality, pneumonia, and the need for tracheostomy (Chapman et al., 2016; Fokin et al., 2018; Maxwell et al., 2012). This poor predictive value is likely due to the lack of detailed fracture pattern information, patient demographics, and patient physiology variables.

Chest trauma score

The chest trauma score (CTS) was developed in 2012 and incorporates patient age, pulmonary contusion, number of rib fractures, and bilateral rib fractures; the score ranges from 0 to 11. Patient cohorts were dichotomized by scores into two groups to determine outcomes. Overall, patients with a score <8 had lower mortality, <7 had fewer admissions to the ICU and were less likely to require mechanical ventilation, and patients with a score <5 had shorter hospital lengths of stay (Pressley et al., 2012). However, what makes this scoring system so unique is that statistic modeling was employed to weigh variables differently. Specifically, the presence of bilateral rib fractures was assigned two points as opposed to one.

Initial validation studies of 1361 CTS demonstrated poor predictive value in detecting pneumonia and only moderate predictive power in detecting mortality and acute respiratory failure (ROC AUC 0.65, 0.71, and 0.72 respectively). The authors also demonstrated that a CTS >4 had a significantly higher rate of tracheostomy, acute respiratory failure, and pneumonia. However, no ROC AUC was calculated for these variables in their study (Chen et al., 2014). However, when further comparisons were made, the AUC ROS was noted to be between 0.62 and 0.67, and overall were poorly predictive of these outcomes (Chapman et al., 2016). The CTS was overall comparable to the RFS and the Chest Wall OIS in the prediction of adverse outcomes.

RibScore

The RibScore was developed to create a completely radiographic score. The impetus behind this was due to the large volume of patients transferred to our institution; CT information was reliable and often present, while clinical information was subjective and often incomplete. Furthermore, our ultimate goal was to predict the need for surgical intervention for rib fractures. Based on both literature and anecdotal evidence of disease severity, we selected 6 radiographic variables: ≥6 total rib fractures; ≥3 fractures with bicortical displacement, flail segment, first rib fracture, bilateral rib fractures, and at least one fracture in each anatomic location (anterior, lateral, posterior; defined by anterior and posterior axillary lines). Overall, the total score ranges from 0 to 6 with each parameter being assigned one point and subsequently added together (Chapman et al., 2016).

In a patient population of 385 at our institution, each of these parameters alone was associated with pneumonia, respiratory failure, and tracheostomy. When looking at the ROC AUC, the RibScore demonstrated moderate predictive outcome (ROC AUC 0.71, 0.69, and 0.75, respectively) in the general population. However, when evaluating patients with isolated rib fractures, the RibScore improved further in regards to predicting pneumonia, respiratory failure, and tracheostomy (ROC AUC: 0.77, 0.83, and 0.85 respectively) (Chapman et al., 2016). When this score was applied to a larger cohort of 1089 patients, the RibScore was relatively poor at predicting mortality and tracheostomy in all patients, non-geriatric, and geriatric patients alone (ROC AUC 0.63–0.68). However, the RibScore was moderate at predicting pneumonia in the geriatric cohort (ROC AUC 0.73) (Fokin et al., 2018). Additionally, for complications (pneumonia, respiratory failure, tracheostomy) in patients ≥55 years of age as a whole, the RibScore faired a little better with good predictive capabilities. In allcomers with isolated rib fractures, however, the RibScore was poorly predictive of respiratory complications (pneumonia, respiratory failure, and empyema) in patients with or without pulmonary contusions (Seok et al., 2019).

The main advantage of the RibScore continues to be the ease and access to the information necessary to calculate the score. It is also the most detailed in regard to the anatomy and fracture pattern of each patient's chest wall trauma. However, the RibScore lacks physiologic findings and fails to incorporate different patient phenotypes. Schmoekel et al. demonstrated that the addition of physiologic (modified frailty index) and laboratory (PaO_2) variables to the RibScore provided much higher predictive capabilities for respiratory complications (Schmoekel et al., 2019). Additionally, equal weighting was assigned to each of the variables while in reality, the presence of any one parameter may have the same risk of pneumonia and is not additive.

Sequential clinical assessment of respiratory function score

Most recently, Hardin et al. created the sequential clinical assessment of respiratory function (SCARF) score, which relies completely upon clinical parameters. The goal of this scoring system was to create a dynamic clinical score that could identify high-risk patients on admission, as well as ICU patients at risk of decompensation during their hospital stay. The score parameters were selected a priori, but were established to be clinically relevant, commonly obtained, and reasonably objective. The parameters included: incentive spirometry <50% predicted, respiratory rate >20 breaths per minute, pain rating of ≥5 based on the numeric rating scale, and weak cough unable to clear secretions (as deemed by the respiratory therapists). The total score ranged from 0 to 4 with each parameter being binary (0 or 1) and equally weighted (Hardin et al., 2019).

During the initial study, the SCARF score was trended throughout the ICU stay. For those who developed a pulmonary complication (pneumonia, empyema, respiratory failure, or tracheostomy), the score was approximately 1 point higher than those who did not develop a complication. Before the development of a complication, the SCARF score was noted to increase one day before the complication arose. In patients whose SCARF score increased, the likelihood of developing pneumonia and remaining in the ICU for 3 days were both increased. When looking at admission SCARF scores, there was moderate predictability for pneumonia (ROC AUC 0.72), prolonged ICU length of stay (ROC AUC 0.70), and the likelihood of needing a FiO_2 >50% (6 L supplemental oxygen) during admission (ROC AUC 0.74). Overall, a SCARF score of ≥3 had at least a 60% probability of developing a pulmonary complication (Hardin et al., 2019).

The SCARF score is unique in that it relies upon complete clinical data and does not utilize radiographic/anatomic factors. It can be easily calculated and is not overly onerous. This score is also able to be easily re-calculated and tracked during a patient's admission and can give providers clues to their patient's condition. Some of the limitations include the use of subjective variables, the small sample cohort during validation, excluding intubated patients, and only tracking the SCARF score in the ICU. Furthermore, the study did not determine if utilizing the SCARF score as a clinical tool would reduce mortality, and if intervention in patients with increasing scores could reduce morbidity.

Chest wall injury scoring systems and surgical stabilization

Newer studies have begun to look at the utilization of scoring systems in the surgical stabilization of rib fractures (SSRF). In a prospective, controlled clinical evaluation of SSRF, 70 patients were followed during their hospitalization. The study was designed with a crossover paradigm with 35 patients being managed nonoperatively. Patients included had: flail chest, ≥3 bicortically displaced fractures, ≥30% volume loss of the hemithorax on chest CT, or any fracture pattern with failure of optimal medical management. Overall, the RibScore was significantly higher in the patients who underwent SSRF (median 4.0 vs 3.0); the RFS and the CTS did not demonstrate significant differences (Pieracci et al., 2016).

Furthermore, a retrospective review was performed to look at different scoring systems in SSRF patients. Utilizing propensity matching to an SSRF cohort, it was noted that there were differences between their two cohorts (the SSRF group had patients with more fractures, more displaced fractures, more pulmonary contusions, and more flail chests). Patients who underwent SSRF had a higher RibScore (2.3 vs 1.7) and a higher CTS (5.7 vs 5.3); however, no differences were observed between cohorts when utilizing RFS. When a ROC AUC was performed for the assignment to SSRF or nonoperative management, the RibScore was most predictive ($R = 0.64$), although this was poor (Wycech et al., 2020).

It is important to note that no studies are yet completed that prospectively look at using a scoring system to determine who benefits from SSRF. The aim of Pieracci et al. was to determine the benefit of utilizing SSRF on a subgroup of severe chest wall injury patients. The difference in RibScore between the two cohorts of patients, despite having similar fracture patterns, was likely incidental (Pieracci et al., 2016). Current systems that have been evaluated appear to have low predictive probability in determining who would benefit from SSRF. There is also an unclear understanding of what the inflection point would be for any of the scoring systems and SSRF.

Future directions

Current strides have been made to aid in the refinement of chest wall scoring systems. CWIS has introduced a nomenclature system using the Delphi method to standardize the language providers use to describe rib fractures. Just before this nomenclature, the AO/OTA also released a taxonomy to further define the fracture patterns. While these systems and organizations are separate, they are symbiotic with some similar definitions (Edwards et al., 2020; Meinberg, Agel, Roberts, et al., 2018). However, they are not the same and thus continued work needs to be done to further standardize the language across multiple organizations. The AAST and other national/international organizations have not uniformly adopted the nomenclature set forth by CWIS. Ultimately, the goal is for a standardized nomenclature to be widely adopted and incorporated into clinical and research practice.

Once a nomenclature is agreed upon, further development of a universal scoring system is necessary. While multiple scoring systems exist, no current system is perfect for predicting complications. Likely, this ideal scoring system would take aspects from each of the current scores to develop an improved model. Ideally, this would utilize a weighted statistical model to include the strength of association with adverse outcomes.

Both a predominantly anatomic and predominantly phenotypic scoring system will likely be required to adequately describe chest wall injuries. A mostly anatomic score would allow for tracking of the overall epidemiology of chest wall injuries and can be readily determined. Additionally, a system utilizing the phenotype of the patient would likely include demographic and physiologic parameters and the relationship to complications. A phenotypically based score would likely be most helpful in defining which patients would benefit from certain therapies after chest wall injury, including locoregional analgesia, as well as the best admission destination.

Additionally, an important task remains in further determining which patients would benefit most from SSRF. Currently, expert opinion remains the mainstay of recommendations for rib plating with the most widespread recommendation being flail chest physiology (Kasotakis et al., 2017; Pieracci et al., 2017). Small randomized controlled trials have demonstrated benefit in both patients with flail chest and also those with severely displaced rib fractures (Marasco et al., 2013; Pieracci et al., 2020). However, further identifying those patients that would benefit from SSRF would be greatly aided by the development of an improved scoring system that utilizes detailed, non-flail fracture patterns.

Lastly, measures in addition to mortality, pneumonia, acute respiratory failure, and tracheostomy should be incorporated as outcomes. Chronic debility, opioid use, and pulmonary function after discharge should all be considered important markers after injury (Fabricant et al., 2013; Pieracci et al., 2021). These should be measured outcomes as newer chest wall injury scoring systems are developed.

Conclusions

Chest wall injuries are a diverse group of pathologies and phenotypes that require dedicated scoring systems to better categorize patients. Scoring systems allow providers to standardize communication, select therapies, and conduct further research. A culture change is needed in regard to rib fractures; it is not just that "the patient has rib fractures," but instead further detailed descriptions are necessary. Recently, improved taxonomy has come into the literature but has not yet been universally adopted. With improved nomenclature, better chest wall scoring systems will be further developed. Currently, multiple chest wall scoring systems are available and should be used despite their limited detail and validation. This structured system will allow for improved patient care, communication, and outcomes research. Outcomes should be further expanded to include disability, opioid use, and long-term function as mortality after rib fractures continues to decline.

References

Baker, J. E., Millar, D. A., Heh, V., Goodman, M. D., Pritts, T. A., & Janowak, C. F. (2020). Does chest wall Organ Injury Scale (OIS) or Abbreviated Injury Scale (AIS) predict outcomes? An analysis of 16,000 consecutive rib fractures. *Surgery, 168*(1), 198−204. Available from https://doi.org/10.1016/j.surg.2020.04.032.

Bulger, E. M., Arneson, M. A., Mock, C. N., & Jurkovich, G. J. (2000). Rib fractures in the elderly. *The Journal of Trauma: Injury, Infection, and Critical Care, 48*(6), 1040–1047. Available from https://doi.org/10.1097/00005373-200006000-00007.

Butts, C. A., Brady, J. J., Wilhelm, S., Castor, L., Sherwood, A., McCall, A., Patch, J., Jones, P., Cortes, V., & Ong, A. W. (2017). Do simple beside lung function tests predict morbidity after rib fractures? *American Journal of Surgery, 213*(3), 473–477. Available from https://doi.org/10.1016/j.amjsurg.2016.11.026.

Carver, T. W., Milia, D. J., Somberg, C., Brasel, K., & Paul, J. (2015). Vital capacity helps predict pulmonary complications after rib fractures. *Journal of Trauma and Acute Care Surgery, 79*(3), 413–416. Available from https://doi.org/10.1097/TA.0000000000000744.

Challener, D. W., Prokop, L. J., & Abu-Saleh, O. (2019). The proliferation of reports on clinical scoring systems: Issues about uptake and clinical utility. *JAMA – Journal of the American Medical Association, 321*(24), 2405–2406. Available from https://doi.org/10.1001/jama.2019.5284.

Chapman, B. C., Herbert, B., Rodil, M., Salotto, J., Stovall, R. T., Biffl, W., Johnson, J., Burlew, C. C., Barnett, C., Fox, C., Moore, E. E., Jurkovich, G. J., & Pieracci, F. M. (2016). RibScore: A novel radiographic score based on fracture pattern that predicts pneumonia, respiratory failure, and tracheostomy. *Journal of Trauma and Acute Care Surgery, 80*(Issue 1), 95–101, Lippincott Williams and Wilkins. Available from https://doi.org/10.1097/TA.0000000000000867.

Chen, J., Jeremitsky, E., Philp, F., Fry, W., & Smith, R. S. (2014). A chest trauma scoring system to predict outcomes. *Surgery, 156*(4), 988–994. Available from https://doi.org/10.1016/j.surg.2014.06.045.

Chien, C. Y., Chen, Y. H., Han, S. T., Blaney, G. N., Huang, T. S., & Chen, K. F. (2017). The number of displaced rib fractures is more predictive for complications in chest trauma patients. *Scandinavian Journal of Trauma, Resuscitation and Emergency Medicine, 25*(1), 19. Available from https://doi.org/10.1186/s13049-017-0368-y.

Clarke, P. T. M., Simpson, R. B., Dorman, J. R., Hunt, W. J., & Edwards, J. G. (2019). Determining the clinical significance of the Chest Wall Injury Society taxonomy for multiple rib fractures. *Journal of Trauma and Acute Care Surgery, 87*(Issue 6), 1282–1288. Available from https://doi.org/10.1097/TA.0000000000002519, Lippincott Williams and Wilkins.

Daurat, A., Millet, I., Roustan, J. P., Maury, C., Taourel, P., Jaber, S., Capdevila, X., & Charbit, J. (2016). Thoracic Trauma Severity score on admission allows to determine the risk of delayed ARDS in trauma patients with pulmonary contusion. *Injury, 47*(1), 147–153. Available from https://doi.org/10.1016/j.injury.2015.08.031.

Easter, A. (2001). Management of patients with multiple rib fractures. *American Journal of Critical Care, 10*(5), 320–327. Available from https://doi.org/10.4037/ajcc2001.10.5.320.

Edwards, J. G., Clarke, P., Pieracci, F. M., Bemelman, M., Black, E. A., Doben, A., Gasparri, M., Gross, R., Jun, W., Long, W. B., Lottenberg, L., Majercik, S., Marasco, S., Mayberry, J., Sarani, B., Schulz-Drost, S., Van Boerum, D., Whitbeck, S. A., & White, T. (2020). Taxonomy of multiple rib fractures: Results of the chest wall injury society international consensus survey. *Journal of Trauma and Acute Care Surgery, 88*(Issue 2), E40–E45. Available from https://doi.org/10.1097/TA.0000000000002282, Lippincott Williams and Wilkins.

Fabricant, L., Ham, B., Mullins, R., & Mayberry, J. (2013). Prolonged pain and disability are common after rib fractures. *American Journal of Surgery, 205*(5), 511–516. Available from https://doi.org/10.1016/j.amjsurg.2012.12.007.

Flagel, B. T., Luchette, F. A., Reed, R. L., Esposito, T. J., Davis, K. A., Santaniello, J. M., & Gamelli, R. L. (2005). Half-a-dozen ribs: The breakpoint for mortality. *Surgery*, *138*(4), 717–725. Available from https://doi.org/10.1016/j.surg.2005.07.022.

Fokin, A., Wycech, J., Crawford, M., & Puente, I. (2018). Quantification of rib fractures by different scoring systems. *Journal of Surgical Research*, *229*, 1–8. Available from https://doi.org/10.1016/j.jss.2018.03.025.

Harden, A., Kang, Y.-S., & Agnew, A. (2019). Rib fractures: Validation of an interdisciplinary classification system. *Forensic Anthropology*, *2*(3), 158–167. Available from https://doi.org/10.5744/fa.2019.1032.

Hardin, K. S., Leasia, K. N., Haenel, J., Moore, E. E., Burlew, C. C., & Pieracci, F. M. (2019). The Sequential Clinical Assessment of Respiratory Function (SCARF) score: A dynamic pulmonary physiologic score that predicts adverse outcomes in critically ill rib fracture patients. *Journal of Trauma and Acute Care Surgery*, *87*(Issue 6), 1260–1268. Available from https://doi.org/10.1097/TA.0000000000002480, Lippincott Williams and Wilkins.

Holcomb, J. B., McMullin, N. R., Kozar, R. A., Lygas, M. H., & Moore, F. A. (2003). Morbidity from rib fractures increases after age 45. *Journal of the American College of Surgeons*, *196*(4), 549–555. Available from https://doi.org/10.1016/S1072-7515(02)01894-X.

Kasotakis, G., Hasenboehler, E. A., Streib, E. W., Patel, N., Patel, M. B., Alarcon, L., Bosarge, P. L., Love, J., Haut, E. R., & Como, J. J. (2017). Operative fixation of rib fractures after blunt trauma: A practice management guideline from the Eastern Association for the Surgery of Trauma. *Journal of Trauma and Acute Care Surgery*, *82*(Issue 3), 618–626. Available from https://doi.org/10.1097/TA.0000000000001350, Lippincott Williams and Wilkins.

Manay, P., Satoskar, R., Karthik, V., & Prajapati, R. (2017). Studying morbidity and predicting mortality in patients with blunt chest trauma using a novel clinical score. *Journal of Emergencies, Trauma and Shock*, *10*(3), 128–133. Available from https://doi.org/10.4103/JETS.JETS_131_16.

Marasco, S. F., Davies, A. R., Cooper, J., Varma, D., Bennett, V., Nevill, R., Lee, G., Bailey, M., & Fitzgerald, M. (2013). Prospective randomized controlled trial of operative rib fixation in traumatic flail chest. *Journal of the American College of Surgeons*, *216*(5), 924–932. Available from https://doi.org/10.1016/j.jamcollsurg.2012.12.024.

Maxwell, C. A., Mion, L. C., & Dietrich, M. S. (2012). Hospitalized injured older adults: Clinical utility of a rib fracture scoring system. *Journal of Trauma Nursing: The Official Journal of the Society of Trauma Nurses*, *19*(3), 168–176. Available from https://doi.org/10.1097/jtn.0b013e318261d201.

Meinberg, E., Agel, J., & Roberts, C. (2018). Thorax. *Journal of Orthopaedic Trauma*, *32* (Suppl. 1), S161–S166. Available from https://doi.org/10.1097/BOT.0000000000001071.

Meinberg, E., Agel, J., Roberts, C., Karam, M., & Kellam, J. (2018). Fracture and dislocation classification compendium—2018. *Journal of Orthopaedic Trauma*, *32*(1), S1–S10. Available from https://doi.org/10.1097/bot.0000000000001063.

Mommsen, P., Zeckey, C., Andruszkow, H., Weidemann, J., Frömke, C., Puljic, P., Van Griensven, M., Frink, M., Krettek, C., & Hildebrand, F. (2012). Comparison of different thoracic trauma scoring systems in regards to prediction of post-traumatic complications and outcome in blunt chest trauma. *Journal of Surgical Research*, *176*(1), 239–247. Available from https://doi.org/10.1016/j.jss.2011.09.018.

Moon, S. H., Kim, J. W., Byun, J. H., Kim, S. H., Choi, J. Y., Jang, I. S., Lee, C. E., Yang, J. H., Kang, D. H., Kim, K. N., & Park, H. O. (2017). The thorax trauma severity score and the trauma and injury severity score: Do they predict in-hospital mortality in patients with severe thoracic trauma? *Medicine (United States), 96*(42). Available from https://doi.org/10.1097/MD.0000000000008317.

Moore, E. E., Cogbill, T. H., Jurkovich, G. J., Mc Aninch, J. W., Champion, H. R., Gennarelli, T. A., Malangoni, M. A., Shackford, S. R., & Trafton, P. G. (1992). Organ injury scaling III: Chest wall, abdominal vascular, ureter, bladder, and urethra. *Journal of Trauma − Injury, Infection and Critical Care, 33*(3), 337−339. Available from https://doi.org/10.1097/00005373-199209000-00001.

Pape, H. C., Remmers, D., Rice, J., Ebisch, M., Krettek, C., & Tscherne, H. (2000). Appraisal of early evaluation of blunt chest trauma: Development of a standardized scoring system for initial clinical decision making. *Journal of Trauma − Injury, Infection and Critical Care, 49* (3), 496−504. Available from https://doi.org/10.1097/00005373-200009000-00018.

Pieracci, F. M., Coleman, J., Ali-Osman, F., Mangram, A., Majercik, S., White, T. W., Jeremitsky, E., & Doben, A. R. (2018). A multicenter evaluation of the optimal timing of surgical stabilization of rib fractures. *Journal of Trauma and Acute Care Surgery, 84*(Issue 1), 1−10. Available from https://doi.org/10.1097/TA.0000000000001729, Lippincott Williams and Wilkins.

Pieracci, F. M., Leasia, K., Bauman, Z., Eriksson, E. A., Lottenberg, L., Majercik, S., Powell, L., Sarani, B., Semon, G., Thomas, B., Zhao, F., Dyke, C., & Doben, A. R. (2020). A multicenter, prospective, controlled clinical trial of surgical stabilization of rib fractures in patients with severe, nonflail fracture patterns (Chest Wall Injury Society NONFLAIL). *Journal of Trauma and Acute Care Surgery, 88*(2), 249−257. Available from https://doi.org/10.1097/TA.0000000000002559.

Pieracci, F. M., Lin, Y., Rodil, M., Synder, M., Herbert, B., Tran, D. K., Stoval, R. T., Johnson, J. L., Biffl, W. L., Barnett, C. C., Cothren-Burlew, C., Fox, C., Jurkovich, G. J., & Moore, E. E. (2016). A prospective, controlled clinical evaluation of surgical stabilization of severe rib fractures. *Journal of Trauma and Acute Care Surgery, 80*(Issue 2), 187−194. Available from https://doi.org/10.1097/TA.0000000000000925, Lippincott Williams and Wilkins.

Pieracci, F. M., Majercik, S., Ali-Osman, F., Ang, D., Doben, A., Edwards, J. G., French, B., Gasparri, M., Marasco, S., Minshall, C., Sarani, B., Tisol, W., VanBoerum, D. H., & White, T. W. (2017). Consensus statement: Surgical stabilization of rib fractures rib fracture colloquium clinical practice guidelines. *Injury, 48*(2), 307−321. Available from https://doi.org/10.1016/j.injury.2016.11.026.

Pieracci, F. M., Schubl, S., Gasparri, M., Delaplain, P., Kirsch, J., Towe, C., White, T. W., Whitbeck, S. A., & Doben, A. R. (2021). The Chest Wall Injury Society recommendations for reporting studies of surgical stabilization of rib fractures. *Injury, 52*(6), 1241−1250. Available from https://doi.org/10.1016/j.injury.2021.02.032.

Pressley, C. M., Fry, W. R., Philp, A. S., Berry, S. D., & Smith, R. S. (2012). Predicting outcome of patients with chest wall injury. *American Journal of Surgery, 204*(6), 910−914. Available from https://doi.org/10.1016/j.amjsurg.2012.05.015.

Schmoekel, N., Berguson, J., Stassinopoulos, J., Karamanos, E., Patton, J., & Johnson, J. L. (2019). Rib fractures in the elderly: Physiology trumps anatomy. *Trauma Surgery & Acute Care Open, 4*(1)e000257. Available from https://doi.org/10.1136/tsaco-2018-000257.

Seok, J., Cho, H. M., Kim, H. H., Kim, J. H., Huh, U., Kim, H. B., Leem, J. H., & Wang, I. J. (2019). Chest trauma scoring systems for predicting respiratory complications in isolated rib fracture. *Journal of Surgical Research, 244*, 84−90. Available from https://doi.org/10.1016/j.jss.2019.06.009.

Tyburski, J. G., Collinge, J. D., Wilson, R. F., & Eachempati, S. R. (1999). Pulmonary contusions: Quantifying the lesions on chest x-ray films and the factors affecting prognosis.

Journal of Trauma – Injury, Infection and Critical Care, 46(Issue 5), 833–838. Available from https://doi.org/10.1097/00005373-199905000-00011, Lippincott Williams and Wilkins.

Van Wijck, S. F. M., Curran, C., Sauaia, A., Van Lieshout, E. M. M., Whitbeck, S. S., Edwards, J. G., Pieracci, F. M., & Wijffels, M. M. E. (2022). Interobserver agreement for the Chest Wall Injury Society taxonomy of rib fractures using computed tomography images. *Journal of Trauma and Acute Care Surgery, 93*(6), 736–742. Available from https://doi.org/10.1097/TA.0000000000003766.

Wycech, J., Fokin, A. A., & Puente, I. (2020). Evaluation of patients with surgically stabilized rib fractures by different scoring systems. *European Journal of Trauma and Emergency Surgery, 46*(2), 441–445. Available from https://doi.org/10.1007/s00068-018-0999-3.

Youden, W. J. (1950). Index for rating diagnostic tests. *Cancer, 3*(1), 32–35. Available from https://doi.org/10.1002/1097-0142(1950)3:1 < 32::AID-CNCR2820030106 > 3.0.CO;2-3.

Chapter 4

Taxonomy of rib fractures and blunt chest wall injury

John G. Edwards
Department of Cardiothoracic Surgery, Sheffield Teaching Hospitals NHS Foundation Trust, Northern General Hospital, Sheffield, United Kingdom

The management plan for patients with rib fractures, whether non-operative or with surgical stabilization of rib fractures, is critically dependent on the description of the injury. A consistent description is also of prime importance in the understanding of studies in the literature, and the collection of data for service evaluation and audit.

> *A classification is only useful if it considers the severity of the bone lesion and serves as a basis for treatment and for the evaluation of results.*
>
> *— Muller 1990* (Müller et al., 1990).

Taxonomy has long been a universal phenomenon across all fields of science and art. Its application to fractures is far from new. The Edwin Smith Papyrus, dating to 1600 BCE, is believed to be the oldest surviving medical text. It contains a rudimentary fracture classification system, dividing fractures into open and closed and deeming the former "an ailment not to be treated." Fractures were characterized to guide treatment: this remains the central aim of modern fracture classification systems.

The taxonomy of rib fractures has been subject to considerable variability, with authors using inconsistent definitions for flail chest and the characterization of displaced fractures. Without the application of consistent taxonomy, the cross-comparison of clinical studies is challenging. The challenge of describing blunt injuries to the skeletal structures of the chest wall is related to the number of ribs, and their relationship to their costal cartilages, the sternum, and the spine, giving rise to an infinite number of unique injuries. While the taxonomy of an individual rib fracture may be relatively straightforward, an account must be taken of where that fracture is situated, and, critically, the relationships with other chest wall skeletal fractures, including those of the costal cartilages, sternum, and spine.

Several scoring systems for polytrauma incorporate assessment of rib fractures, such as the Abbreviated Injury Scale (Gennarelli & Wodzin, 2005), Organ Injury Scale (Moore et al., 1994), and Injury Severity Score (Osler et al., 1997), but the injury characterization is limited to the number of fractures, bilaterality and the presence of "flail chest." Several specific rib fracture scoring systems have been proposed, which similarly do not address the degree of displacement of individual fractures, nor the co-location of rib fractures. Furthermore, the thoracic skeleton includes costal cartilage and the sternum, and the combination of injuries of these components with individual and collective rib fractures deserves acknowledgment in thoracic skeleton trauma taxonomy.

AO Foundation/Orthopaedic Trauma Association fracture and dislocation classification compendium

The AO Foundation/Orthopaedic Trauma Association (AO/OTA) fracture classification system is the most widely used generic method addressing fractures of the limbs, spine, and thoracic skeleton, having first been published in 1996 (Orthopaedic Trauma Association Committee for Coding & Classification, 1996). This used the principles of the Comprehensive Classification of Fractures of the Long Bones developed by Müller and collaborators (Müller et al., 1987), and the OTA classification committee classified and coded the remaining bones. A revision in 2007 revision (Marsh et al., 2007) standardized the two different alphanumeric codes assigned to each fracture into one agreed scheme, hence developing an internationally recognized uniform system for clinical research on fractures and dislocations. There was a further comprehensive review of the Compendium in 2018, with the inclusion of a preliminary classification of rib and sternal fractures for the first time (Meinberg et al., 2018).

The AO/OTA classification uses an alphanumeric structure to indicate the localization (the bone and location) and fracture morphology (type, group, subgroup, qualifications, and modifiers) to assign a unique code to each fracture. The anatomical location for the Thorax is 16, with the next identifier being the side (right = 1, left = 2), followed by the rib number (1–12). The location on each rib is described next. A fracture between the costovertebral joint and the lateral tip of the vertebral transverse process is within the "posterior end segment" (assigned the numeric 1), between the transverse process and the costochondral junction is the "shaft" (assigned 2), and of the costochondral cartilage the "anterior end segment" (assigned 3). Location qualifications are defined for complete and partial costotransverse disruption, where the posterior end segment is fractured in association with a transverse process fracture (1C), or not (1B) respectively, or an extra-articular fracture (1 A). The next letter characterizes the type of fractures of the "shaft," being simple (2 A), multifragmentary wedge (2B), or multifragmentary segmental

(2 C). Anterior end segments are similarly subclassified as simple (3 A), wedge (3B), or multifragmentary (3 C). Notably, while the anatomical site and location of every rib fracture can be assigned an alphanumeric code, there is no definition offered for the differentiation of undisplaced or displaced fractures, nor to characterize any radiological or functional relationship between fractures on neighboring ribs, nor to specify the location of fractures with any more detail than lying between the spine transverse process and the costochondral junction. For example, a displaced flail chest is not coded by the AO/OTA system. The practicality of the AO/OTA system for rib fractures in clinical and research settings is unknown, as few publications are demonstrating its value.

Chest Wall Injury Society Taxonomy

The nascent Chest Wall Injury Society (CWIS) initiated a Delphi Consensus process in 2017, to establish an opinion-based taxonomy for rib fractures (Edwards et al., 2020). An international review committee of surgeons experienced in the management of rib fractures was established to coordinate the exercise. A literature search was performed of published manuscripts, trial protocols, research theses, conference proceedings, and personal correspondence to identify all possible rib fracture definitions in terms of location, displacement, and patterns of injury. An international group of 113 surgeons from 18 countries with expertise in rib fracture management were included in the exercise. First and lead authors of all rib fractures surgery papers from 2006 to 2017, faculty members of educational courses in the United States of America and Europe, clinical trial chief investigators, and the mailing list of the CWIS membership were included in the process. Three rounds of questioning were performed, each round designed to capture opinions about common rib fracture definitions for which a lack of consensus was evident. Definitions for displacement, fracture line characterization, fragmentation, associations between neighboring ribs, anatomical sectors, flail chest, and anterior (sternal) flail chest were included. The results from each round were analyzed by the international review committee and assessed to determine whether consensus was reached, determined by a level of 80% agreement of each individual definition. Where agreement was not reached, questioning was revised and repeated, according to standard Delphi methods.

A summary of the results is shown in Table 4.1. The consensus was reached for the use of three categories of displacement, namely "Undisplaced," "Offset," and "Displaced." The use of "Simple," "Wedge," and "Complex" as the descriptions for individual fracture characterization was agreed upon, terms which are in keeping with the AO/OTA taxonomy. Following numerous potential options, in round 1, for the description of functionally and anatomically associated fractures on neighboring ribs, the term "series" of fractures reached the threshold for consensus.

TABLE 4.1 Rib Fracture Taxonomy proposed by the Chest Wall Injury Society (Edwards et al., 2020).

Category	Definitions	Description	Level of agreement
Displacement	Undisplaced	Greater than 90% cortical contact	87%
	Offset	90% to 0% cortical contact	
	Displaced	No cortical contact	
Character of fracture	Simple	One fracture line traversing the rib, resulting	87%
	Wedge	Two fracture lines traversing the rib, resulting in one wedge-shaped fragment	
	Complex	Multiple fracture lines	
Association between fractures	Series	Functionally and anatomically associated fractures on neighboring ribs	88%
Location (sectors)	Three sectors	There should be three sectors between the costochondral joint and the transverse process of the spine: anterior, lateral, and posterior	82%
	Costochondral sector	There should be a costochondral sector, between the costochondral joint and the sternocostal joint	91%
	Paravertebral sector	There should be a paravertebral sector, between the costotransverse joint and the costovertebral joint	77%
	Boundaries between sectors	The boundaries between the anterior and lateral, and the lateral and posterior sectors should be the anterior and posterior axillary lines, respectively	72%
Flail	Flail segment	A flail segment describes the radiographic appearance of three or more sequential ribs, each fractured in two or more separate places	88%

(Continued)

TABLE 4.1 (Continued)

Category	Definitions	Description	Level of agreement
	Flail chest	A flail chest describes the clinical scenario of paradoxical motion on clinical examination, most often (but not exclusively) in association with a flail segment	
	Displaced flail chest	There was no consensus on the number of displaced rib fractures required to define a flail chest	48%
	Inclusion of paravertebral sector factors in a flail segment	There was no consensus as to whether fractures in the paravertebral sector should be included in a flail segment	60%
Anterior flail	Anterior flail segment	There should be a minimum of 3 rib or costal cartilage fractures on both sides to define an anterior flail segment	94%
	Anterior flail segment with a sternal fracture	When associated with a transverse sternal fracture, how many ribs or costal cartilage fractures should be to give an anterior flail segment	51%

With regards to the location of rib fractures around the chest wall, the concept of three sectors (anterior, lateral, and posterior) gained consensus, together with the additional inclusion of a separate costal cartilage sector. The adoption of a paravertebral sector (equivalent to the AO/OTA "posterior end segment") achieved a level of 77% agreement, which did not quite reach the threshold for consensus. Nevertheless, this sector has gained traction in conference and clinical discussions since the publication of the manuscript. The determination of the boundaries between the anterior and lateral, and the lateral and posterior sectors, generated significant discussion, however. Although the use of anterior and posterior axillary lines was popular, there is no agreement on the definition of these lines. It is not easy to identify axillary lines on Computed Tomography (CT) scans, and not all definitions would give easily reproducible sectors. Axillary lines can be defined by vertical lines, according to the muscle borders, giving equal-sized sectors between the costotransverse joint and costochondral junction as a "clock face" angle from the mid-thoracic point (Simpson et al., 2020). There was no consensus achieved in the second Delphi round, and, when further explicit descriptions of the advantages and disadvantages of each were offered in the third round, there was a roughly even distribution of opinion between the four options.

The definition of a flail segment was proposed as the radiographic appearance of three sequential ribs, where there were two distinct fractures on each. It was agreed that a "flail segment" should describe the radiographic appearance, whereas a "flail chest" is the clinical scenario where paradoxical motion is present. Definitions concerning the number of rib fractures required for a flail chest did not gain consensus, nor did consideration of the inclusion of paravertebral sector fractures in a flail segment. While it was felt that there should be a minimum of three rib or costal cartilage fractures on both sides of the sternum for the definition of an anterior flail segment, there was no consensus when there was a sternal fracture present.

There have been several studies that have challenged the CWIS Taxonomy, in terms of clinical applicability, interobserver agreement, refining definitions, and setting up large datasets to allow future evolution of the taxonomy. Recognizing that the CWIS Taxonomy was derived from expert opinion, rather than a data-driven exercise, there have been efforts to validate the categories derived. Clarke et al assessed the outcomes of the different CWIS categories in a study using 539 cases, demonstrating clinical significance for several of the proposals (Clarke et al., 2019). Van Wijck et al. investigated the interobserver agreement of the CWIS taxonomy, with 76 respondents categorizing CT scan images of 11 different fracture types (Van Wijck et al., 2022). While there was strong agreement regarding fracture location, it was assessed as moderate for the fracture type and degree of displacement.

CWIS/ASER working group

In 2023, a collaboration between CWIS and the American Society of Emergency Radiology (ASER) undertook an initiative to review the CWIS Taxonomy, through a series of meetings, to consider each CWIS consensus statement (Nguyen et al., 2023). It was recognized that there were some ambiguities and incomplete concepts which required further development. A group of radiologists and surgeons from these organizations held a series of meetings to develop and propose more precise definitions (Table 4.2). These were focused on the degree of displacement, cartilage fractures, anatomic sector boundaries, and the definitions of flail segment/chest. The CWIS/ASER working group considered that the descriptor for the degree of displacement should not be "cortical overlap" but "cross-sectional overlap." Greater precision was felt to be offered using the term "costal cartilage" rather than "costochondral cartilage," and clear definitions of the junctions

TABLE 4.2 Improved CWIS Rib Fracture Taxonomy proposed by the Chest Wall Injury Society and American Society of Emergency Radiology collaboration (Nguyen et al., 2023).

Category	CWIS/ASER improved taxonomy and recommendations
Degree of displacement	The descriptor should be "cross-sectional overlap," rather than "cortical contact"
Cartilage fractures	"Costal cartilage" refers to the cartilage itself The term "costochondral" should not be used in isolation "Costochondral junction" refers to the transition between the rib and the cartilage "Chondrosternal junction" refers to the transition between the costal cartilage and sternum Costal cartilage fractures should only refer to the "true" ribs, where ribs do not share costal cartilages, not the "false ribs" (typically 8th–10th)
Sector boundaries	Challenges were recognized in determining a consensus for the boundaries between sectors. A recommendation was made to identify landmarks for each rib in robust prospective studies, with those landmarks easy to identify and clinically relevant
Flail segment/chest	The distinction between a flail segment and a flail chest was acknowledged, with a recommendation for disseminated education
	Costal cartilage fractures or junction disruptions should contribute to a flail segment
	A flail segment is defined as "ipsilateral segmental fracture of the ribs, cartilage, costochondral junction, or chondrosternal junctions at three or more contiguous levels"
Anterior flail segment	Further multidisciplinary collaborative clarification is required for the definition of an anterior flail segment

ASER, American Society of Emergency Radiology; *CWIS*, Chest Wall Injury Society.

between costal cartilages and ribs/sternum were proposed. Although not expressly stated, there is the inference of a separate definition for the costal margin, the arch of cartilage and ligaments composed primarily from the false ribs. There remains a need for clarity in the definitions of costal cartilage anatomy and injury taxonomy. The uncertainties of sector boundaries were discussed, with no resolution achieved. Challenges remain, as outlined above. The group recommended identifying, in robust prospective studies, landmarks relevant to each rib that are easy to identify and clinically relevant. With regards to a flail segment/chest, the definition was tightened by including not just the "ribs" but the costal cartilages. A definition was proposed: "ipsilateral segmental fracture of the ribs, cartilage, costochondral junction, or chondrosternal junctions at three or more contiguous levels."

Much of the discussion regarding an anterior flail segment/chest centered on the apparent lack of familiarity of the radiologists of the clinical entity. Certainly, clinical teams are familiar with the paradoxical motion that can occur following crush injury (e.g., the classical steering wheel injury, or after cardiopulmonary resuscitation), but it was felt that further multidisciplinary work was required to refine the definition of an anterior flail segment. Despite the work of CWIS and the CWIS/ASER working group, there remain sub-categories of displacement that have thus far not been addressed. For example, within the undisplaced category, fractures could potentially be categorized as unicortical or bicortical, as these patterns are seen on CT scans. Furthermore, "buckle" type fractures can occur, with angulation between the rib fragments, despite maintained cross-sectional overlap. The crux of any category of displacement is to be able to determine the clinical feature of instability, with the resulting clinical implications, and yet instability (manifest by progressive deformity with time) may be seen with fractures that are initially categorized as undisplaced or even "incomplete." The clinical relevance of these — and indeed any — categories of displacement requires validation in clinical studies.

Chest Injuries International Database

A significant initiative from the Chest Wall Injury Society has been the genesis of the Chest Injuries International Database (CIID) (Chest Wall Injury Society, 2024). Collaborators enter patient-level data pertaining to individual rib fractures according to the CWIS Taxonomy, with the dataset including outcome measures (for both operated and non-operative cases). Currently, there are over 3000 cases entered from 14 different centers. Using a similar methodology to that employed by the International Association for the Study of Lung Cancer for the rolling revisions of Tumour Nodes Metastasis (TNM) staging of lung cancer, for example, (Rami-Porta et al., 2014) CWIS will be able to propose outcome-based revisions to the CWIS Taxonomy based on the analyzed data from the CIID.

Other rib fracture classifications

A challenge of the AO/OTA and CWIS taxonomy is that, while they may provide an accurate characterization of every rib fracture, neither system groups multiple fractures into different patterns of injury (aside from the assignment of "a flail segment" by CWIS taxonomy). It may be found useful to plot the character and location of individual fractures onto a Rib Fracture Scoring Form (Fig. 4.1). This is a diagrammatic table, which allows the reader to begin to build up a mental picture of the pattern of injury, improving the planning of clinical management.

Mark the Location of each fracture using:

Individual fracture description:
  "S" – Simple fracture with a single fracture line.
  "W" – Wedge fracture with 2 fracture lines creating a butterfly fragment.
  "C" – Complex fracture with ≥2 fracture lines where ≥1 lines span the whole rib.

Categorisation of fracture displacement:
  "U" - Simple/ minimal/ Undisplaced – cross-sectional overlap >90%
  "O" - Offset – cross-sectional overlap 0-90%
  "D" - Displaced – no cross-sectional overlap: overriding, overlapped, distracted.

	Para-vertebral Medial to CTJ	Post Post to PAxL	Lat AAxL to PAxL	Ant Ant. To AAxL	Costal Cartilage	Sternum	Costal Cartilage	Ant Ant. To AAxL	Lat AAxL to PAxL	Post Post to PAxL	Para-vertebral Medial to CTJ	
1												1
2												2
3												3
4												4
5												5
6												6
7												7
8												8
9												9
10												10
11												11
12												12

FIGURE 4.1 Extract of a rib fracture record form. Completion of this form using the Chest Wall Injury Society Taxonomy allows a visual representation of the anatomy and severity of the injury, aiding communication and management planning.

A dozen or so thoracic injury scoring classifications have been proposed, which can be used to characterize the injury further. As discussed in Chapter 10, these include an assessment of the anatomical injury, all assessing the number of fractures, and variably anatomical features such as the presence of flail chest, sternal fractures, thoracic spine fractures, bilaterality, displacement, and the presence of pulmonary contusions, hemothorax or pneumothorax. Additionally, several of the systems incorporate non-anatomical, biochemical, and physiological parameters, such as age, comorbidities, hypoalbuminemia, hypoxia, and the presence of a tube thoracostomy tube. The undertaking of external validation of these classifications within large datasets has been variable.

Some of these classifications, such as the Abbreviated Injury Scale and the Injury Severity Score, are in widespread use, for example in the assessment and clinical coding of injury severity in polytrauma patients. However, arguably, these do not drive the clinical management of the individual patient. Of the other scores, generated by clinicians specifically for the assessment of chest wall injury arguably RibScore (Chapman et al., 2016) has generated the most attention, having been validated in other studies (Wycech et al., 2018) and being incorporated into clinical management protocols. While rib fracture scores are not strictly taxonomy, they do have potential value in the characterization of injury, the assessment of risk, and the determination of management.

SMuRFS

Despite the ability to derive a detailed classification of individual fractures, the only descriptors to address the combination of multiple fractures in the CWIS/ASER Taxonomy are the assignment of a series of fractures or a flail segment. While the various rib fracture scoring systems may assign a degree of severity to a pattern of injury, they do not allow the assessor to comprehend that pattern of injury.

The Sheffield Multiple Rib Fractures Study (SMuRFS) classification was derived to allow a rapid visualization of the pattern of injury (Simpson et al., 2020). The authors described a series of fractures, the term which was adopted by the CWIS taxonomy, but added a measure of the extent by defining a series as "three or more neighboring ribs each fractured at a similar location along their length (each fracture within 20° of each other, measured from the mid-thoracic point)." Furthermore, between three and five such fractures comprised a "short" series, whereas six or more formed a "long" series. Similarly, the degree of overlap between series was assessed, allowing the discrimination between a "short flail segment," where the longest series was three to five rib fractures, versus a "long flail segment," with the longest series being six or more rib fractures. Thus, with the inclusion of "no series," five different categories, A through E, were proposed, which were then subdivided according to whether there were two or more displaced rib fractures present. A pilot agreement study was performed, to assess interobserver variation. Ensuring low variability between observers is an essential step in the proposal of any classification system, and the consistency seen with the SMuRFS Classification, which was within the limits of widely adopted classifications for other fracture types, demonstrates its potential utility as a communication tool: an important aspect of any taxonomy system. Ultimately, this classification adds further detail to a statement describing, for example, a "four segment flail."

Costal cartilage

Taxonomy of rib fractures would be incomplete without addressing injuries to the costal cartilage. While the AO/OTA Fracture Compendium and the CWIS included injuries to the costal cartilages as an anatomic sector, neither takes into account the anatomy of the costal cartilages themselves, not important associated injury patterns (Meinberg et al., 2018).

The "true ribs" are typically defined as those where each rib joins, through its own specific continuous costal cartilage, to the sternum, using costochondral and sternocostal joints. The former are primary cartilaginous joints. The first sternocostal joint is also a primary cartilaginous joint, whereas the other sternocostal joints are synovial. The first seven ribs are typically "true ribs".

"False ribs" are typically those from the 8th to the 10th, whereby the 8th rib may have a synovial joint or joints receiving the tip of the 9th cartilage: an "end to side" interchondral joint. Similarly, the tip of the 10th may articulate with the side of the 9th costal cartilage. The anatomy of these costal cartilages, which form the "costal margin," is highly variable (Briscoe, 1925; Laswi et al., 2022; Patel et al., 2023). This variation has long been recognized, with the concept of a "costal arch" consisting of continuous cartilage running from the 10th rib to the sternum perhaps misguided and only occuring in about 5% of patients. The 11th and 12th ribs are classically described as "Floating Ribs," although frequently the 10th rib may also be "floating," having no synovial or ligamentous articulation with the 9th costal cartilage. A second type of interchondral joint is the "side to side" interchondral joint which may also produce synovial articulations between neighboring costal cartilage of "true" costal cartilages, such as commonly seen between the 5th and 4th costal cartilages.

The anatomical taxonomy of the interchondral joints deserves further clarification and consensus, given the functional and injury-related differences between "side to side" interchondral and end-to-side interchondral joints.

An understanding of the costal cartilage and costal margin anatomy in the individual patient is critical in the characterization of conditions such as slipping rib syndrome and injuries to the costal margin.

Injuries to the costal margin

Injuries to the costal margin can be complex and are not adequately classified by the current AO/OTA Fracture Compendium or the CWIS Taxonomy. It is important to note that the costal margin has anatomical relationships with the intercostal spaces, the diaphragm, and the abdominal wall muscles. Hence injuries to the costal margin can involve the muscles of the diaphragm, the intercostal, and the abdominal wall. We applied "sequential segmental analysis" of the costal margin and the associated muscles in the proposal of the Sheffield Classification of injuries associated with the costal margin (Gooseman et al., 2019) (Fig. 4.2). Seven different categories of injury are described, according to the presence or absence of costal margin rupture, intercostal herniation, and diaphragm rupture. Considering the variability of terminology cited in the large body of case reports, the use of the categorical description afforded by the Sheffield Classification is to be

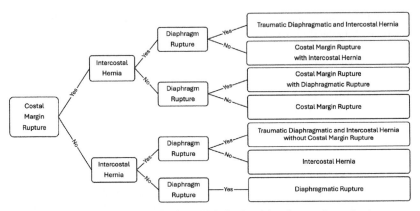

FIGURE 4.2 The Sheffield Classification of injuries involving the costal margin. Assessment of the state of costal margin rupture, intercostal herniation, and diaphragmatic rupture leads to seven distinct categories. In over eighty cases categorized in this way, the entity of "Traumatic Diaphragmatic and Intercostal Hernia without Costal Margin Rupture" has not yet been encountered at the author's institution, confirming the pivotal role of Costal Margin Rupture in these injuries. This classification facilitates accurate analysis of imaging to assign a categorical diagnosis and hence guide logical clinical decision-making.

encouraged, as it relies upon and recognizes precise assessment and consistent description of the anatomy.

In a prospective study, we have found that the Sheffield Classification has value in the description of injury and the determination of management options (Byers et al., 2023; Wijerathne et al., 2024). Of specific note is that the term "costal margin rupture" is preferred: we have noted several cases, typically with cough, sneeze, or retch associated "expulsive" injuries, where there is no "fracture" of costal cartilage, but rupture of the ligaments between a tip of costal cartilage and its cephalad neighbor, resulting in intercostal herniation.

Injuries to the sternum

The AO/OTA Fracture Compendium described useful fracture patterns applied to the sternum, with individual codes for the manubrium, body, and xiphoid process. Arguably, the CWIS Taxonomy allows a consensus for a description of the degree of displacement of sternal as well as rib fractures. However, manubriosternal joint dislocations may not fall within these descriptive boundaries. Thirupathi and Husted described two types of manubriosternal dislocations, where the sternal body may be dislocated posteriorly (type I) or anteriorly (type II) to the manubrium: this classification is frequently quoted (Thirupathi & Husted, 1982). A third type of manubriosternal joint dislocation may be seen, where there is cephalocaudal separation of the manubrium from the body of the sternum, but no anterior or posterior dislocation.

Summary

The adoption of taxonomy for fractures to the ribs, sternum, and costal cartilages is important, for the following reasons:

1. to enable clinical team members to communicate accurately
2. to ensure optimal clinical management
3. to collect meaningful data for service evaluations and clinical audit
4. to select appropriate cases for clinical trials and allow the correct interpretation of trial outcome data

No single taxonomy system yet fulfills all these objectives, while many have merits. The AO/OTA Fracture Compendium collects detailed data about different fractures. The CWIS Taxonomy (with CWIS/ASER revision) provides for a more detailed record of individual fractures. The SMuRFS classification helps summarize the extent of injury. Certain rib fracture scoring systems assist in the risk prediction of outcomes. The Sheffield Classification defines injuries related to the costal margin injuries. However, large-scale studies, such as the CWIS Chest Injuries International Database, with periodic iterative data-based revision, are required to refine the better aspects of the various options and to determine the most appropriate taxonomy for widespread utilization.

References

Briscoe, C. (1925). The interchondral joints of the human thorax. *Journal of Anatomy, 59,* 432−437, 1925.

Byers, J. L., Rao, J. N., Socci, L., Hopkinson, D. N., Tenconi, S., & Edwards, J. G. (2023). Costal margin injuries and trans-diaphragmatic intercostal hernia: Presentation, management and outcomes according to the Sheffield classification. *Journal of Trauma and Acute Care Surgery, 95*(6), 839−845. Available from https://doi.org/10.1097/TA.0000000000004068, http://journals.lww.com/jtrauma.

Chapman, B. C., Herbert, B., Rodil, M., Salotto, J., Stovall, R. T., Biffl, W., Johnson, J., Burlew, C. C., Barnett, C., Fox, C., Moore, E. E., Jurkovich, G. J., & Pieracci, F. M. (2016). RibScore: A novel radiographic score based on fracture pattern that predicts pneumonia, respiratory failure, and tracheostomy. *Journal of Trauma and Acute Care Surgery, 80*(1), 95−101. Available from https://doi.org/10.1097/TA.0000000000000867, http://journals.lww.com/jtrauma.

Chest Wall Injury Society. (2024). Chest Injuries International Database. Available from https://cwisociety/ciid. (2024) Accessed 15.11.24.

Clarke, P. T. M., Simpson, R. B., Dorman, J. R., Hunt, W. J., & Edwards, J. G. (2019). Determining the clinical significance of the Chest Wall Injury Society taxonomy for multiple rib fractures. *Journal of Trauma and Acute Care Surgery, 87*(6), 1282−1288. Available from https://doi.org/10.1097/TA.0000000000002519, http://journals.lww.com/jtrauma.

Edwards, J. G., Clarke, P., Pieracci, F. M., Bemelman, M., Black, E. A., Doben, A., Gasparri, M., Gross, R., Jun, W., Long, W. B., Lottenberg, L., Majercik, S., Marasco, S., Mayberry, J., Sarani, B., Schulz-Drost, S., Van Boerum, D., Whitbeck, S. A., & White, T. (2020). Taxonomy of multiple rib fractures: Results of the chest wall injury society international consensus survey. *Journal of Trauma and Acute Care Surgery*, 88(2), E40−E45. Available from https://doi.org/10.1097/TA.0000000000002282, http://journals.lww.com/jtrauma.

Gennarelli, T. A., & Wodzin, E. (2005). *AIS - Abbreviated Injury Scale*. Barrington IL: Association for the Advancement of Automotive Medicine (AAAM).

Gooseman, M. R., Rawashdeh, M., Mattam, K., Rao, J. N., Vaughan, P. R., & Edwards, J. G. (2019). Unifying classification for transdiaphragmatic intercostal hernia and other costal margin injuries. *European Journal of Cardio-thoracic Surgery*, 56(1), 150−158. Available from https://doi.org/10.1093/ejcts/ezz020, http://ejcts.oxfordjournals.org/.

Laswi, M., Lesperance, R., Kaye, A., Bauman, Z., Hansen, A., Achay, J., Kubalak, S., & Eriksson, E. (2022). Redefining the costal margin: A pilot study. *Journal of Trauma and Acute Care Surgery*, 93(6), 762−766. Available from https://doi.org/10.1097/TA.0000000000003792, http://journals.lww.com/jtrauma.

Marsh, J. L., Slongo, T. F., Agel, J., Broderick, J. S., Creevey, W., DeCoster, T. A., Prokuski, L., Sirkin, M. S., Ziran, B., Henley, B., & Audigé, L. (2007). Fracture and dislocation classification compendium − 2007: Orthopaedic Trauma Association Classification, Database and Outcomes Committee. *Journal of Orthopaedic Trauma*, 21(10), S1. Available from https://doi.org/10.1097/00005131-200711101-00001, http://journals.lww.com/jorthotrauma.

Meinberg, E. G., Agel, J., Roberts, C. S., Karam, M. D., & Kellam, J. F. (2018). Fracture and dislocation classification compendium-2018. *Journal of Orthopaedic Trauma*, 32, S1. Available from https://doi.org/10.1097/BOT.0000000000001063.

Moore, E. E., Malangoni, M. A., Cogbill, T. H., Shackford, S. R., Champion, H. R., Jurkovich, G. J., McAninch, J. W., & Trafton, P. G. (1994). Organ injury scaling IV: Thoracic vascular, lung, cardiac, and diaphragm. *Journal of Trauma − Injury, Infection and Critical Care*, 36(3), 299−300. Available from https://doi.org/10.1097/00005373-199403000-00002.

Müller, M. E., Nazarian, S., & Koch, P. (1987). *Classification AO des fractures*. Springer-Verlag, Tome I. Les os longs 1987.

Müller., Nazarian, S., & Koch, P. (1990). *The comprehensive classification of fractures of long bones*. Springer-Verlag, 1990.

Nguyen, J., Archer-Arroyo, K., Gross, J. A., Steenburg, S. D., Sliker, C. W., Meyer, C. H., Nummela, M. T., Pieracci, F. M., & Kaye, A. J. (2023). Improved chest wall trauma taxonomy: An interdisciplinary CWIS and ASER collaboration. *Emergency Radiology*, 30(5), 637−645. Available from https://doi.org/10.1007/s10140-023-02171-4, https://www.springer.com/journal/10140.

Orthopaedic Trauma Association Committee for Coding and Classification. (1996). *Journal of Orthopaedic Trauma*, 10, 1−154, 1996.

Osler, T., Baker, S. P., & Long, W. (1997). A modification of the injury severity score that both improves accuracy and simplifies scoring. *Journal of Trauma − Injury, Infection and Critical Care*, 43(6), 922−926. Available from https://doi.org/10.1097/00005373-199712000-00009, http://www.jtrauma.com.

Patel, A., Privette, A., Bauman, Z., Hansen, A., Kubalak, S., & Eriksson, E. (2023). Anatomy of the anterior ribs and the composition of the costal margin: A cadaver study. *Journal of Trauma and Acute Care Surgery*, 95(6), 875−879. Available from https://doi.org/10.1097/TA.0000000000004115, http://journals.lww.com/jtrauma.

Rami-Porta, R., Bolejack, V., Giroux, D. J., Chansky, K., Crowley, J., Asamura, H., & Goldstraw, P. (2014). The IASLC lung cancer staging project: The new database to inform the eighth edition of the TNM classification of lung cancer. *Journal of Thoracic Oncology*, 9(11), 1618−1624. Available from https://doi.org/10.1097/JTO.0000000000000334, https://www.journals.elsevier.com/journal-of-thoracic-oncology/.

Simpson, R. B., Dorman, J. R., Hunt, W. J., & Edwards, J. G. (2020). Multiple rib fractures: A novel and prognostic CT-based classification system. *Trauma*, 22(4), 265−272. Available from https://doi.org/10.1177/1460408619895683.

Thirupathi, R., & Husted, C. (1982). Traumatic disruption of the manubriosternal joint. A case report. *Bulletin of the Hospital for Joint Diseases Orthopaedic Institute*, 42(2), 242−247.

Van Wijck, S. F. M., Curran, C., Sauaia, A., Van Lieshout, E. M. M., Whitbeck, S. S., Edwards, J. G., Pieracci, F. M., & Wijffels, M. M. E. (2022). Interobserver agreement for the Chest Wall Injury Society taxonomy of rib fractures using computed tomography images. *Journal of Trauma and Acute Care Surgery*, 93(6), 736−742. Available from https://doi.org/10.1097/TA.0000000000003766, http://journals.lww.com/jtrauma.

Wijerathne., Rao, J. N., Wijffels, M. M. E., Tamburrini, A., Tenconi., & Edwards, J. G. (2024). Surgical management of costal margin rupture associated with intercostal hernia: Evolution of techniques. *Journal of Trauma and Acute Care Surgery*. Available from https://doi.org/10.1097/TA.0000000000004440.

Wycech, J., Fokin., & Puente, I. (2018). Evaluation of patients with surgically stabilized rib fractures by different scoring systems. *European Journal of Trauma and Emergency Surgery*, 46(2). Available from https://doi.org/10.1007/s00068-018-0999-3, Epub2018.

Chapter 5

Rib fractures management algorithm

Didier Lardinois
Department of Thoracic Surgery, University Hospital Basel, Basel, Switzerland

Introduction

Rib fractures are one of the most common injuries observed in approximately 10%–20% of all trauma patients and about 40%–50% of those sustaining blunt chest trauma (Lardinois, 2018). The treatment of rib fractures associated with blunt chest wall trauma is still commonly conservative. Usual therapy consists of pain treatment (peridural catheter, patient-controlled analgesia, oral therapy), mobilization (physiotherapy), and eventually bronchoscopy (bronchial toilette) and antibiotics. However, conservative management can be associated with complications in the acute phase following the trauma, due to its duration or inefficacity (Bhatnagar et al., 2012; Lardinois et al., 2001). Most usual complications consisted of infections and prolonged respiratory failure, leading to extended hospitalization and a high mortality rate. It should be mentioned that the mortality rate is not only associated with the thorax trauma itself but is also strongly dependent on other injuries, age, and comorbidities. Several reports have shown that conservatively treated rib fractures can often be accompanied by impairment of daily activity and of the quality of life, due to persistent pain or chest wall deformity, fracture nonunion, and chest tightness (Bhatnagar et al., 2012; Landercasper et al., 1984; Lardinois et al., 2001). Bhatnagar observed a loss of an average of 70 days of work during recovery after conservative therapy of the flail chest (Bhatnagar et al., 2012), Landerscaper described a long-term disability in up to 60% of the patients (Landercasper et al., 1984).

In recent years, an increasing interest in stabilization of the chest wall in flail chest patients or rib fixation has considerably raised. This was underlined by the development of new surgical fixation devices and the creation of several courses on chest wall surgery to make the indications and techniques more familiar to the surgical community. This is a very important point since rib fractures are still often banalized and their long-term consequences are underestimated. Indeed, many surgeons do not pay much attention to rib fractures and are unfamiliar with the indications and fixation techniques (Lardinois, 2018). The level of evidence of the published data is indeed relatively low, due to the heterogeneity of the indications for surgery, of the techniques used, and the relatively small number of patients included in most of the trials. An often-mentioned argument to favorize conservative management of rib fractures is the absence of solid data from prospective, randomized trials showing clear advantages for the patients after surgical fixation. The question is: do we need such prospective randomized trials? Although prospective, randomized trials should always be attempted because of their strong scientific value, such trials for rib fractures would require large multicentric studies including a very high number of patients (results difficult to interpret due to the high heterogeneity), and would be associated with important costs. Moreover, for a surgeon who is experienced with the procedure and who is convinced of the potential positive results of the operation in selected patients with rib flail chest, the process of randomization would probably be not possible in many cases for ethical reasons. Additionally, there is more and more data clearly showing an advantage of rib fixation in comparison with conservative therapy in the long-term results, functional outcome, pain control, quality of life, and survival.

The selection of the patients who should undergo rib fixation is an extremely important step. This selection requires well-established cooperation between the emergency station, traumatology, and thoracic surgery units. It is a considerable advantage if thoracic surgeons are early involved, due to the relatively high incidence of associated intrathoracic injuries but also to reach a rapid consensus immediately at admission at the emergency unit or latest at the intensive care unit. In most of the cases, the situation will be discussed 1 or 2 days after admission again. An interdisciplinary plan for the management of the patients should be given, according to the associated injuries.

It is usual to differentiate between patients with or without flail chest and between patients with a series of fractures or isolated fractures in the management of rib fractures Fig. 5.1.

Management in flail chest patients

Flail chest is a severe injury of which the mortality rate can reach 30%—40% depending on the number of fractured ribs and the age of the patient. Flail chest belongs to the lethal six defined by the American College of Surgeons Committee on Trauma (Van Olden et al., 2004). Small prospective, randomized trials, several systematic reviews and meta-analyses, and other studies have shown that early stabilization can improve the incidence of pneumonia, the need for tracheostomy, duration of intubation, length of intensive care unit (ICU) stay, duration of hospital stay (Coughlin et al., 2016; Granetzny et al., 2005; Kasotakis et al., 2017; Leinicke et al., 2013; Liu et al., 2019;

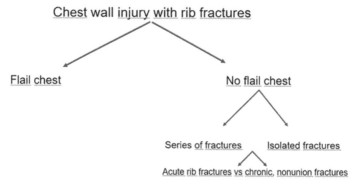

FIGURE 5.1 Types of rib fractures following blunt chest trauma.

Majercik et al., 2014; Marasco et al., 2013; Pieracci et al., 2016; Slobogean et al., 2013; Tanaka et al., 2002). Some prospective monocentric studies have described a positive long-term outcome after rib fixation in patients with flail chest with a return-to-work rate at 2 months of 100% with a 95%–100% working capacity, no restriction in lung function test at 6 months (Mouton et al., 1997; Qiu et al., 2016). Majercik observed no residual pain after fixation already 5.4 weeks postoperatively with a mean satisfaction score of 9.2 on a visual scale (Majercik et al., 2014). Two retrospective studies including 1022 and 5293 patients with severe chest blunt injury observed that the mortality rate was significantly lower in the fixation group after matching. On multivariate analysis, rib fixation was associated with improved mortality (Craxford et al., 2022; Gerakopoulos et al., 2019; Owattanapanich et al., 2022). In one of these studies, early fixation ($\leq$72 h) was associated with a significant reduction of prolonged ventilation (7 days) (Owattanapanich et al., 2022). The role of timing in chest wall stabilization in patients with flail chest was also investigated in a study involving 20,457 patients from the Trauma Quality Improvement Program. Patients undergoing operation had a reduced mortality (2.9% vs 11.7%) in comparison with the patients undergoing conservative therapy (Patel et al., 2024). Multivariate logistic regression identified operation as the only risk factor associated with decreased mortality. The cut-off was 4 days. When looking at management costs, operative therapy seems to be less expensive (between 10,000 USD and 15,000 USD) than conservative therapy (Marasco et al., 2013; Tanaka et al., 2002).

Summary

Early chest wall stabilization should be considered in all patients with flail chests whenever possible.

Practically, different situations can be pointed out (Fig. 5.2).

74 Chest Blunt Trauma

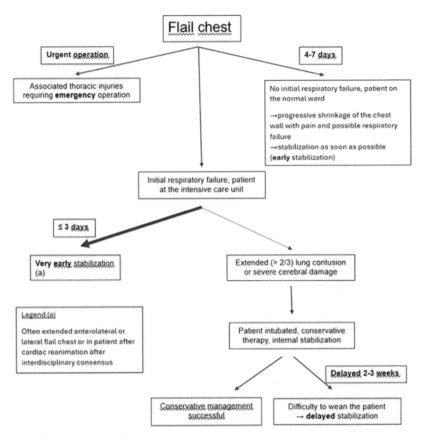

FIGURE 5.2 Algorithm for the management of patients with flail chest.

1. **Urgent operation**: Patients with flail chest presenting with associated thoracic injuries requiring immediate surgery (hemothorax with persistent bleeding and hemodynamically unstable patient, diaphragm, or heart injuries). In this situation, the operation will be performed urgently, including the chest wall stabilization.
2. **Very early operation (≤3 days)**: Patients with severe extended flail chest (mostly anterolateral) with respiratory failure and intubation. These patients should be operated on as soon as possible, generally 2 or 3 days after admission to avoid pneumonia, which could ruin the potential advantages of the operation. In most of the cases (90%), extubation is possible immediately after operation or in the first 24 h (Lardinois et al., 2001) (Fig. 5.3A and B).
3. **Very early operation (≤3 days)**: A special absolute indication for stabilization consists of patients presenting with bilateral parasternal fractures (costochondral separation), often resulting from cardiac reanimation.

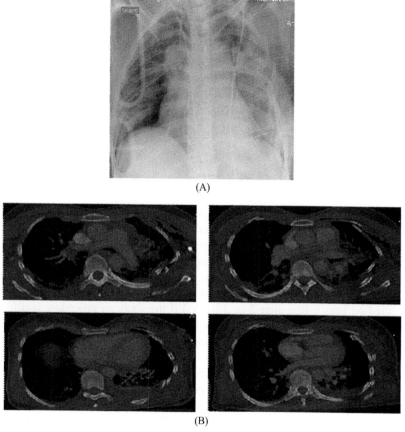

FIGURE 5.3 (A) Chest X-ray showing bilateral dislocated rib fractures in a patient with a flail chest. (B) Chest computerized tomography (CT)-scan with complicated multiple plurifragmentary rib fractures, hemothorax, and moderate lung contusion.

The sternum itself is the unstable segment of the chest wall. In this situation, chest wall stabilization should also be performed very early if the cardiac situation is stable enough.

4. **Early operation (4–7 days)**: Another indication is observed in patients without initial respiratory failure. Most of these patients have been transferred to the normal ward with an adapted pain therapy. After a few days to 7 days, a progressive shrinkage of the chest wall is observed with an increase in pain or respiratory failure. Stabilization should be performed as soon as dislocation of the fractures has been documented to avoid pneumonia (Fig. 5.4).

5. **Delayed operation (2–3 weeks)**: Special situations are observed in patients with lung contusion and cerebral injuries requiring a prolonged intubation time. Although lung contusion per se is not a contraindication

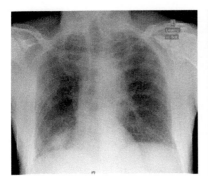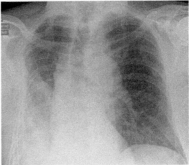

FIGURE 5.4 Chest X-ray with a series of dislocated rib fractures on the right side.

for operation, Voggenreiter has shown in a small series of patients that complication rate and mortality were higher than in the group conservative therapy in patients with extended (more than 2/3 of the volume) bilateral lung contusion (Voggenreiter et al., 1996). In this particular situation, the positive ventilatory pressure used to treat the lung contusion can be sufficient to achieve internal fixation. These patients can be operated on later (often after 2 or 3 weeks) if the weaning procedure is difficult. In the case of moderate lung contusion, an operation should be performed in the first days following admission (early operation). In the situation of cerebral injuries requiring intubation, the operation can be a little bit delayed and performed a few days after trauma (4–7 days).

6. **Special situation in patients with pneumonia or empyema**: There is not much data in the literature discussing this topic. Pneumonia and empyema are not absolute contraindications for operation, but it is advisable to wait a few days until the infection is under control with the antibiotics (decrease of C-reactive protein (CRP) and leucocytes) before performing the stabilization. Longer treatment with antibiotics should also be indicated.

Management in no flail chest patients

The most common situation consists of patients with a series of rib fractures (Fig. 5.5). For multiple rib fractures without a flail chest, the benefit of rib fixation remains a topic of debate and the situation is less clear than in flail chest patients. Standard treatment involves adequate pain therapy, mainly by use of a peridural catheter, and physiotherapy with mobilization. Usually, symptoms progressively improve within 4–7 days and patients can recover. However, it is not rare to observe that pain does not decrease, leading to prolonged hospital stays and to higher risk of complications like pneumonia that can be lethal in the elderly. Moreover, a part of these patients will develop chronic pain with long-term impairment of quality of life. Conservative

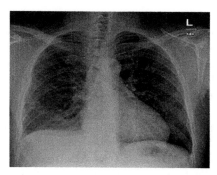

FIGURE 5.5 Chest X-ray with a series of dislocated rib fractures on the right side.

therapy can be associated with high rates of morbidity and low quality of life (chronic pain up to 60%, disability, loss of performance) at long-term follow-up (Fabricant et al., 2013; Marasco et al., 2015).

Due to the positive results of chest wall stabilization in flail chest patients regarding the prevention of complications and functional outcomes, the indication for rib fixation was also extended in selected patients without flail chests in the last few years.

Several single-center studies including a limited number of patients and mostly retrospective showed an improvement in the pain and the functional outcome after rib fixation (Akil et al., 2019; De Moya et al., 2011; Khandelwal et al., 2011; Majercik et al., 2015; Pieracci et al., 2016). In most of these studies, the operation was performed relatively early (4−7 days after trauma). A relatively recent study investigated the long-term outcome after rib fixation for a series of dislocated and painful rib fractures in terms of pain and quality of life assessed by the short-form-12 (SF12) questionnaire (short form SF36) (Hojski et al., 2022). About 50 patients suffered from a series of fractures in this study. The authors observed a significant decrease in the pain assessed by the visual analogue scale (VAS) score (0−10) immediately after the operation but also in long-term follow-up up to 1 year. The score was 8.3 preoperatively, 3.6 at discharge from the hospital, 2.7 at 1 month, 2 at 3 months, 1.4 at 6 months, and 1.5 at 1 year after rib fixation. Additionally, there was a significant improvement in the physical score measured by the SF-12 questionnaire at 6 and at 12 months postoperative in comparison with the preoperative situation (Hojski et al., 2022).

There is one interesting multicenter prospective controlled trial comparing rib fixation with conservative therapy in 110 patients. The operation was performed within 72 h after trauma (Pieracci et al., 2020). At 2 weeks, the pain score was significantly lower (2.9 vs 4.5, $P < .01$), the quality of life significantly improved (disability score, 21 vs 25, $P = .03$), the narcotic consumption also trended toward being lower (0.5 vs 1.2 narcotic equivalents, $P = .05$), and the pleural space complications significantly lower in the operative group (0% vs 10.2%, $P=.02$). All these data support the role of surgical stabilization of rib fractures (SSRF) in patients without flail chests.

However, in many centers, there is still a certain unwillingness to fix the rib in non-flail patients. The main argument is the absence of prospective, randomized trials demonstrating an advantage of the operative treatment over the conservative one.

The results of a multicenter randomized controlled trial from The Netherlands comparing early fixation versus conservative therapy of multiple, simple rib fractures should soon be published (Wijffels et al., 2019). This study should definitely bring more scientific value to the medical community. On the other hand, is it absolutely necessary to demonstrate that surgical treatment is superior to conservative one to fix rib fractures? There is enough data in the literature showing that rib fixation is an effective, easy, safe treatment, with acceptable morbidity and that this treatment can significantly improve the symptoms and the functional outcomes of the patients. It is an alternative to conservative therapy. The true and main question is to be able to select patients who can benefit from an operation. Also if the prospective, randomized trial does not show a significant advantage of operative treatment in the collective of patients included in this study, it should not be a reason to stop to fix ribs in a series of fractures but to better select the patients who will benefit from the operation. The goal of the operation is to decrease pain and to give the patient the same quality of life as before trauma as fast as possible, according to age, own physical characteristics, physical activities in all-day life, hobbies, and profession. It is therefore justifiable to offer the possibility of rib fixation in each patient with a painful series of fractures if there is no tendency to decrease (particularly when coughing) and if the mobilization of the patient remains difficult after 3 or 4 days of conservative therapy. The patient should be aware that the possibility of the operation exists, and he should be correctly informed on the modalities of the operation and the experience in the literature. The patient will then make the decision himself without any pressure. The decision to operate should be taken case-by-case, based on daily clinical examination and physical activity level before trauma. There are 3 subgroups of patients who could benefit from a very early operation (within 72 h). First, young active, and sporty patients who wish to be fit as soon as possible again; second, older patients who are living alone, independently, and are still active with a good quality of life. Third, patients with multiple, severe dislocations of the fractures, since the complication rate is higher (chronic pain, fracture nonunion). Prevention is the best therapy for fracture nonunion in very dislocated fractures. In such patients, operations should be proposed early at admission or after some days (within 72 h).

Special attention should be given during the initial physical examination to correctly diagnose multiple rib fractures associated with dislocation of the costal arch. This dislocation is not always easy to notice and if several ribs are fractured, this situation leads to instability. If only the ribs are fixed without reparation of the costal arch, shearing forces with rotation components will usually result in persisting pain, deformity, and eventually fractures or dislocation of the material. Reconstruction of the costal arch should be always performed, avoiding injury of small branches of intercostal nerves which could lead to an invalidating relaxation of a part of the superficial abdominal musculature.

Is it necessary to fix all the fractured ribs? Fixation of all the fractures is probably not necessary. However, care should be taken to avoid persistent deformity of the chest wall, leading to possible functional restriction. Generally, after repositioning the most dislocated fractures, the other fractures automatically take a normal alignment. Some surgeons try to define which rib fractures are most painful by daily palpation to fix these fractures only, but experience shows that it is often impossible to palpate the fractures separately due to the diffuse pain. Other tools like 3-dimensional reconstruction images from chest CT scans and thoracoscopy can help to better define the most dislocated ribs to be fixed. Additionally, thoracoscopy often plays an important role in evacuating the hemothorax, rinsing the cavity, and not seldom diagnosing other injuries in the thorax like diaphragmatic injuries, lung lesions, aortic contusion, and pericardial lesions.

Summary

Rib fixation in patients with a series of rib fractures should/could be considered (Fig. 5.6).

1. **Urgent operation**: Patients with a series of rib fractures presenting with associated thoracic injuries requiring immediate surgery (hemothorax with persistent bleeding and hemodynamically unstable patient, diaphragm

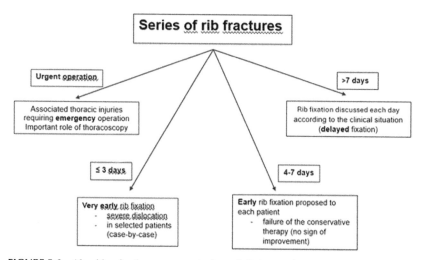

FIGURE 5.6 Algorithm for the management of non-flail chest patients.

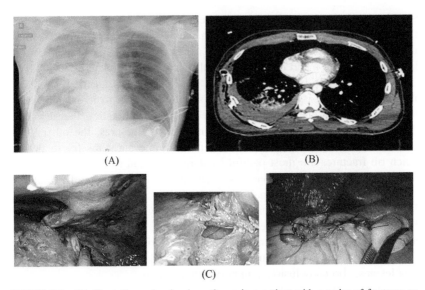

FIGURE 5.7 (A) Chest X-ray showing hemothorax in a patient with a series of fractures on the right side and persisting bleeding requiring urgent operation. (B) Chest CT scan demonstrating the dislocated fractures and the hemothorax. (C) Intraoperative view with a lesion of the diaphragm responsible for the bleeding. Minimal invasive repair was possible.

injuries). In this situation, the operation will be performed urgently, including the rib fixation (Fig. 5.7A−C).

2. **Very early operation (≤3 days)**: The operation should be proposed and discussed in patients with radiological severe dislocation of the fractures or special situations (young sporty patients, old patients living independently and still active).
3. **Early operation (4−7 days)**: The operation should be early proposed and discussed in each patient with failure of the initial conservative therapy when there is no sign of clinical improvement.
4. **Delayed operation (>7 days)**: The operation is proposed and discussed each day with the patient according to the clinical situation in patients who did not want early rib fixation or who required more time for a decision.

Management of rib fracture nonunion

Fracture nonunion is defined by a radiological absence of bridging callus between the 2 fragments of the fracture later than 6 months after trauma (Fig. 5.8). It can be hypertrophic but most of the time the nonunion is pseudarthrotic. Up to 5%−10% of all rib fractures treated conservatively go on to nonunion (Buehler et al., 2020; de Jong et al., 2018; DeGenova et al., 2023; Fabricant et al., 2014; Van Wijck et al., 2022). Symptoms mainly

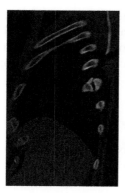 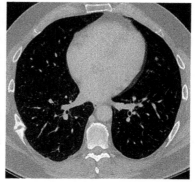

FIGURE 5.8 Chest CT scan illustrating a fracture nonunion on the right side.

consist of chronic painkiller refractory pain, clicking sensation or motion at the fracture site, restrictive range of motion of the ipsilateral extremity, or more rarely dyspnea. Risk factors for fracture nonunion include distraction between fractures, excessive displacement and dislocation, infection, soft tissue interposition, fracture location, osteoporosis, and other factors like nicotine use, diabetes mellitus, the use of nonsteroidal anti-inflammatory drugs (DeGenova et al., 2024).

According to the management of fracture nonunion, there is a paucity of literature, including case reports or small case series with a low number of patients. In these reports, the definition of rib nonunion is also variable, either already later than 3 months or later than 6 months after injury. A recently published multicenter prospective cohort study by Van Wijck including 68 patients found nonunions in 12% of the rib fractures in 43% of the patients treated conservatively on chest CT scans conducted 6 months after trauma (Van Wijck et al., 2024). The incidence of fracture nonunion was particularly high in fractures located at the costal cartilages or in the posterior and lateral parts of the ribs and dislocated fractures. Nonunion was most commonly observed in ribs 7–10 (Van Wijck et al., 2024). Most of the studies involved in the systematic review of the literature by DeGenova including between 6 and 36 patients showed a significant improvement in patients' outcomes in terms of pain, return to normal activity, and decrease in the use of narcotic pain medication (DeGenova et al., 2024). In 4 of the 9 studies included in the systematic review, the authors observed up to 50%–58% of the patients with persistent pain postoperatively, in one report like preoperatively in 32% of the patients (Buehler et al., 2020; de Jong et al., 2018; Van Wijck et al., 2022, 2024). Fabricant described that the percentage of patients taking opioids did not change from preoperative to postoperative (Fabricant et al., 2014). Possible explanations for the persistence of chronic pain can result from intercostal nerve damage from the injury or nerve entrapment in the healing efforts (scar) across the fracture. In this situation, the simple elimination of motion across the non-healed fracture by fixing the fracture would have a lesser impact on the symptoms. Another explanation is that the patients are «fixed» on the symptoms so it takes longer to observe an improvement of the symptoms despite correct operation. In some of these patients, the operation could potentially worsen the symptomatology.

Summary

Fixation of rib fracture nonunions appears to be an appropriate treatment for persistent disabling pain refractory to conservative treatment methods. The pressure of suffering is very important in decision-making.

When is rib fixation in non-flail patients usually not necessary?

- simple, isolated rib fracture or series of fractures under control with pain therapy and not very dislocated.
- fracture of the first and second ribs with or without fracture of the clavicle. In the case of a simultaneous fracture of the clavicle, osteosynthesis of the clavicle should be performed to ensure good stability of the shoulder girdle.
- fracture of ribs 11 and 12 (contribute little to breathing).

References

Akil, A., Ziegeler, S., Reichelt, J., Semik, M., Müller, M. C., & Fischer, S. (2019). Rib osteosynthesis is a safe and effective treatment and leads to a significant reduction of trauma associated pain. *European Journal of Trauma and Emergency Surgery, 45*(4), 623−630. Available from https://doi.org/10.1007/s00068-018-01062-5, http://link.springer.com/journal/68.

Bhatnagar, A., Mayberry, J., & Nirula, R. (2012). Rib fracture fixation for flail chest: What Is the benefit? *Journal of the American College of Surgeons, 215*(2), 201−205. Available from https://doi.org/10.1016/j.jamcollsurg.2012.02.023.

Buehler, K. E., Wilshire, C. L., Bograd, A. J., & Vallières, E. (2020). Rib plating offers favorable outcomes in patients with chronic nonunion of prior rib fractures. *Annals of Thoracic Surgery, 110*(3), 993−997, Elsevier USA United States. Available from https://doi.org/10.1016/j.athoracsur.2020.03.075, www.elsevier.com/locate/athoracsur.

Coughlin, T. A., Ng, J. W. G., Rollins, K. E., Forward, D. P., & Ollivere, B. J. (2016). Management of rib fractures in traumatic flail chest: A meta-analysis of randomised controlled trials. *Bone and Joint Journal, 98-B*(8), 1119−1125. Available from https://doi.org/10.1302/0301-620X.98B8.37282, http://www.bjj.boneandjoint.org.uk/content/jbjsbr/98-B/8/1119.full.pdf.

Craxford, S., Marson, B. A., Nightingale, J., Forward, D. P., Taylor, A., & Ollivere, B. (2022). Surgical fixation of rib fractures improves 30-day survival after significant chest injury an analysis of ten years of prospective registry data from england and Wales. British Editorial Society of Bone and Joint Surgery, United Kingdom. *Bone and Joint Journal, 104*(6),

729–735. Available from https://doi.org/10.1302/0301-620X.104B6.BJJ-2021-1502.R1, https://online.boneandjoint.org.uk/doi/full/10.1302/0301-620X.104B6.BJJ-2021-1502.R1.

de Jong, M. B., Houwert, R. M., van Heerde, S., de Steenwinkel, M., Hietbrink, F., & Leenen, L. P. H. (2018). Surgical treatment of rib fracture nonunion: A single center experience. *Injury, 49*(3), 599–603. Available from https://doi.org/10.1016/j.injury.2018.01.004.

De Moya, M., Bramos, T., Agarwal, S., Fikry, K., Janjua, S., King, D. R., Alam, H. B., Velmahos, G. C., Burke, P., & Tobler, W. (2011). Pain as an indication for rib fixation: A bi-institutional pilot study. *Journal of Trauma – Injury, Infection and Critical Care, 71*(6), 1750–1754. Available from https://doi.org/10.1097/TA.0b013e31823c85e9, United States.

DeGenova, D. T., Miller, K. B., McClure, T. T., Schuette, H. B., French, B. G., & Taylor, B. C. (2023). Operative fixation of rib fracture nonunions. *Archives of Orthopaedic and Trauma Surgery, 143*(6), 3047–3054. Available from https://doi.org/10.1007/s00402-022-04540-z, https://www.springer.com/journal/402.

DeGenova, D. T., Peabody, J. T., Schrock, J. B., Homan, M. D., Peguero, E. S., & Taylor, B. C. (2024). Symptomatic rib fracture nonunion: A systematic review of the literature. *Archives of Orthopaedic and Trauma Surgery, 144*(5), 1917–1924. Available from https://doi.org/10.1007/s00402-024-05264-y, https://www.springer.com/journal/402.

Fabricant, L., Ham, B., Mullins, R., & Mayberry, J. (2013). Prolonged pain and disability are common after rib fractures. *The American Journal of Surgery, 205*(5), 511–516. Available from https://doi.org/10.1016/j.amjsurg.2012.12.007.

Fabricant, L., Ham, B., Mullins, R., & Mayberry, J. (2014). Prospective clinical trial of surgical intervention for painful rib fracture nonunion. *The American Surgeon, 80*(6), 580–586. Available from https://doi.org/10.1177/000313481408000622.

Gerakopoulos, E., Walker, L., Melling, D., Scott, S., & Scott, S. (2019). Surgical management of multiple rib fractures reduces the hospital length of stay and the mortality rate in major trauma patients: A comparative study in a UK major trauma center. *Journal of Orthopaedic Trauma, 33*(1), 9–14. Available from https://doi.org/10.1097/bot.0000000000001264.

Granetzny, A., El-Aal, M. A., Emam, E. R., Shalaby, A., & Boseila, A. (2005). Surgical versus conservative treatment of flail chest. Evaluation of the pulmonary status. *Interactive Cardiovascular and Thoracic Surgery, 4*(6), 583–587. Available from https://doi.org/10.1510/icvts.2005.111807 Germany, http://icvts.ctsnetjournals.org/cgi/reprint/4/6/583?ck = nck, Germany.

Hojski, A., Xhambazi, A., Wiese, M. N., Subotic, D., Bachmann, H., & Lardinois, D. (2022). Chest wall stabilization and rib fixation using a nitinol screwless system in selected patients after blunt trauma: Long-term results in a single-centre experience. *Interactive Cardiovascular and Thoracic Surgery, 34*(3), 386–392. Available from https://doi.org/10.1093/icvts/ivab278, http://icvts.oxfordjournals.org/.

Kasotakis, G., Hasenboehler, E. A., Streib, E. W., Patel, N., Patel, M. B., Alarcon, L., Bosarge, P. L., Love, J., Haut, E. R., & Como, J. J. (2017). Operative fixation of rib fractures after blunt trauma: A practice management guideline from the Eastern Association for the Surgery of Trauma. *Journal of Trauma and Acute Care Surgery, 82*(3), 618–626, Lippincott Williams and Wilkins United States. Available from https://doi.org/10.1097/TA.0000000000001350, http://journals.lww.com/jtrauma.

Khandelwal, G., Mathur, R. K., Shukla, S., & Maheshwari, A. (2011). A prospective single center study to assess the impact of surgical stabilization in patients with rib fracture. *International Journal of Surgery, 9*(6), 478–481. Available from https://doi.org/10.1016/j.ijsu.2011.06.003.

Landercasper, J., Cogbill, T. H., & Lindesmith, L. A. (1984). Long-term disability after flail chest injury. *Journal of Trauma – Injury, Infection and Critical Care, 24*(5), 410–414. Available from https://doi.org/10.1097/00005373-198405000-00007.

Lardinois, D., Krueger, T., Dusmet, M., Ghisletta, N., Gugger, M., & Ris, H.-B. (2001). Pulmonary function testing after operative stabilisation of the chest wall for flail chest. *European Journal of Cardio-Thoracic Surgery*, *20*(3), 496–501. Available from https://doi.org/10.1016/S1010-7940(01)00818-1.

Lardinois, D. (2018). General considerations, indications, and potential advantages of chest wall stabilization, rib fixation, and sternum osteosynthesis in selected patients after blunt trauma. *Shanghai Chest*, *2*(8), 62. Available from https://doi.org/10.21037/shc.2018.06.11.

Leinicke, J. A., Elmore, L., Freeman, B. D., & Colditz, G. A. (2013). Operative management of Rib fractures in the setting of flail chest: A systematic review and meta-analysis. *Annals of Surgery*, *258*(6), 914–921. Available from https://doi.org/10.1097/SLA.0b013e3182895bb0.

Liu, T., Liu, P., Chen, J., Xie, J., Yang, F., & Liao, Y. (2019). A randomized controlled trial of surgical rib fixation in polytrauma patients with flail chest. *Journal of Surgical Research*, *242*, 223–230. Available from https://doi.org/10.1016/j.jss.2019.04.005.

Majercik, S., Wilson, E., Gardner, S., Granger, S., Van, B. D. H., & White, T. W. (2015). In-hospital outcomes and costs of surgical stabilization versus nonoperative management of severe rib fractures. *Journal of Trauma and Acute Care Surgery*, *79*(4), 533–539. Available from https://doi.org/10.1097/TA.0000000000000820, http://journals.lww.com/jtrauma.

Majercik, S., Cannon, Q., Granger, S. R., VanBoerum, D. H., & White, T. W. (2014). Long-term patient outcomes after surgical stabilization of rib fractures. *The American Journal of Surgery*, *208*(1), 88–92. Available from https://doi.org/10.1016/j.amjsurg.2013.08.051.

Marasco, S. F., Davies, A. R., Cooper, J., Varma, D., Bennett, V., Nevill, R., Lee, G., Bailey, M., & Fitzgerald, M. (2013). Prospective randomized controlled trial of operative rib fixation in traumatic flail chest. *Journal of the American College of Surgeons*, *216*(5), 924–932. Available from https://doi.org/10.1016/j.jamcollsurg.2012.12.024.

Marasco, S., Lee, G., Summerhayes, R., Fitzgerald, M., & Bailey, M. (2015). Quality of life after major trauma with multiple rib fractures. *Injury*, *46*(1), 61–65. Available from https://doi.org/10.1016/j.injury.2014.06.014.

Mouton, W., Lardinois, D., Furrer, M., Regli, B., & Ris, H. (1997). Long-term follow-up of patients with operative stabilisation of a flail chest. *The Thoracic and Cardiovascular Surgeon*, *45*(05), 242–244. Available from https://doi.org/10.1055/s-2007-1013735.

Owattanapanich, N., Lewis, M. R., Benjamin, E. R., Jakob, D. A., & Demetriades, D. (2022). Surgical rib fixation in isolated flail chest improves survival, Elsevier Inc., undefined*Annals of Thoracic Surgery*, *113*(6), 1859–1865, Elsevier Inc., undefined. Available from https://doi.org/10.1016/j.athoracsur.2021.05.085, www.elsevier.com/locate/athoracsur.

Patel, D. D., Zambetti, B. R., & Magnotti, L. J. (2024). Timing to rib fixation in patients with flail chest. *Journal of Surgical Research*, *294*, 93–98. Available from https://doi.org/10.1016/j.jss.2023.09.057, https://www.sciencedirect.com/science/journal/00224804.

Pieracci, F. M., Leasia, K., Bauman, Z., Eriksson, E. A., Lottenberg, L., Majercik, S., Powell, L., Sarani, B., Semon, G., Thomas, B., Zhao, F., Dyke, C., & Doben, A. R. (2020). A multi-center, prospective, controlled clinical trial of surgical stabilization of rib fractures in patients with severe, nonflail fracture patterns (Chest Wall Injury Society NONFLAIL). *Journal of Trauma and Acute Care Surgery*, *88*(2), 249–257. Available from https://doi.org/10.1097/TA.0000000000002559, http://journals.lww.com/jtrauma.

Pieracci, F. M., Lin, Y., Rodil, M., Synder, M., Herbert, B., Tran, D. K., Stoval, R. T., Johnson, J. L., Biffl, W. L., Barnett, C. C., Cothren-Burlew, C., Fox, C., Jurkovich, G. J., & Moore, E. E. (2016). A prospective, controlled clinical evaluation of surgical stabilization of severe rib fractures. *Journal of Trauma and Acute Care Surgery*, *80*(2), 187–194. Available from https://doi.org/10.1097/TA.0000000000000925, http://journals.lww.com/jtrauma.

Qiu, M., Shi, Z., Xiao, J., Zhang, X., Ling, S., & Ling, H. (2016). Potential benefits of rib fracture fixation in patients with flail chest and multiple non-flail rib fractures. *Indian Journal of Surgery*, *78*(6), 458−463. Available from https://doi.org/10.1007/s12262-015-1409-2.

Slobogean, G. P., MacPherson, C. A., Sun, T., Pelletier, M. E., & Hameed, S. M. (2013). Surgical fixation vs nonoperative management of flail chest: A meta-analysis. *Journal of the American College of Surgeons*, *216*(2), 302. Available from https://doi.org/10.1016/j.jamcollsurg.2012.10.010, www.elsevier.com/locate/jamcollsurg.

Tanaka, H., Yukioka, T., Yamaguti, Y., Shimizu, S., Goto, H., Matsuda, H., & Shimazaki, S. (2002). Surgical stabilization of internal pneumatic stabilization? A prospective randomized study of management of severe flail chest patients. *The Journal of Trauma: Injury, Infection, and Critical Care*, *52*(4), 727−732. Available from https://doi.org/10.1097/00005373-200204000-00020.

Van Olden, G. D. J., Dik Meeuwis, J., Bolhuis, H. W., Boxma, H., & Goris, R. J. A. (2004). Clinical impact of advanced trauma life support. *American Journal of Emergency Medicine*, *22*(7), 522−525. Available from https://doi.org/10.1016/j.ajem.2004.08.013.

Van Wijck, S. F. M., Van Diepen, M. R., Prins, J. T. H., Verhofstad, M. H. J., Wijffels, M. M. E., Van Lieshout, E. M. M., Blokhuis, T. J., Boersma, D., De Loos, E. R., Flikweert, E. R., IJpma, F. F. A., Kleinveld, S., Knops, S. P., Pull ter Gunne, A. F., Spanjersberg, W. R., Van der Bij, G., Van Eijck, F. C., Van Huijstee, P. J., Van Montfort, G., ... Vos, D. I. (2024). Radiographic rib fracture nonunion and association with fracture classification in adults with multiple rib fractures without flail segment: A multicenter prospective cohort study. *Injury*, *55*(5). Available from https://doi.org/10.1016/j.injury.2024.111335, https://www.sciencedirect.com/science/journal/00201383.

Van Wijck, S. F. M., Van Lieshout, E. M. M., Prins, J. T. H., Verhofstad, M. H. J., Van Huijstee, P. J., Vermeulen, J., & Wijffels, M. M. E. (2022). Outcome after surgical stabilization of symptomatic rib fracture nonunion: A multicenter retrospective case series. *European Journal of Trauma and Emergency Surgery*, *48*(4), 2783−2793. Available from https://doi.org/10.1007/s00068-021-01867-x, http://link.springer.com/journal/68.

Voggenreiter, G., Neudeck, F., Aufmkolk, M., Obertacke, U., & Schmit-Neuerburg, K. P. (1996). Behandlungsergebnisse der operativen Thoraxwandstabilisierung bei instabilem Thorax mit und ohne Lungenkontusion. *Der Unfallchirurg*, *99*(6), 425−434.

Wijffels, M. M. E., Prins, J. T. H., Polinder, S., Blokhuis, T. J., De Loos, E. R., Den Boer, R. H., Flikweert, E. R., Pull Ter Gunne, A. F., Ringburg, A. N., Spanjersberg, W. R., Van Huijstee, P. J., Van Montfort, G., Vermeulen, J., Vos, D. I., Verhofstad, M. H. J., & Van Lieshout, E. M. M. (2019). Early fixation versus conservative therapy of multiple, simple rib fractures (FixCon): Protocol for a multicenter randomized controlled trial. *World Journal of Emergency Surgery*, *14*(1). Available from https://doi.org/10.1186/s13017-019-0258-x, http://www.wjes.org/.

Chapter 6

Early rib fixation: the evidence, the trials

Namariq Abbaker, Ahmed Hamada and May Al-Sahaf
Imperial College NHS Healthcare Trust and National Heart and Lung Institute, Division of Thoracic Surgery, London, United Kingdom

Introduction

Rib fractures, frequently resulting from blunt chest trauma, are a significant medical concern due to their associated morbidity and mortality (Ziegler & Agarwal, 1994). The management of these fractures has undergone considerable evolution over the years, with early surgical stabilisation of rib fractures (SSRF) gaining prominence as a viable approach to enhance patient outcomes. This strategy, primarily focused on restoring chest wall integrity, aims to reduce pain, optimise respiratory function, and expedite the recovery process.

The increasing adoption of SSRF in clinical practice has been influenced by a growing body of research exploring its potential benefits. Recent studies and clinical trials have delved into various aspects of SSRF, including the determination of the optimal timing for surgical intervention, identification of suitable patient populations, and evaluation of its impact on key clinical outcomes. These outcomes encompass variables such as hospital and intensive care unit (ICU) length of stay (LOS), duration of mechanical ventilation, rates of complications, and mortality.

This chapter presents a detailed and comprehensive review of the current clinical guidelines for early rib fixation. It examines recommendations from leading medical societies and organisations to present a comprehensive insight into the current practices for utilising early rib fixation in clinical practice. Following this, the chapter transitions into an in-depth analysis of trials focusing on early surgical stabilisation. It reviews the existing literature, which includes retrospective studies, prospective trials, and meta-analyses, to assess the overall impact of early surgical stabilisation on patient recovery trajectories. In doing so, the chapter not only highlights the strengths and benefits of early SSRF but also critically appraises the limitations and gaps present in the current body of evidence.

Evidence synthesis: guidelines and recommendations for early rib fixation

Historically, the stance of preeminent medical societies on thoracic trauma has skewed toward conservative management for rib fractures. However, exceptions were made predominantly in cases of severe trauma, such as complicated and extensive multiple rib fractures or flail chest, where surgical intervention was deemed necessary (Pieracci, Leasia, et al., 2021). Yet, the landscape of clinical practice is ever-evolving, as evidenced by a discernible trend in recent guidelines, which now delineate a more proactive approach to surgical intervention. This includes establishing clear criteria for the types of fractures that warrant surgery, alongside specific indications, contraindications, and the timing of these procedures. This shift is substantiated by a growing corpus of research suggesting superior outcomes with surgical intervention over non-operative strategies. A synthesised overview of these nuanced guidelines is presented in Table 6.1.

Diving deeper into these recommendations, authoritative entities such as the National Institute for Health and Care Excellence (NICE), The Society of Thoracic Surgeons, and the European Society of Thoracic Surgeons have been vocal advocates for surgical stabilisation in instances of flail chest and respiratory compromise (1Guidance et al., n.d.; Bedetti et al., 2018). Moreover, these bodies propose that surgical fixation should be considered for patients suffering from multiple rib fractures coupled with severe pain or functional impairment. NICE, in particular, underscores the necessity importance of judicious patient selection, the provision of multidisciplinary care, and timely interventions, emphasising a practice built around safety and accountability (Insertion of metal rib reinforcements to stabilize a flail chest wall, National Institute for Health and Clinical Excellence, 2010).

Conversely, The American Association for the Surgery of Trauma maintains that rib fractures, by themselves, do not warrant a unique treatment algorithm, instead endorsing supportive measures such as pain management and incentive spirometry, with surgical fixation reserved for complicated cases of flail chest (Rib Fractures, 2013).

Meanwhile, the Chest Wall Injury Society (CWIS) has articulated a comprehensive set of indications and contraindications for surgical stabilisation, recommending intervention for clinical signs of chest wall instability, paradoxical motion, or patient-reported instability, particularly in non-ventilated individuals. The society further advocates for prompt surgical action within 24 hours for non-ventilated patients when practical, extending the window to 72 hours post-injury for those with non-flail indications who require mechanical ventilation. These recommendations are tailored by considering the potential presence of other critical injuries (Kasotakis et al., 2017; Pieracci, Schubl, et al., 2021).

TABLE 6.1 Summary table for organisation stands on surgical stabilisation of rib fractures.

Organization/Society	Final recommendation on rib fracture management	Position on early surgical fixation	Recommended time frame for surgery
NICE	Surgical stabilization in cases of flail chest and respiratory compromise; consideration for surgical fixation in multiple rib fractures with severe pain or functional impairment. Emphasises patient selection and early intervention.	For (in specific cases)	Not specifically mentioned
STS (The Society of Thoracic Surgeons) and ESTS (European Society of Thoracic Surgeons)	Advocate surgical stabilization in patients with flail chest and respiratory compromise; recommend considering surgical fixation in patients with multiple rib fractures and severe pain or functional impairment.	For (in specific cases)	Not specifically mentioned
AAST (American Association for the Surgery of Trauma)	Suggests no specific treatment for rib fractures but supports supportive measures and surgical fixation of flail chest if necessary. No clear recommendation for non-flail rib fractures.	Neutral/Conditional	Not specifically mentioned
CWIS (Chest Wall Injury Society)	Detailed indications for surgical intervention, including chest wall instability, paradoxical motion in non-ventilated patients, and specific contraindications. Recommends surgery within specific time frames based on the patient's condition.	For (with detailed indications)	Within 24 h for non-ventilated patients; within 72 h of injury for ventilated patients
Eastern Association of the Surgery of Trauma	Conditional recommendations for operative rib ORIF in adult patients with flail chest; notes limited and low-quality evidence for these recommendations.	Conditional	Not specifically mentioned

NICE, National Institute for Health and Care Excellence; *ORIF*, open reduction and internal fixation.

The Eastern Association for the Surgery of Trauma, through a systematic review and meta-analysis classified as Level III evidence, cautiously suggests that operative rib open reduction and internal fixation (ORIF) in adult patients with flail chest may improve a variety of clinical outcomes. This recommendation, however, is made with an understanding of the limitations inherent in the current evidence base, which includes a scarcity of high-quality randomised controlled trials and an overall assessment of the data as being of moderate quality. This perspective acknowledges the potential benefits of ORIF while also underscoring the need for further research to strengthen the evidence base (Pieracci et al., 2017; Kasotakis et al., 2017).

In essence, the trajectory of rib fracture management is progressively gravitating toward evidence-based interventions, with surgical options gaining favour, particularly for more grievous injuries. Despite a consensus among leading medical associations on the need for patient-specific care and interdisciplinary collaboration, the variability in guidelines between societies points to prevailing uncertainties and a dearth of evidence, most notably for less critical cases. This variability serves to show the evolving nature of the field and is a clarion call for the execution of high-quality randomised controlled trials to forge more universally applicable guidelines and solidify the best practices for rib fracture management.

Trial analysis and discussion: understanding early surgical stabilisation

Movement toward surgical intervention

Our systematic literature search involved searching PubMed, Ovid Medline, Embase, and Cochrane databases for studies published between 2010 and 2024 that investigated the timing of early rib fixation. We have used search terms such as 'rib fracture,' 'early fixation,' and 'SSRF.' This refined approach led us to identify nine pivotal studies, which were selected based on inclusion criteria emphasising research specifically exploring the efficacy and outcomes of early versus delayed SSRF. See Table 6.2 for a summary of the studies.

The non-operative management of rib fractures was commonplace, but this approach is increasingly being scrutinised due to its associated risks and potential for escalating complications. In recent years, there has been a discernible shift toward early surgical management for rib fractures driven by the demonstrated efficacy of SSRF in enhancing patient outcomes. For example, studies conducted by Su et al., (2019) and Wang et al., (2023) underscore the benefits of early SSRF, including reduced ICU and hospital stay, and a lowered incidence of complications such as pneumonia and tracheostomy. As such, contemporary research is shining a light on the benefits of early surgical intervention, advocating a more proactive approach to managing these injuries.

TABLE 6.2 Summary of studies looking at early versus late surgical fixation of rib fractures.

Study	Period	Design	SSRF indicators	Intervention window	Participants	Follow-up	Key findings	Study constraints
Iqbal et al. (2018)	Mar 2015 – May 2016	Retrospective cohort, single center	≥ 3 rib fractures displaced	≤ 48 h (early) vs >48 h (late)	65 (early), 37 (late)	3 months	Shorter hospital and ICU stays in early SSRF, less pneumonia and tracheostomy	Retrospective nature, limited follow-up, single center
Harrell et al. (2020)	Jan 2007 – Jan 2018	Retrospective cohort, single center	Rib displacement ≥3, additional complexities	0–2 days (early) vs >2 days (late)	Not specified	Not specified	Reduced hospital stays with early SSRF; increased length of stay with delayed SSRF	Retrospective, limited sample size, brief follow-up
Pieracci et al. (2018)	Jan 2006 – Jan 2017	Retrospective cohort, multicenter	Flail chest, displaced rib fractures ≥ 3, other criteria	≤ 24 h (early); 1–2 days (mid); 3–10 days (late)	207 (early), 168 (mid), 176 (late)	Not specified	Shortest hospital and ICU stays with earliest SSRF; each day's delay increased pneumonia, tracheostomy, and DMV likelihood	Retrospective, short follow-up, lack of stratification details
Su et al. (2019)	Not specified	Retrospective cohort, single center	≥ 4 displaced rib fractures, respiratory failure, etc.	≤ 3 days (early) vs >3 days (late)	16 (early), 17 (late)	Not specified	Early SSRF led to shorter hospital, and ICU stays, and DMV	Single centre, small cohort, retrospective, brief follow-up

(Continued)

TABLE 6.2 (Continued)

Study	Period	Design	SSRF indicators	Intervention window	Participants	Follow-up	Key findings	Study constraints
Otaka et al. (2020)	Jul 2010 – Apr 2018	Retrospective, national database	Multiple rib fractures needing mechanical ventilation	≤3 days (early); ≤6 days (mid); ≤10 days (late)	62 (early), 113 (mid), 162 (late)	Not specified	Early SSRF linked to shorter hospital stays and DMV; mid and late SSRF had outcomes like nonoperative management	Retrospective, no detailed injury severity, no stratification insight
Zhu et al. (2020)	Jan 2016 – Jan 2018	Retrospective, national database	Elderly patients with flail chest or multiple rib fractures	≤3 days (early) vs >3 days (late)	366 (early), 375 (late)	Not specified	Early SSRF resulted in lower pneumonia and tracheostomy rates, shorter hospital and ICU stays, and DMV	Retrospective, lacks details on injury severity
Pieracci et al. (2021)	Jan 2015 – Apr 2020	Retrospective cohort, multicenter	Flail chest, ≥3 ipsilateral displaced rib fractures	<3 days (early) vs ≥3 days (late)	63 (early), 70 (late)	Not specified	Early SSRF linked to lower mortality but higher pneumonia risk and longer ICU stays	Retrospective, brief follow-up, no stratification rationale
Wang et al. (2023)	Jan 2018 – Dec 2021	Randomised controlled trial, multicenter	Patients with multiple rib fractures	≤48 h (early) vs >48 h (delayed)	201 (early), 202 (delayed)	30 days	Early SSRF is associated with shorter ICU and hospital stays, less ventilation required, lower costs; reduced pneumonia and atelectasis rates	Limited quality of life evaluation, no detailed fracture morphology, subjective parameters like pain not fully assessed

DMV, Duration of mechanical ventilation; *ICU*, intensive care unit; *SSRF*, surgical stabilisation of rib fractures.

Elderly versus young

The shift in focus on rib fractures is driven by the changing demographic profile of the population, with an increasing prevalence of an aging demographic. This is particularly relevant as the primary cause of blunt trauma has shifted from vehicular accidents to falls, making the elderly more susceptible to rib fractures due to decreased bone density. Studies such as those by Bulger et al., (2000), Van Vledder et al., (2019), and Kelley et al., (2019) converge on the critical understanding that rib fractures significantly impact the elderly, underscoring a compounded vulnerability due to age. Van Vledder's et al., (2019) research specifically profiles the elderly with rib fractures, indicating a median age of 76 that suffers an elevated risk of complications and mortality. Kelley's et al., (2019) work, through the lens of a proposed early chest trauma protocol, demonstrates the potential of structured interventions to reduce adverse outcomes, reinforcing the importance of early tailored care strategies. In patients aged 65 and above, thoracic injuries, predominantly rib fractures, are a major cause of mortality due to higher rates of complications following rib fractures compared to younger cohorts. The mechanism of injury, including motor vehicle crashes and falls, plays a crucial role in the patterns and severity of these fractures (Bulger et al., 2000).

In contrast, younger patients, often victims of severe trauma like visceral thoracic injuries, face different challenges. As highlighted by Prins et al., (2020) initial medical attention in these cases focuses on life-threatening conditions, potentially delaying rib fracture management. This necessitates a balance between addressing immediate critical injuries and managing rib fractures to prevent complications such as flail chest and pneumonia. Once critical injuries are stabilised, the focus shifts to SSRF to enhance pain control, pulmonary function, and recovery, with the timing of SSRF being crucial and dependent on the patient's overall injury profile and recovery trajectory (Chou et al., 2015). The timing of SSRF in younger patients thus becomes a critical decision, influenced by the overall injury profile and the patient's recovery path.

Indications for surgical approach

In examining trials focused on early surgical intervention for rib fractures, a key finding is the consistent identification of flail chest as an indication for operative surgery. However, the criteria for other indications, such as the number of displaced ribs, show a lack of uniformity. For example, Iqbal et al., (2018) and Harrell et al., (2020) used a threshold of ≥ 3 rib displacement alone as an indication, while Pieracci et al., (2018) added the condition of the fractures being ipsilaterally displaced to this number. This hints at identifying patients with more severe localised chest trauma with potentially significant respiratory impairment rather than a more spread trauma. Su et al., (2019) opted for a higher threshold of ≥ 4 displacement as an indication for fixation. Other indications noted in these studies include respiratory failure, mechanical ventilatory requirements within one day of admission, flail sternum, pulmonary hernia, and significant volume loss in haemothorax (Harrell et al., 2020; Otaka et al., 2020; Pieracci et al., 2018; Su et al., 2019).

It is critical to note that in some studies, patients who have failed nonoperative approaches, particularly in achieving pain control, are considered for surgical intervention at a later stage (Pieracci et al., 2018; Su et al., 2019). This situation often results in these patients being grouped into the late surgical intervention category. This can potentially skew the outcomes of surgical intervention studies. These patients may present with more severe conditions due to the delay, thereby impacting the perceived effectiveness of surgical approaches when compared to early intervention.

Timing to surgical intervention

The optimal timing for SSRF poses a significant challenge to determine the ultimate time frame. There is a growing consensus in the surgical community about the benefits of early intervention, typically within 48 hours to 3 days post-injury. A range of studies have shown early SSRF within this time frame to be associated with shorter hospital and ICU stays, a reduced need for mechanical ventilation, mortality, and lower rates of pneumonia and tracheostomy (Iqbal et al., 2018; Otaka et al., 2020; Zhu et al., 2020). In line with this, a recent study conducted across multiple centres in Jiangsu, China, compared early (≤48 h) and delayed (>48 h) rib fixation in patients with multiple rib fractures, reinforcing the benefits of early intervention with significant reductions in ICU and hospital stays, ventilation days, and costs (Patel et al., 2024).

The trend toward earlier intervention is further supported by literature from other medical specialities. For example, Chou et al., (2015) discussed the use of guided loco-regional anaesthesia and early management of pleural space issues in mitigating the severity of complications such as pneumonia and respiratory failure. This shift toward earlier intervention is corroborated by the application of principles from orthopaedic literature which suggests that early fixation of long bone fractures improves patient outcomes in the management of fractures (Dabezies & D'Ambrosia, 1986). Another influence is the changing practice patterns among trauma surgeons. Pieracci et al., (2018) analysed data from four high-volume centers, which demonstrated a significant trend toward early SSRF over the past decade, underscoring the evolving understanding and preference for early surgical intervention in the management of rib fractures. However, it's important to note that the definition of 'early' intervention varies across studies, reflecting the need for a more standardised approach.

Although the body of studies supports early SSRF, these studies also highlight the complexity of determining the optimal timing for surgery. For example, Pieracci et al., (2021) identified potential risks associated with early SSRF in the retrospective cohort of rib fracture patients aged 80 years or older, such as an increased risk of pneumonia and prolonged ICU-LOS. Fig. 6.1 shows the benefits and risk outcomes in studies that looked into early SSRF. In addition, a study published in the *Journal of Personalized Medicine* in 2022 examined the clinical impact and proper surgical timing of rib fixation in geriatric patients. It found that rib fixation complications were significantly related to injury severity scores but not associated with age or surgical timing. This study suggests that rib fixation can be delayed for geriatric patients until they are stabilised, and other critical injuries are managed (Chen et al., 2022). These findings challenge the prevailing view and suggest a more nuanced approach may be necessary, particularly for specific patient populations. It also demonstrates the need to balance the immediate surgical risks with the potential long-term benefits of early intervention. For instance, performing SSRF within the first 24 h can be challenging due to increased bleeding risks, as the inflammatory response and edema typically peak around 72 h post-injury. Thus, the decision-making process for the timing of rib fixation in trauma patients involves

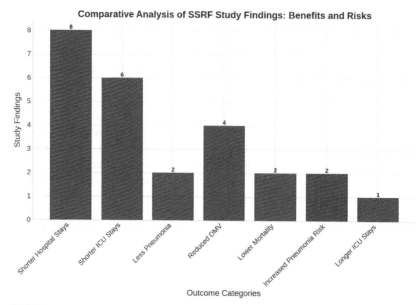

FIGURE 6.1 Benefits and risks of early rib fixation. The bar graph compares the benefits (in blue) and risks (in red) of early rib fixation in surgical stabilisation studies. The height of each bar indicates the number of studies that reported each specific outcome.

considering a multitude of factors, including patient-specific physiology, the nature of the fractures, the body's response following the injury, and potentially the age of the patients.

Patient selection and allocation challenges

The methodological complexities inherent in interpreting trial data on SSRF timing, such as selection and attrition bias, highlight the need for standardised guidelines to enable consistent decision-making in surgical interventions. Selection bias emerges from the tendency to choose patients with less severe injuries for early SSRF. As such, the observed benefits may be attributed to factors other than the rib fracture treatments themselves. Attrition bias, on the other hand, may arise when patients initially considered for SSRF subsequently improve without the need for surgery, leading to their exclusion from comparative analyses. Conversely, those who deteriorate during observation and later undergo surgery represent a subgroup whose delayed intervention leads to significant heterogeneity in sample allocation and potentially skewing results in studies comparing different surgical timings. One notable gap in the SSRF literature is the lack of comprehensive information on the reasons behind patients undergoing "late" surgery. While injury severity is one potential factor, other elements such as surgeon and operating room availability, competing operations, and patient conditions may also contribute to the decision. For example, surgeons' experience in rib fixation, higher priority injuries, or comorbidities including cardiac problems, medication use such as anticoagulation might confound time from admission to the theatre. This variability in surgical intervention criteria underscores the urgent need for further research to highlight these factors.

Multi-disciplinary approach

The multidisciplinary approach to early rib fixation, particularly in the treatment of non-flail rib fractures following trauma, represents a complex yet potentially transformative strategy in trauma care, as evidenced by recent studies. This approach, underscored by the NICE guidelines, emphasises the importance of early surgical stabilisation to optimise patient outcomes, particularly with the incorporation of other specialities such as orthopaedic and trauma surgery teams, as shown in Bauer's study (Bauer et al., 2023). The results demonstrate reduced complications and mortality rates with surgical fixation compared to non-operative management, supporting the growing body of evidence in favour of the surgical approach. Additionally, the study observes a decrease in the duration of ICU stays, reinforcing the benefits of early surgical intervention. Collaboration between specialities enables the undertaking of high-risk cases with varying complexities, reducing the need for patient transfer and enhancing overall efficiency. In line with this, similar studies by Pieracci et al., (2020) and Marasco et al., (2022) have shown the tangible benefits of a tandem approach that combines the expertise of orthopaedic and trauma surgeons. Pieracci's et al., (2020) study revealed significantly lower pain scores and fewer pleural space complications in surgically managed patients, advocating for the efficacy of leveraging diverse surgical skills. However, in this study, along with Marasco's et al., (2022) work, the suggested improvement in return-to-work rates in selected patients is limited by potential allocation bias and high crossover rates, raising questions about their generalizability.

Furthermore, research efforts like those of Li et al., (2020) and Zhang et al., (2019) have demonstrated superior outcomes in pain control and quality of life in surgical groups. Yet, these findings are tempered by their non-randomized designs and small sample sizes, which limit broader applicability. Additionally, the variation in analgesic regimens noted in Zhang's et al., (2019) study adds complexity to the assessment of surgical interventions, compounded by the lack of standardised pain measurement tools. The success of this team approach underscores the potential cost savings, particularly in terms of ICU stay duration, and suggests the need for further multi-institutional studies to explore the utilisation of specialised teams.

Quality of life

The surgical option is promising in terms of improving patient outcomes. However, Delaplain (Pieracci et al., 2021) and Raza & Eckhaus, (2022) raised concerns about the longevity of benefits post-surgery. Girsowicz et al., (2012) reviewed nine studies, concluding that surgical fixation is safe and effective for pain and disability relief, but noted that evidence is primarily from cohort studies and case reports. Building on this, Li et al., (2020) and Zhang et al., (2019) found that patients undergoing surgical treatment reported lower pain scores and improved quality of life, suggesting more immediate postoperative benefits.

In a similar vein, Qiu et al., (2016) observed that surgical patients experienced shorter hospital stays and a faster return to normal activities, reinforcing the notion of short-term advantages. Pieracci et al., (2020) extended this perspective in their multicentre trial, noting improved pain scores and respiratory quality of life in surgically managed patients, adding another layer to the growing evidence of the benefits of surgical intervention.

However, the study by Marasco et al., (2022) introduces a counterpoint to this narrative. Their randomised trial revealed no significant differences in pain scores between patients receiving surgical and conservative treatments at a 3-month follow-up. Interestingly, they noted that a greater proportion of patients in the surgical group returned to work by six months, hinting at possible long-term functional benefits that might not be immediately apparent in using pain scores alone. Their study, however, suggests that the benefits of surgical fixation may be limited to specific patient populations. One explanation could be related to the different techniques and analgesia regimens, potential allocation biases, and loss of follow-up. Furthermore, the use of one-dimensional pain indices in many of the cited studies limits their applicability to real-world patients. Despite the overall lean toward the potential benefits of surgical fixation over conservative management, further research is needed to establish more robust recommendations.

The ongoing FixCon trial which is a multicentric randomised control trial looking to study the outcome after operative versus nonoperative treatment of multiple simple rib fractures, could provide further insight into this topic by examining thoracic pain during different activities and analysing complications in detail (Wijffels et al., 2019).

The CWIS recommends that outpatient follow-up visits for patients with rib fractures include the reporting of analgesic modalities, pain scores, and pulmonary function. This can be assessed using bedside spirometry or formal pulmonary function tests. It is important to note that there is currently no validated quality-of-life metric specific to rib fractures. If other validated scores are utilised to assess quality of life, it is recommended to explicitly mention and reference the specific score being used (Pieracci et al., 2021).

The multicentred randomised controlled trial provides strong evidence for the benefits of early intervention in reducing short-term adverse outcomes. Future research should aim for inclusivity in patient demographics and assess long-term outcomes.

Cost-effectiveness

In a comprehensive retrospective study using data from the National Inpatient Sample, Choi et al., (2021) conducted a detailed investigation into the cost-effectiveness of SSRF in adults. Analysing three years of data, the study offered an extensive overview of hospital discharges across the United States, with a focus on patients of varying ages and the presence or absence of a flail chest, using propensity score matching for accurate comparison (Choi et al., 2021).

The study's findings strengthen the argument for the cost-effectiveness of SSRF, particularly in patients with flail chest. It showed that SSRF is a financially viable treatment option, being cost-effective at a threshold of US $150,000/QALY, and balancing patient benefits with healthcare costs. However, the study also uncovered a notable underutilisation of SSRF despite its proven clinical and economic benefits, pointing to systemic issues in healthcare, such as provider awareness or insurance coverage hesitations (Chou et al., 2015).

The research particularly highlights the importance of considering patient demographics in SSRF discussions. For older patients, SSRF not only aids immediate recovery but also significantly enhances long-term quality of life, making it a valuable investment. In younger patients, the benefits of SSRF are even more pronounced, offering substantial long-term health advantages. These findings call for a re-evaluation of the healthcare system, suggesting a need to expand SSRF use in these specific patient groups.

Summary

The increasing focus on early SSRF is supported by medical society guidelines, showing benefits like reduced hospital stay and ventilation duration. However, potential complications necessitate caution. The need for standardised guidelines for patient selection and surgery timing is critical, along with comprehensive research into long-term outcomes, quality of life, and cost-effectiveness, to optimise patient care in rib fracture management.

References

1Guidance. (n.d.). *Insertion of metal rib reinforcements to stabilize a flail chest wall*. Guidance, NICE.

Bauer, F., Haag, S., Najafi, K., Miller, B., & Kepros, J. (2023). Surgical stabilization of rib fracture patients versus nonoperative controls treated by a multidisciplinary team in a single institution. *Heliyon*, 9(4), e15205. Available from https://doi.org/10.1016/j.heliyon.2023.e15205.

Bedetti, B., Patrini, D., Bertolaccini, L., Crisci, R., Solli, P., Schmidt, J., & Scarci, M. (2018). Focus on specific disease-part 2: The European Society of Thoracic Surgery chest wall database. *Journal of Thoracic Disease*, 10(S29), S3500. Available from https://doi.org/10.21037/jtd.2018.05.115.

Bulger, E. M., Arneson, M. A., Mock, C. N., & Jurkovich, G. J. (2000). Rib fractures in the elderly. *The Journal of Trauma: Injury, Infection, and Critical Care*, 48(6), 1040–1047. Available from https://doi.org/10.1097/00005373-200006000-00007.

Chen, S. A., Liao, C. A., Kuo, L. W., Hsu, C. P., Ouyang, C. H., & Cheng, C. T. (2022). The surgical timing and complications of rib fixation for rib fractures in geriatric patients. *Journal of Personalized Medicine*, 12(10), 1567. Available from https://doi.org/10.3390/jpm12101567.

Choi, J., Mulaney, B., Laohavinij, W., Trimble, R., Tennakoon, L., Spain, D. A., Salomon, J. A., Goldhaber-Fiebert, J. D., & Forrester, J. D. (2021). Nationwide cost-effectiveness analysis of surgical stabilization of rib fractures by flail chest status and age groups. *Journal of Trauma and Acute Care Surgery*, 90(3), 451–458. Available from https://doi.org/10.1097/TA.0000000000003021, http://journals.lww.com/jtrauma.

Chou, Y. P., Lin, H. L., & Wu, T. C. (2015). Video-assisted thoracoscopic surgery for retained hemothorax in blunt chest trauma. *Current Opinion in Pulmonary Medicine*, 21(4), 393–398. Available from https://doi.org/10.1097/mcp.0000000000000173.

Dabezies, E. J., & D'Ambrosia, R. (1986). Fracture treatment for the multiply injured patient. *Instructional Course Lectures*, 35, 13–21.

Girsowicz, E., Falcoz, P.-E., Santelmo, N., & Massard, G. (2012). Does surgical stabilization improve outcomes in patients with isolated multiple distracted and painful non-flail rib fractures? *Interactive Cardiovascular and Thoracic Surgery*, 14(3), 312–315. Available from https://doi.org/10.1093/icvts/ivr028.

Harrell, K. N., Jean, R. J., Dave Bhattacharya, S., Hunt, D. J., Barker, D. E., & Maxwell, R. A. (2020). Late operative rib fixation is inferior to nonoperative management. *American Surgeon*, *86*(8), 944−949. Available from https://doi.org/10.1177/0003134820942185, https://journals.sagepub.com/home/asua.

Insertion of metal rib reinforcements to stabilise a flail chest wall. (2010). *National Institute for Health and Clinical Excellence*.

Iqbal, H. J., Alsousou, J., Shah, S., Jayatilaka, L., Scott, S., Scott, S., & Melling, D. (2018). Early surgical stabilization of complex chest wall injuries improves short-term patient outcomes. *Journal of Bone and Joint Surgery - American*, *100*(15), 1298−1308. Available from https://doi.org/10.2106/JBJS.17.01215, http://jbjs.org/issues.aspx.

Kasotakis, G., Hasenboehler, E. A., Streib, E. W., Patel, N., Patel, M. B., Alarcon, L., Bosarge, P. L., Love, J., Haut, E. R., & Como, J. J. (2017). Operative fixation of rib fractures after blunt trauma: A practice management guideline from the Eastern Association for the Surgery of Trauma. *Journal of Trauma and Acute Care Surgery*, *82*(3), 618−626. Available from https://doi.org/10.1097/TA.0000000000001350, http://journals.lww.com/jtrauma.

Kelley, K. M., Burgess, J., Weireter, L., Novosel, T. J., Parks, K., Aseuga, M., & Collins, J. (2019). Early use of a chest trauma protocol in elderly patients with rib fractures improves pulmonary outcomes. *American Surgeon*, *85*(3), 288−291. Available from https://www.ingentaconnect.com/contentone/sesc/tas/2019/00000085/00000003/art00034.

Li, Y., Gao, E., Yang, Y., Gao, Z., He, W., Zhao, Y., Wu, W., Zhao, T., & Guo, X. (2020). Comparison of minimally invasive surgery for non-flail chest rib fractures: A prospective cohort study. *Journal of Thoracic Disease*, *12*(7), 3706−3714. Available from https://doi.org/10.21037/jtd-19-2586.

Marasco, S. F., Balogh, Z. J., Wullschleger, M. E., Hsu, J., Patel, B., Fitzgerald, M., Martin, K., Summerhayes, R., & Bailey, M. (2022). Rib fixation in non-ventilator-dependent chest wall injuries: A prospective randomized trial. *Journal of Trauma and Acute Care Surgery*, *92*(6), 1047−1053. Available from https://doi.org/10.1097/TA.0000000000003549, http://journals.lww.com/jtrauma.

Otaka, S., Aso, S., Matsui, H., Fushimi, K., & Yasunaga, H. (2020). Effectiveness of surgical fixation for rib fractures in relation to its timing: A retrospective Japanese nationwide study. *European Journal of Trauma and Emergency Surgery*, *2020*, 1−8.

Patel, D. D., Zambetti, B. R., & Magnotti, L. J. (2024). Timing to rib fixation in patients with flail chest. *Journal of Surgical Research*, *294*, 93−98. Available from https://doi.org/10.1016/j.jss.2023.09.057, http://www.elsevier.com/inca/publications/store/6/2/2/9/0/1/index.htt.

Pieracci, F. M., Coleman, J., Ali-Osman, F., Mangram, A., Majercik, S., White, T. W., Jeremitsky, E., & Doben, A. R. (2018). A multicenter evaluation of the optimal timing of surgical stabilization of rib fractures. *Journal of Trauma and Acute Care Surgery*, *84*(1), 1−10. Available from https://doi.org/10.1097/TA.0000000000001729, http://journals.lww.com/jtrauma.

Pieracci, F. M., Leasia, K., Bauman, Z., Eriksson, E. A., Lottenberg, L., Majercik, S., Powell, L., Sarani, B., Semon, G., Thomas, B., Zhao, F., Dyke, C., & Doben, A. R. (2020). A multicenter, prospective, controlled clinical trial of surgical stabilization of rib fractures in patients with severe, nonflail fracture patterns (Chest Wall Injury Society NONFLAIL). *Journal of Trauma and Acute Care Surgery*, *88*(2), 249−257. Available from https://doi.org/10.1097/TA.0000000000002559, http://journals.lww.com/jtrauma.

Pieracci, F. M., Leasia, K., Hernandez, M. C., Kim, B., Cantrell, E., Bauman, Z., Gardner, S., Majercik, S., White, T., Dieffenbaugher, S., Eriksson, E., Barns, M., Christie, D. B., Lasso, E. T., Schubl, S., Sauaia, A., & Doben, A. R. (2021). Surgical stabilization of rib fractures in octogenarians and beyond-what are the outcomes? *Journal of Trauma and Acute Care Surgery*, *90*(6), 1014−1021. Available from http://journals.lww.com/jtrauma.

Pieracci, F. M., Majercik, S., Ali-Osman, F., Ang, D., Doben, A., Edwards, J. G., French, B., Gasparri, M., Marasco, S., Minshall, C., Sarani, B., Tisol, W., VanBoerum, D. H., & White, T. W. (2017). Consensus statement: Surgical stabilization of rib fractures rib fracture colloquium clinical practice guidelines. *Injury*, *48*(2), 307−321. Available from https://doi.org/10.1016/j.injury.2016.11.026, www.elsevier.com/locate/injury.

Pieracci, F. M., Schubl, S., Gasparri, M., Delaplain, P., Kirsch, J., Towe, C., White, T. W., Whitbeck, S. A., & Doben, A. R. (2021). The Chest Wall Injury Society recommendations for reporting studies of surgical stabilization of rib fractures. *Injury*, *52*(6), 1241−1250. Available from https://doi.org/10.1016/j.injury.2021.02.032, www.elsevier.com/locate/injury.

Prins, J. T. H., Van Lieshout, E. M. M., Reijnders, M. R. L., Verhofstad, M. H. J., & Wijffels, M. M. E. (2020). Rib fractures after blunt thoracic trauma in patients with normal versus diminished bone mineral density: A retrospective cohort study. *Osteoporosis International*, *31*(2), 225−231. Available from https://doi.org/10.1007/s00198-019-05219-9.

Qiu, M., Shi, Z., Xiao, J., Zhang, X., Ling, S., & Ling, H. (2016). Potential benefits of rib fracture fixation in patients with flail chest and multiple non-flail rib fractures. *Indian Journal of Surgery*, *78*(6), 458−463. Available from https://doi.org/10.1007/s12262-015-1409-2.

Raza, S., & Eckhaus, J. (2022). Does surgical fixation improve pain and quality of life in patients with non-flail rib fractures? A best evidence topic review. *Interactive Cardiovascular and Thoracic Surgery*, *35*(3), 1569−9285. Available from https://doi.org/10.1093/icvts/ivac214.

Rib Fractures, 2013. https://www.aast.org/resources-detail/rib-fractures [Accessed Jul 16, 2023].

Su, Y.-H., Yang, S.-M., Huang, C.-H., Ko, H.-J., & Lee, H.-S. (2019). Early versus late surgical stabilization of severe rib fractures in patients with respiratory failure: A retrospective study. *PLoS One*, *14*(4), e0216170. Available from https://doi.org/10.1371/journal.pone.0216170.

Van Vledder, M. G., Kwakernaak, V., Hagenaars, T., Van Lieshout, E. M. M., Verhofstad, M. H. J., Boonstra, O., den Hoed, P. T., Jakma, T., van Niekerk, J. L. M., de Rijcke, P. A. R., Roukema, G. R., de Ridder, V. A., Schmidt, G. B., & Waleboer, M. (2019). Patterns of injury and outcomes in the elderly patient with rib fractures: A multicenter observational study. *European Journal of Trauma and Emergency Surgery*, *45*(4), 575−583. Available from https://doi.org/10.1007/s00068-018-0969-9, http://link.springer.com/journal/68.

Wang, Z., Jia, Y., & Li, M. (2023). The effectiveness of early surgical stabilization for multiple rib fractures: A multicenter randomized controlled trial. *Journal of Cardiothoracic Surgery*, *18*(1). Available from https://doi.org/10.1186/s13019-023-02203-7.

Wijffels, M. M. E., Prins, J. T. H., Polinder, S., Blokhuis, T. J., De Loos, E. R., Den Boer, R. H., Flikweert, E. R., Pull Ter Gunne, A. F., Ringburg, A. N., Spanjersberg, W. R., Van Huijstee, P. J., Van Montfort, G., Vermeulen, J., Vos, D. I., Verhofstad, M. H. J., & Van Lieshout, E. M. M. (2019). Early fixation versus conservative therapy of multiple, simple rib fractures (FixCon): Protocol for a multicenter randomized controlled trial. *World Journal of Emergency Surgery*, *14*(1). Available from https://doi.org/10.1186/s13017-019-0258-x, http://www.wjes.org/.

Zhang, J. P., Sun, L., Li, W. Q., Wang, Y. Y., Li, X. Z., & Liu, Y. (2019). Surgical treatment of patients with severe non-flail chest rib fractures. *World Journal of Clinical Cases*, *7*(22), 3718−3727. Available from https://doi.org/10.12998/wjcc.v7.i22.3718, https://www.wjgnet.com/2307-8960/about.htm.

Zhu, R. C., de Roulet, A., Ogami, T., & Khariton, K. (2020). Rib fixation in geriatric trauma: Mortality benefits for the most vulnerable patients. *Journal of Trauma and Acute Care Surgery*, *89*(1), 103−110. Available from https://doi.org/10.1097/ta.0000000000002666.

Ziegler, D. W., & Agarwal, N. N. (1994). The morbidity and mortality of RIB fractures. *Journal of Trauma - Injury, Infection and Critical Care, 37*(6), 975–979. Available from https://doi.org/10.1097/00005373-199412000-00018.

Kasotakis, G., Hasenboehler, E. A., Streib, E. W., Patel, N., Patel, M. B., Alarcon, L., Bosarge, P. L., Love, J., Haut, E. R., Como, J. J. (2017). *Rib fractures, open reduction and internal fixation of (update in process) – Practice management guideline.*

Chapter 7

Modern approach to rib fixation: surgical techniques

Federico Raveglia[1], Riccardo Orlandi[1], Federica Danuzzo[1], Ugo Cioffi[2] and Marco Scarci[3,4]

[1]*San Gerardo Hospital, Thoracic Surgery Department, Monza, Italy,* [2]*Department of Surgery, University of Milan, Milan, Italy,* [3]*Department of Thoracic Surgery, Imperial College NHS Healthcare Trust, London, United Kingdom,* [4]*National Heart and Lung Institute, Imperial College, London, United Kingdom*

Introduction

Rib fractures are recorded in at least 10%–20% of chest trauma (Ziegler & Agarwal, 1994) representing the most frequent injury. Their management varies from conservative treatment to surgical repair based on symptom severity. The conservative approach is mainly based on pain control and positive pressure ventilation whereas surgery involves fracture reduction and rib fixation by prosthetic devices.

Flail chest, the contiguous segment of three or more ribs with plurime fractures supporting chest wall paradoxical movement during respirations, represents the most dramatic scenario, associated with respiratory failure and high mortality. Surgical repair is commonly accepted by most thoracic surgeons in the case of chest flail; whereas in all the other cases consensus regarding indications and timing is still poor (Sawyer et al., 2022).

Surgical rib reduction and internal fixation have been first described in the 1940–1950s, but as soon as positive pressure ventilation technology had improved it became the preferred strategy even for flail chest. Then, in the 1980s, surgical repairing hardware systems were developed and the advantages of fixation on respiration have been better understood (Mohr et al., 2007; Slobogean et al., 2015). Therefore, surgeons have focused once again their attention on open reduction and internal repair; literature confirms this growing trend so much, that it has increased 10-fold over the past decade. Despite ongoing guidelines (Chest Wall Injury Society "Guideline for surgical stabilization of rib fractures (SSRF) indications, contraindications and timing"), it is a common opinion that the surgical approach is still underused (Richardson et al., 2007) and a 10-year ago survey revealed that only 26% of surgeons who managed patients deserving surgery had experience with the topic (Mayberry et al., 2009). This concern mainly originates from the idea that the disadvantages of a thoracotomy fixation may be greater than the advantages of the fixation itself. Indeed, this surgery usually involves multiple ribs or multiple rib segments and requires good chest wall exposure by large skin incisions and muscle disruption. That could potentially further worsen respiratory mechanics. Overcoming these points would reduce the surgical impact and enlarge indications including fewer severe cases that would still benefit in terms of chronic pain. Fortunately, new approaches are emerging and thanks to advances in technology and VATS experience, minimally invasive techniques are now available also for chest wall. In this chapter, our attention has been focused on a tree among the most promising advances: minimally invasive surgical rib fixation, 3D printing technology, and evolution of the surgical approaches.

Minimal invasive surgical rib fixation

Recent studies have definitively confirmed the advantages of surgical ribs stabilization over no surgical approach in the chest blunt trauma especially in case of flail chest or chronic pain (Granetzny et al., 2005; Tanaka et al., 2002). However, despite much evidence in favor of surgery even in no flail chest cases, concerns still exist; mainly because of thoracotomy invasiveness and possible side effects (Pieracci et al., 2020). Indeed, the traditional technique is based on extra thoracic fixation systems applied through large skin incisions and muscle dissection or cutting. With advances in technology and the advent of minimally invasive surgery, VATS seems to be the next natural step to look forward to. Thoracoscopy for rib fracture stabilization may present many logical advantages. First of all, the intra pleural vision allows an easier and more accurate fracture localization. Indeed, despite ultra sound and 3D computed tomography (CT) scan chest wall reconstruction advent, correct intraoperative fracture localization is always one of the most challenging issues. Second, VATS allows a comprehensive pleural cavity exploration that could help diagnose and treat concurrent post-traumatic intrathoracic injuries. Then, VATS enlarges the fracture fixation range (for example in case of retro scapular ribs involvement) and makes stabilization safer avoiding intercostal vessels, nerves, and lung parenchyma injuries through small incisions and direct intra-thoracic visualization. Moreover, this approach adds the advantages of minimally invasive surgery: faster recovery and esthetic outcomes. Lastly, VATS may avoid scapular distraction and eliminate the touch ability of the bone fracture plate to the scapula through medial cortical fixation (Zens et al., 2018; Zhang et al., 2019).

Despite these possible advantages, VATS advent has been limited by two issues: poor practical experience and the absence of tailored instruments (Pierracci, 2019). Surgeons are also discouraged by the following challenges:

(1) potential longer operative time, (2) potential plates dislodging into the pleural cavity. The dislodgment is probably one of the more terrific complications requiring a new VATS, or thoracotomy, to remove and replace plates but also an exploration of the wound to remove nuts and washers.

To date, minimally invasive techniques can be basically separated into two groups. One is based on the classic internal rib fixation system but supported by thoracoscope assistance; the other is performed by an intrapleural thoracoscopic approach. The first involves plate placement on the outer face of the ribs, whereas the second is on the inner one.

Unfortunately, a disturbing confusion between the two video-assisted techniques distinctions is still present in the literature. Summarizing, the main distinctive criteria are (1) the plate's position on the inner or the outer rib face and (2) the extra versus intra pleural cavity approach. When the thoracoscope does not enter the pleural cavity and the plates are positioned on the outer rib surface, the technique is called "internal rib fixation," whereas when the plates are on the inner face and the thoracoscope enters the pleural cavity, the technique is called "intrathoracic rib fixation."

Internal rib fixation

Internal rib fixation is characterized by fracture repair through an extra-thoracic VATS incision for rib plate placement. This approach avoids the classic thoracotomy exposition and related muscle disruption; in fact, thanks to the thoracoscope assistance, thoracotomy has been replaced by single or multiple small skin incisions.

In 2018, Merchant and Onugha (2018) described their innovative approach to internal rib fixation. This consists of (1) elevating sub-muscular, extra-thoracic flaps using a balloon dilator (2) insufflation of this space (3) video-assisted visualization to repair the rib fractures using extra-thoracic plates.

After chest palpation to detect the wall instability and optimal incision point, an extra-thoracic space between the chest wall muscles and the outer face of the ribs has been obtained using a "space maker balloon" and wound retractor. This artificial space guarantees optimal surgical field exposure and allows camera vision. After fracture identification, stabilization of the titanium plate/screw system has been obtained by placing bilateral sutures around the flail segments. These maneuvers should be easily achieved through optimal exposure and a combination of right-angle and straight-angle electric drills. Unfortunately, the author's experience consists of a single case report and, despite it being safe and effective, there is a lack of evidence.

In the same period, Xia et al. (2018) reported a retrospective study of 84 cases describing another similar approach. Patients were divided into groups (minimally invasive internal fixation and traditional incision surgery group) to compare the techniques. Indeed, the authors have developed an internal support system (also called internal support system of chest wall [ISSW]) capable of distracting the muscular chest wall internally to successfully achieve a thoracoscope-assisted minimally invasive open reduction and internal rib fracture fixation. Follows a step-by-step technical description:

- fracture localization based on physical examination and 3D-CT;
- minimal skin incision over the midpoint of upper and lower fractures edges;
- chest wall muscles dissociation and cut off of the attachment points;
- ISSW is used to move part of the dissociated osseous and muscular thoraxes to create the space to insert into a thoracoscope;
- exposure of the fractures;
- anatomical reduction of the broken edges sparing intercostal muscles, nerves, and vascular structures;
- memory alloy rib bone plates (ice immersion) placement on both sides of the broken ends using the special rib plate holding the clamp in ISSW;
- sterile saline gauze at $\sim 60°C$ on the surface of the rib plate to reshape the jaw to hold the rib and fix fractures.

Results showed in favor of the novel vs the traditional approach in terms of operating time, total drainage volume, tardive chest numbness, postoperative pain, C-reactive protein levels, and pulmonary functions. Despite good results, further studies and more data are needed before to definitively support it.

Intrathoracic rib fixation

The intrathoracic technique involves (1) the introduction of the thoracoscope in the pleural cavity and (2) plate positioning on the inner rib face. This, in turn, provides for two different approaches: (1) full VATS, without any further skin incision apart from the uni o multiports thoracostomies, and (2) partial VATS, with the use of utility skin incisions over the fractures.

Full VATS approach

In 2019, Pieracci et al. (2020) introduced the full intra-thoracic approach by the use of traditional rib plates. The technique is performed using traditional plates usually intended to be placed on the outer rib surface. This is the first weakness, since they need to be reshaped during surgery; plates contoured to match the inner surface are not still available. The second weakness is that traditional plates need special articulating drilling and screws tailored for thoracoscopy but such technology is rare and still rudimentary.

Unfortunately, despite the Authors have well described the technique's theoretical advantages versus traditional thoracotomy, there is not any comparative study. Therefore, they concluded that this approach can not still be considered a standard of care.

Zhang et al. (2022), Xia et al. (2018) and Yang et al. (2023) improved the technique. They introduced a complete uniportal approach to fixing ribs using an innovative dedicated memory alloy rib coaptation board instead of traditional plates. Follows a step-by-step technical description:

- fractured ribs found by thoracoscope and preoperative chest 3D CT reconstruction;
- fractures reduction using oval forceps with an elongated notch and teeth;
- rib coaptation board selection and cooling preparation (0°C ice sterile saline to open the arms);
- rib coaptation board connection to the clip head of the detachable clip applier; (Fig. 7.1).
- assembly connection to the detachable clip applier;
- board application to the target rib; (Fig. 7.2).
- rib coaptation board washing with 40°C–45°C sterile saline to obtain its original shape to clasp and fix fractures.

Authors reported their experience with 35 cases concluding that full video-assisted thoracic surgery (VATS) guidance for internal fixation of rib coaptation boards is easy to perform and may guarantee good results.

In our opinion the main issues in favor of this approach are: (1) no drilling or binding is required (2) board dynamic and continuous self-pressurization function (3) intercostal blood vessels and nerves damage sparing (4) absence of discomfort caused by the traditional plates on the chest wall surface.

On the contrary, the approach described by Pieracci et al. (2020) presents some disadvantages. The first is represented by the mechanical limits of today's fixation devices, such as the 90-degree screwdriver to thoracoscopic ribs repair; the second is that intrathoracic rib plates and screws may dislodge into the chest.

Partial VATS approach

This technique is based on a thoracoscopic approach combined with small utility skin incisions for plate anchoring. The approach involves a selling system. Eight French catheters are fed through drill holes on each side of the fracture. Guide wires are then fed through the catheters and pulled to place plates under the ribs to reduce fractures. The plates are stabilized with 2 bolts to keep the reduction. Intrathoracic anatomically contoured titanium plates, combined intra and extra-thoracic passage guide wires and bolts are part of an innovative system recently approved for use by the Food and Drug Administration (SIG Medical wins food and drug authority (FDA) nod for Advantage Rib fracture repair device. Available online: https://www.massdevice.com/sig-medical-wins-fda-nod-for-advantage-rib-fracture-repair-device/).

In 2021 for the first time in literature, Bauman et al. described the case of a 48-year-old male with four right-sided fractured ribs after a ground-level

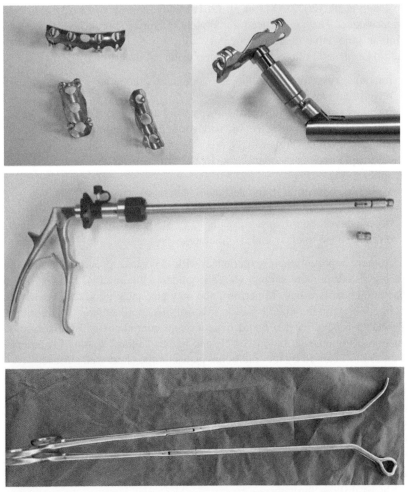

FIGURE 7.1 Rib coaptation board and its detachable clip applier. The picture shows a rib coaptation board, its applier, and the board connection to the clip head of the detachable clip applier. Oval forceps for fracture reduction have also been reported. *From Yang, Z., Wen, M., Kong, W., Li, X., Liu, Z., & Liu, X. (2023). Complete uni-port video-assisted thoracoscopic surgery for surgical stabilization of rib fractures: A case report. Journal of Cardiothoracic Surgery, 18(1), 61.*

fall and presenting poor pain control and breathing mechanics (Bauman et al., 2021), managed to adopt this new minimally invasive technique. Follow a step-by-step technical description:

- small incision/incisions in the middle of any palpable rib fractures for external exposure;
- pleural space entering through a posterior incision;

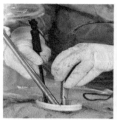

FIGURE 7.2 The fixation procedure. The picture represents some steps of the fixation procedures. In particular, board application to the target rib is well shown. *From Yang, Z., Wen, M., Kong, W., Li, X., Liu, Z., & Liu, X. (2023). Complete uni-port video-assisted thoracoscopic surgery for surgical stabilization of rib fractures: A case report.* Journal of Cardiothoracic Surgery, 18(1), 61.

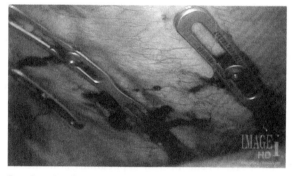

FIGURE 7.3 Intrathoracic plates. Intrathoracic plates with bicortical fixation and adequate reduction of rib fractures are shown. *From Bauman, Z. M., Beard, R., & Cemaj, S. (2021). When less is more: A minimally invasive, intrathoracic approach to surgical stabilization of rib fractures.* Trauma Case Reports, 32, 100452. https://doi.org/10.1016/j.tcr.2021.100452. Erratum in: Trauma Case Rep. 2023 Feb 17;45:100795. Erratum in: Trauma Case Rep. 2023 Mar 01;45:100811.

- holes drilling through non-fractured ribs at least 1 cm on either edge of the fracture to allow for rubber guide tubes passage from the inside of the chest cavity outward;
- plate insertion into the chest through the thoracostomy and its guide to the fractures via cables fed through the guide tubes.
- fractures reduction, using the cables attached to the plate;
- plates fixation from the outside using bolts and nuts (Fig. 7.3).

The authors observed excellent outcomes in terms of pain control and breathing mechanics restoration; therefore, concluded in favor of further studies to support its use.

In 2022, Castater et al. published the first case series of 10 patients managed with this new system (Castater et al., 2022), describing their learning curve moving from a hybrid technique where both internal anterior and intrathoracic plates were adopted to a complete use of intrathoracic plates with smaller and smaller skin incisions. Outcomes were encouraging despite population heterogeneity; therefore, they supported the potential advantage of this system that guarantees smaller incisions, less pain, hemothorax evacuation, and fewer wound complications. Both intrathoracic VATS approaches are aimed to preserve the chest wall integrity but only the full VATS one avoids any skin incision and muscle disruption. To conclude, all the techniques described in this chapter are innovative and potentially effective but more data are still needed before to be considered standard of care.

The application of 3D-printing technology in rib fracture fixation

3D printing technology

Three-dimensional printing (3D-P), also known as additive manufacturing, gained an important role in the medical world in the last decades.

3D-P creates three-dimensional objects by layering several materials starting from two-dimensional images which are usually obtained from CT or magnetic resonance scans. It represents an innovative and useful tool in the medical field for different specialties thanks to its ability to produce automatically complex geometries with precision and customization (Tack et al., 2016).

One of the most significant medical applications is in the surgical field: 3D-P technology has changed and improved the way of teaching trainees, the surgical approach to challenging cases, and the availability of new customized prostheses and materials (Pontiki et al., 2023).

Nowadays, thoracic anatomic models have been used to plan or simulate surgery in cases of complex anatomy, risky tumor resections, chest wall reconstruction, and complex segmentectomy planning (Kwok et al., 2018).

In the context of rib fracture fixation, 3D-P technology allows the creation of patient-specific models reproducing accurately the size and shape of thorax and rib fractures: this helps in preoperative planning, fabrication of personalized implants, and shaping of the locking plates.

Preoperative planning with 3D printing models

Surgery of highly complex rib fractures presents unique challenges due to the complex anatomy of the rib cage, the difficult position of fractures, multiple fracture segments, the locking plate shaping, and its placement.

Criteria for high complex rib fractures definition are one to three fractures in the II-IV ribs with ≥ 3 fracture segments for each rib; the length of the middle fracture segment ≤ 5 cm; costal cartilage fracture (Zhou, Zhang, Xie, et al., 2019).

According to CT scans and data from patients, 3D-P provides anatomically and customized precise models.

These models are a tangible representation of the patient's rib cage anatomy, allowing surgeons to locate the rib fracture site, assess the fracture pattern, plan sites and length of incisions, simulate the reduction and fixation techniques, shape fixation plates (Kwok et al., 2018; Zhang et al., 2019).

This detailed and accurate preoperative planning enhances surgical procedures, reduces operation time and risk of complications, and improves clinical outcomes and aesthetical results.

Customized implants, surgical guides and templates, minimally invasive SSRF

One of the specific advantages of 3D-P technology in rib fracture fixation is the possibility of fabricating customized implants. Nowadays, standard fixation tools such as plates and screws are routinely used, but in some demanding cases, they may not provide an ideal fit for the patient. This level of customization ensures a surgical approach tailored to the specific needs of the patient, improving the accuracy of reduction and fixation and the rib cage stability.

In addition to patient-specific models and implants, 3D-P technology can also be used to create surgical guides and templates, which provide precise guidance for screw placement and fixation. By using these tools, surgeons can achieve greater accuracy, ensure precise screw trajectory and optimal screw placement, minimizing the risk of screw misplacement, nerve and vascular injury, or damage to adjacent structures. The guides are designed preoperatively with patient-specific models.

Furthermore, the use of surgical guides and templates can play a great role in minimally invasive rib fracture fixation. The limited visibility and access can hinder accurate screw placement: surgical guides and templates provide a guide for the insertion of screws through small incisions used in this minimally invasive approach.

This personalized and precise procedure can lead to improved outcomes, reduced operative times, postoperative pain and complications, and shorter recovery times (Zhou et al., 2021; Zhou, Zhang, Yang, et al., 2019).

Potential for training and education

3D-P models can also be used in training programs to teach surgeons and residents about rib fracture fixation and screw placement techniques. In this way, trainees can practice on patient-specific models and simulate the surgical procedure to develop their skills and gain confidence before performing the procedure on actual patients.

Future directions and challenges

As technology continues to develop and become more accessible, it is expected that 3D-P will play an increasingly significant role in rib fracture fixation. Even though several challenges still need to be addressed: the cost for 3D printers and materials can be significant, limiting its widespread adoption; the integration of 3D- into existing surgical pathways and the standardization of techniques may be difficult; this technology is not suitable for emergency conditions because 3D-P is time-consuming (Chen et al., 2019); additionally, long-term studies are needed to evaluate the durability and clinical outcomes associated with 3D-printed implants.

Evolution of the surgical approaches

Surgical approaches

Rib fixation could be performed through several different approaches, which have gradually developed throughout the decades. Each of them has a range of applications, depending on the location of fractured ribs. Conventionally, rib fractures are grouped in three zones: anterior, from sternum to anterior axillary line, middle, from anterior to the posterior axillary line, and posterior, from posterior axillary line to backbone. Recently, further subdivision in six zones has been proposed (Zhang et al., 2022): zone 1 (cartilage zone) containing costal cartilages and arch area, zone 2 (chest zone) containing pectoralis major and minor muscles until clavicle midline, zone 3 (lateral costal zone) containing middle area from posterior axillary line to subscapular back-midline, zone 4 (high posterior costal zone) containing the back area of the first four ribs until caput costae, zone 5 (low posterior subscapular coastal zone) containing the back area of 5th–12th ribs from subscapular back-midline to the lateral edge of erector spinae muscles, zone 6 (low posterior costal paraspinal zone) containing the back area of 5th–12th ribs from the lateral edge of erector spinae muscles to caput costae. Each zone has its anatomical landmarks guiding the proper incision, based on rib fracture lines. Whatever the involved zone, each approach has its pros and cons. Eventually, deep knowledge of surface topography, together with thorough preoperative CT scan study, preferably by 3-dimensional reconstructions, and intraoperative ultrasound aid are highly recommended to get the most out of each approach minimizing risks of iatrogenic injuries.

Standard open approach

Traditionally, conventional postero-lateral thoracotomy has been the approach of choice until the early 2000s (Fig. 7.4).

Based on lung surgery experience, it ensures great exposure of intrathoracic structures, allowing for a satisfactory field of view of the thoracic wall, too. Nonetheless, it is burdened by a high morbidity rate (Taylor et al., 2013;

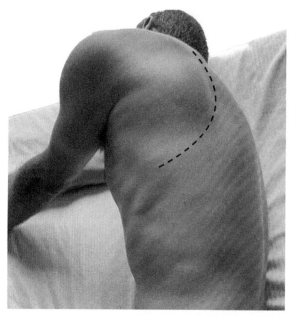

FIGURE 7.4 Postero-lateral thoracotomy. Traditional postero-lateral thoracotomy.

Zhang et al., 2019) due to transection of the trapezius, rhomboid major, and latissimus dorsi muscles as well as extensive damage to local blood vessels and nerves. Usually, it provides great access to lateral and posterior zones, although apical and anterior areas are hardly accessible. The patient is on lateral decubitus with the ipsilateral arm hanged overhead or prepped to be intraoperatively manipulated to lateralize the scapula allowing for better access to some fractures. Since this traumatic incision often harmed more than the fracture itself, it has been abandoned and limited to a few urgent or emergency settings. Another traditional approach is the anterolateral longitudinal thoracotomy (Fig. 7.5).

Fig. 7.5, which gains access to the entire anterolateral ribcage (Zhang et al., 2019). The patient is in the supine position, variably inclined up to 45 degrees towards the healthy side, and the arm is abducted to 90 degrees. After dissecting subcutaneous tissues, the attachments of the serratus anterior, minor, and major pectoralis muscles are divided and retracted to reach the ribcage. Albeit providing an excellent view of the anterolateral portion of ribs, this access doesn't allow to reach the posterolateral and posterior zones, besides being burdened by high morbidity and functional disability.

Modified open approaches

According to principle of broad exposure but without compromising on principles of minimal injury, functionality-saving and aesthetic appearance, modified

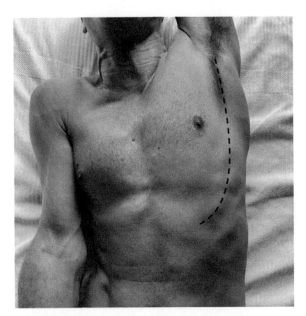

FIGURE 7.5 Antero-lateral thoracotomy. Traditional anterolateral thoracotomy.

patterns of standard open access rapidly spread throughout the world (de Campos & White, 2018; Greiffenstein et al., 2019; Pieracci et al., 2017; Taylor et al., 2013; Xia et al., 2020; Zhang et al., 2019): modified postero-lateral thoracotomy, muscle-sparing lateral thoracotomy, axillary approach, inframammary approach, posterior approach. The tenets of these approaches are to avoid muscle transection and nervous injury. Indeed, muscles are lifted and retracted after dissecting subcutaneous tissue, whereas thoraco-dorsal, long thoracic, intercostal, and inter-costobrachial nerves are protected. The modified postero-lateral thoracotomy includes the same positioning as described, but trapezius, rhomboid major, and latissimus dorsi muscles are preserved (Fig. 7.6).

It ensures good access to lateral, posterolateral, and posterior ribs. Lateral thoracotomy overlaps the postero-lateral one without reaching the medial board of the scapula: latissimus dorsi muscle is preserved and retracted whereas the serratus anterior muscle is split along its fibers (Fig. 7.7).

It allows access to ribs 4th—8th in their lateral area; potentially, the incision could be extended posteriorly or anteriorly to have access to the posterior or anterior area, respectively. The axillary approach ensures large exposure of anterolateral ribs, though without gaining access to the chondral zone and to the posterior area (Fig. 7.8).

The patient is on lateral decubitus, and the incision is carried vertically along the mid-axillary line; again, the serratus anterior muscle can be split in line with its fibers. The inframammary approach allows great access to anterior ribs, comprising the cartilage segment (Fig. 7.9).

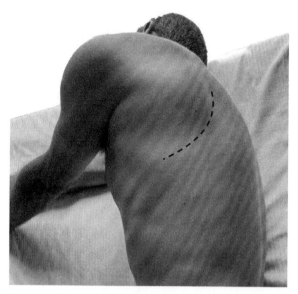

FIGURE 7.6 Muscle-sparing postero-lateral thoracotomy.

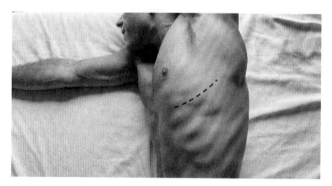

FIGURE 7.7 Muscle-sparing lateral thoracotomy.

The patient is on supine decubitus and the incision is carried along the mammillary crease and the pectoralis major is lifted, together with the breast. The field of view is limited to the anterior zone, but if needed the incision can be extended laterally into antero-lateral thoracotomy or axillary approach. In case of high anterior fractures, an incision directly over the fractured ribs is recommended, by dividing major and minor pectoral muscles along their fibers (Fig. 7.10).

The posterior approach is performed in the auscultation triangle, within the medial border of the scapula, latissimus dorsi, and trapezius muscles (Fig. 7.11).

116 Chest Blunt Trauma

FIGURE 7.8 Muscle-sparing axillary thoracotomy.

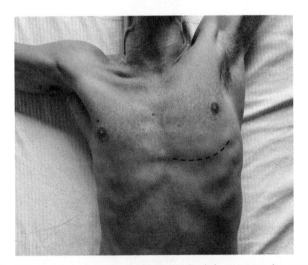

FIGURE 7.9 Inframammary thoracotomy. Muscle-sparing inframammary thoracotomy.

The patient is in the prone position. This incision provides good exposure to the rear infra-scapular chest wall. Generally, the first two (1st−2nd) and last two (11th−12th) ribs as well as posterior fractures within 2 cm from the transverse process are not accessible through muscle-sparing approaches; however, as a matter of fact, the target ribs are usually 3rd−10th, since it is better to repair easily accessible rib fractures through muscle-sparing approaches within the exposed surgical field than repairing all the fractured ribs by extensive and traumatic incisions (de Campos & White, 2018).

Tunnel-based open approaches

Generally, the specific position of incision in open approaches is determined according to the location and number of fractures, attempting to choose the

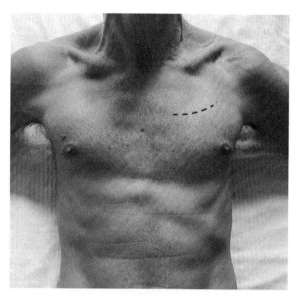

FIGURE 7.10 Pectoral thoracotomy. Muscle-sparing pectoral thoracotomy.

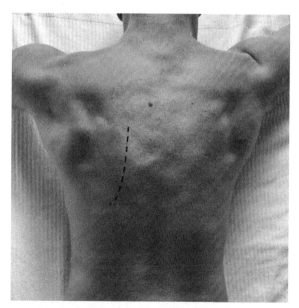

FIGURE 7.11 Posterior thoracotomy. Muscle-sparing posterior thoracotomy.

center of multiple consecutive fractures as a guideline. In the case of solitary rib fractures away from the incision, the tunnel-based technique could be adopted (de Campos & White, 2018; Taylor et al., 2013; Xia et al., 2020).

Recently developed right-angled drilling and screw-driving tools have simplified this technique. Instead of massively dissecting tissue or extremely extending the incision, a thin tunnel over the periosteum of the fractured rib can be created from the incision towards the fracture, a plate is run along the tunnel and is fixed by specific locking pin through percutaneous trocar ancillary access.

Video-assisted thoracoscopic hybrid approaches

Hybrid VATS approaches have been developed from the practice of using a thoracoscope to explore the pleural cavity during open repair of rib fractures, allowing for hemothorax evacuation and hemostasis, identification, and repair of pulmonary or diaphragmatic injuries, as well as accurate placement of chest drains or loco-regional anesthesia devices (de Campos & White, 2018; Pieracci et al., 2017; Xia et al., 2020). Furthermore, the thoracoscope could allow us to identify and reach rib fractures otherwise inaccessible, such as those beneath the scapula (Butterfield et al., 2023). By precisely locating rib fractures, VATS can intraoperatively help plan the best incision to reach each fracture (Bae et al., 2023). Another application of the hybrid technique is exemplified by the extra-thoracic VATS approach (Merchant & Onugha, 2018): through minimal incision, a chest wall expander is inserted beneath muscles creating an operative space to introduce a thoracoscope, which can allow realize endoscopic-guided tunnels for rib plating.

Video-assisted thoracoscopic approaches

This approach represents the implementation of the VATS hybrid technique. Thanks to the development of specific instrumentations, a totally VATS approach allows rib fixation from inside the chest wall without extensively dissecting tissues or damaging critical vascular and nervous structures (Merchant & Onugha, 2018; Mischler et al., 2022; Pieracci et al., 2017; Xia et al., 2020). Additionally, besides all the above-cited advantages, totally VATS approach allows internal plate placement, avoiding uncomfortable sensations of palpable or dislodged plates and ensuring better biomechanical performance. Indeed, this endoscopic technique is based on reducing and drilling the fractured ribs from inside and placing plates or coaptation board to the inner cortex of the rib under direct view, minimizing trauma and injuries to extra-thoracic structures. Furthermore, a totally VATS technique could allow repair of those fractures traditionally deemed inaccessible, such as 1st–2nd, extremely posterior, and subscapular ribs. Naturally, this approach requires specific equipment to ensure reliable results, which is still under implementation, justifying its lack of employment on a large scale. Surgical rib fracture repair has been demonstrated to improve clinical outcomes in selected patients (Craxford et al., 2022), explaining the increasing procedures of rib fixation observed in recent years. Technological advancements, technical improvements, and instrumental developments have allowed the implementation of minimally invasive approaches, which are going to replace traditional open accesses, by ensuring lower morbidity and mortality rates together with enhanced results.

References

Bae, C. M., Son, S. A., Lee, Y. J., & Lee, S. C. (2023). Clinical outcomes of minimally invasive surgical stabilization of rib fractures using video-assisted thoracoscopic surgery. *Journal of Chest Surgery*, 56(2), 120–125. Available from https://doi.org/10.5090/jcs.22.119, http://www.jchestsurg.org/main.html.

Bauman, Z. M., Beard, R., & Cemaj, S. (2021). When less is more: A minimally invasive, intrathoracic approach to surgical stabilization of rib fractures. *Trauma Case Reports*, 32, 100452. Available from https://doi.org/10.1016/j.tcr.2021.100452.

Butterfield, J. H., Hessey, J. A., & Reparaz, L. (2023). Trans-scapular approach to intrathoracic rib plating of upper rib fractures: An innovative technique. *Trauma Case Reports*, 46, 100840. Available from https://doi.org/10.1016/j.tcr.2023.100840.

de Campos, J. R. M., & White, T. W. (2018). Chest wall stabilization in trauma patients: Why, when, and how? *Journal of Thoracic Disease*, 10, S951. Available from https://doi.org/10.21037/jtd.2018.04.69, http://jtd.amegroups.com/article/download/20411/pdf.

Castater, C., Hazen, B., Davis, C., Hoppe, S., Butler, C., Grant, A., Archer-Arroyo, K., Maceroli, M., Todd, S. R., & Nguyen, J. (2022). Video-assisted thoracoscopic internal rib fixation. *American Surgeon*, 88(5), 994–996. Available from https://doi.org/10.1177/00031348211060450, https://journals.sagepub.com/home/ASU.

Chen, Y. Y., Lin, K. H., Huang, H. K., Chang, H., Lee, S. C., & Huang, T. W. (2019). The beneficial application of preoperative 3D printing for surgical stabilization of rib fractures. *PLoS One*, 13(10). Available from https://doi.org/10.1371/journal.pone.0204652, https://journals.plos.org/plosone/article/file?id = 10.1371/journal.pone.0204649&type = printable.

Craxford, S., Owyang, D., Marson, B., Rowlins, K., Coughlin, T., Forward, D., & Ollivere, B. (2022). Surgical management of rib fractures after blunt trauma: A systematic review and meta-analysis of randomised controlled trials. *Annals of the Royal College of Surgeons of England*, 104(4), 249–256. Available from https://doi.org/10.1308/rcsann.2021.0148, https://publishing.rcseng.ac.uk/doi/abs/10.1308/rcsann.2021.0148.

Granetzny, A., El-Aal, M. A., Emam, E. R., Shalaby, A., & Boseila, A. (2005). Surgical versus conservative treatment of flail chest. Evaluation of the pulmonary status. *Interactive Cardiovascular and Thoracic Surgery*, 4(6), 583–587. Available from https://doi.org/10.1510/icvts.2005.111807, http://icvts.ctsnetjournals.org/cgi/reprint/4/6/583?ck = nck.

Greiffenstein, P., Tran, M. Q., & Campeau, L. (2019). Three common exposures of the chest wall for rib fixation: Anatomical considerations. *Journal of Thoracic Disease*, 11, S1034. Available from https://doi.org/10.21037/jtd.2019.03.33, http://www.jthoracdis.com/.

Kwok, J. K. S., Lau, R. W. H., Zhao, Z. R., Yu, P. S. Y., Ho, J. Y. K., Chow, S. C. Y., Wan, I. Y. P., & Ng, C. S. H. (2018). Multi-dimensional printing in thoracic surgery: Current and future applications. *Journal of Thoracic Disease*, 10, S756. Available from https://doi.org/10.21037/jtd.2018.02.91, http://jtd.amegroups.com/article/download/19968/pdf.

Mayberry, J. C., Ham, L. B., Schipper, P. H., Ellis, T. J., & Mullins, R. J. (2009). Surveyed opinion of American trauma, orthopedic, and thoracic surgeons on rib and sternal fracture repair. *Journal of Trauma – Injury, Infection and Critical Care*, *66*(3), 875–879. Available from https://doi.org/10.1097/TA.0b013e318190c3d3.

Merchant, N. N., & Onugha, O. (2018). Extra-thoracic video-assisted thoracoscopic surgery rib plating and intra-thoracic VATS decortication of retained hemothorax. *Surgical Technology International*, *33*, 251–254.

Mischler, D., Schopper, C., Gasparri, M., Schulz-Drost, S., Brace, M., & Gueorguiev, B. (2022). Is intrathoracic rib plate fixation advantageous over extrathoracic plating? A biomechanical cadaveric study. *Journal of Trauma and Acute Care Surgery*, *92*(3), 574–580. Available from https://doi.org/10.1097/TA.0000000000003443, http://journals.lww.com/jtrauma.

Mohr, M., Abrams, E., Engel, C., Long, W. B., & Bottlang, M. (2007). Geometry of human ribs pertinent to orthopedic chest-wall reconstruction. *Journal of Biomechanics*, *40*(6), 1310–1317. Available from https://doi.org/10.1016/j.jbiomech.2006.05.017.

Pieracci, F. M., Leasia, K., Bauman, Z., Eriksson, E. A., Lottenberg, L., Majercik, S., Powell, L., Sarani, B., Semon, G., Thomas, B., Zhao, F., Dyke, C., & Doben, A. R. (2020). A multicenter, prospective, controlled clinical trial of surgical stabilization of rib fractures in patients with severe, nonflail fracture patterns (Chest Wall Injury Society NONFLAIL). *Journal of Trauma and Acute Care Surgery*, *88*(2), 249–257. Available from https://doi.org/10.1097/TA.0000000000002559, http://journals.lww.com/jtrauma.

Pieracci, F. M., Majercik, S., Ali-Osman, F., Ang, D., Doben, A., Edwards, J. G., French, B., Gasparri, M., Marasco, S., Minshall, C., Sarani, B., Tisol, W., VanBoerum, D. H., & White, T. W. (2017). Consensus statement: Surgical stabilization of rib fractures rib fracture colloquium clinical practice guidelines. *Injury*, *48*(2), 307–321. Available from https://doi.org/10.1016/j.injury.2016.11.026, http://www.elsevier.com/locate/injury.

Pieracci, F. M. (2019). Completely thoracoscopic surgical stabilization of rib fractures: Can it be done and is it worth it? *Journal of Thoracic Disease*, *11*.

Pontiki, A. A., Rhode, K., Lampridis, S., & Bille, A. (2023). Three-dimensional printing applications in thoracic surgery. *Thoracic Surgery Clinics*, *33*(3), 273–281. Available from https://doi.org/10.1016/j.thorsurg.2023.04.012.

Richardson, J. D., Franklin, G. A., Heffley, S., & Seligson, D. (2007). Operative fixation of chest wall fractures: An underused procedure? *American Surgeon*, *73*(6), 591–596.

Sawyer, E., Wullschleger, M., Muller, N., & Muller, M. (2022). Surgical rib fixation of multiple rib fractures and flail chest: A systematic review and meta-analysis. *Journal of Surgical Research*, *27*, 221–234. Available from https://doi.org/10.1016/j.jss.2022.02.055, http://www.elsevier.com/inca/publications/store/6/2/2/9/0/1/index.htt.

Slobogean, G. P., Kim, H., Russell, J. P., Stockton, D. J., Hsieh, A. H., & O'Toole, R. V. (2015). Rib fracture fixation restores inspiratory volume and peak flow in a full thorax human cadaveric breathing model. *Archives of Trauma Research*, *4*(4). Available from https://doi.org/10.5812/atr.28018.

Tack, P., Victor, J., Gemmel, P., & Annemans, L. (2016). 3D-printing techniques in a medical setting: A systematic literature review. *Biomedical Engineering Online*, *15*(1). Available from https://doi.org/10.1186/s12938-016-0236-4, http://www.biomedical-engineering-online.com/start.asp.

Tanaka, H., Yukioka, T., Yamaguti, Y., Shimizu, S., Goto, H., Matsuda, H., & Shimazaki, S. (2002). Surgical stabilization of internal pneumatic stabilization? A prospective randomized study of management of severe flail chest patients. *Journal of Trauma*, *52*(4), 727–732. Available from https://doi.org/10.1097/00005373-200204000-00020.

Taylor, B. C., French, B. G., & Fowler, T. T. (2013). Surgical approaches for rib fracture fixation. *Journal of Orthopaedic Trauma*, *27*(7), e168. Available from https://doi.org/10.1097/BOT.0b013e318283fa2d.

Xia, H., Zhu, D., Li, J., Sun, Z., Deng, L., Zhu, P., Zhang, Y., Li, X., & Wang. (2020). Current status and research progress of minimally invasive surgery for flail chest. *Experimental and Therapeutic Medicine*, *19*(1), 421−427.

Xia, H., Zhu, P., Li, J., Zhu, D., Sun, Z., Deng, L., Zhang, Y., & Wang, D. (2018). Thoracoscope combined with internal support system of chest wall in open reduction and internal fixation for multiple rib fractures. *Experimental and Therapeutic Medicine*, *16*(6), 4650−4654. Available from https://doi.org/10.3892/etm.2018.6817, http://www.spandidos-publications.com/etm/16/6/4650/download.

Yang, Z., Wen, M., Kong, W., Li, X., Liu, Z., & Liu, X. (2023). Complete uni-port video-assisted thoracoscopic surgery for surgical stabilization of rib fractures: A case report. *Journal of Cardiothoracic Surgery*, *18*(1). Available from https://doi.org/10.1186/s13019-023-02167-8, https://cardiothoracicsurgery.biomedcentral.com.

Zens, T., Beems, M. V., & Agarwal, S. (2018). Thoracoscopic, minimally invasive rib fixation after trauma. *Trauma (United Kingdom)*, *20*(2), 142−146. Available from https://doi.org/10.1177/1460408616681600, https://journals.sagepub.com/home/TRA.

Zhang, D., Zhou, X., Yang, Y., Xie, Z., Chen, M., Liang, Z., & Zhang, G. (2022). Minimally invasive surgery rib fracture fixation based on location and anatomical landmarks. *European Journal of Trauma and Emergency Surgery*, *48*(5), 3613−3622. Available from https://doi.org/10.1007/s00068-021-01676-2, http://link.springer.com/journal/68.

Zhang, Q., Song, L., Ning, S., Xie, H., Li, N., & Wang, Y. (2019). Recent advances in rib fracture fixation. *Journal of Thoracic Disease*, *11*, S1070. Available from https://doi.org/10.21037/jtd.2019.04.99, http://www.jthoracdis.com/.

Zhou, X., Zhang, D., Xie, Z., Chen, M., Yang, Y., Liang, Z., & Zhang, G. (2019). 3D printing and thoracoscopy assisted MIPO in treatment of long-range comminuted rib fractures, a case report. *Journal of Cardiothoracic Surgery*, *14*(1). Available from https://doi.org/10.1186/s13019-019-0892-0, http://www.cardiothoracicsurgery.org/.

Zhou, X., Zhang, D., Xie, Z., Yang, Y., Chen, M., Liang, Z., Zhang, G., & Li, S. (2021). Application of 3D printing and framework internal fixation technology for high complex rib fractures. *Journal of Cardiothoracic Surgery*, *16*(1). Available from https://doi.org/10.1186/s13019-020-01377-8, http://www.cardiothoracicsurgery.org/.

Zhou, X. T., Zhang, D. S., Yang, Y., Zhang, G. L., Xie, Z. X., Chen, M. H., & Liang, Z. (2019). Analysis of the advantages of 3D printing in the surgical treatment of multiple rib fractures: 5 cases report. *Journal of Cardiothoracic Surgery*, *14*(1). Available from https://doi.org/10.1186/s13019-019-0930-y, http://www.cardiothoracicsurgery.org/.

Ziegler, D. W., & Agarwal, N. N. (1994). The morbidity and mortality of RIB fractures. *Journal of Trauma − Injury, Infection and Critical Care*, *37*(6), 975−979. Available from https://doi.org/10.1097/00005373-199412000-00018.

Chapter 8

Postoperative complications

Rachel Chubsey[1], Savannah Gysling[2] and Edward J. Caruana[3]
[1]Nottingham City Hospital, Nottingham, United Kingdom, [2]Queen Elizabeth Hospital Birmingham, Birmingham, United Kingdom, [3]Glenfield Hospital, Leicester, United Kingdom

Introduction

Chest trauma is often managed nonsurgically, with many patients recovering with observation alone. Surgery may be indicated at different time points after injury and for various indications, but is associated with its own risk of complications. It is often challenging to distinguish between complications of the original injury and those related to a superimposed surgical intervention (Nirula, 2018).

Postoperative complications have an important impact on survival, in-hospital recovery, postdischarge rehabilitation, the ability and timing of return to activities of daily living, and quality of life. The effect of potential postoperative complications on the patient's return to health is critical to the process of patient selection for surgery and during shared decision-making, particularly where nonsurgical options exist.

This chapter focuses on complications associated with rib fixation for traumatic rib fractures.

Preexisting patient factors, including advancing age, malnutrition, smoking, diabetes, peripheral vascular disease, rheumatoid arthritis, and vitamin D deficiency are strong determinants of the incidence and outcome of postoperative complications (Chen et al., 2022). Due consideration is required when weighing the role of surgery in individual patients. Although several prediction models have been developed to quantify the risk of poor outcomes in patients with major trauma across a variety of anatomical regions, such as the Injury Severity Score (ISS) (Baker et al., 1974), New Injury Severity Score (NISS) (Osler et al., 1997), and Trauma Injury Severity Score (TRISS) (Boyd et al., 1987), these have not been shown to correlate with postoperative complications after rib fracture fixation (Chen et al., 2022).

Similarly, the more specific scoring systems developed for use in patients with chest trauma, such as the Rib Fracture Score (RFS) (Fokin et al., 2018), the Chest Trauma Score (CTS) (Chen et al., 2014), and the RibScore (Chapman et al., 2016), have not been designed to differentiate between patients who will or will not benefit from surgery. At present, no universally accepted risk stratification tool for patients with traumatic rib fractures reliably predicts the failure of conservative management or the risk of postoperative complications for the selected patients who undergo open reduction and internal fixation (ORIF) (Callisto et al., 2022; Williams et al., 2020; Wycech et al., 2020).

There is a high incidence of admission to the intensive care unit (ICU) postsurgery ($>75\%$) with long lengths of stay (Bae et al., 2023; Hoepelman et al., 2023). It is not always clear if this is a direct result of the initial injury or secondary to the surgical intervention. Several studies have demonstrated longer hospital and ICU lengths of stay following fixation (Kheirbek et al., 2022; Martin et al., 2023) However, a large Systematic Review demonstrated a reduced ICU length of stay in those under 60 years, who underwent ORIF compared with conservative management ($P < .03$). This benefit however was not seen in those over 60 years (Sawyer et al., 2022). There was however a significant survival benefit seen across all age groups in those who underwent fixation, compared to conservative treatment ($P < .04$) (Sawyer et al., 2022). Of note, patients in the original studies were highly selected.

Quality-of-life measures have been used to evaluate the long-term outcome following rib fixation after chest trauma. Results are varied, with some studies suggesting no benefit and even a worse outcome following fixation, in areas such as mobility and ability to self-care in the first month, plus lower EuroQol-5D scores compared to the general population (Beks et al., 2019; Marasco et al., 2015; Meyer et al., 2023; Walters et al., 2019). However, a systematic review in 2020 did identify four studies reporting a weighted mean EQ-5D index of 0.80 indicating good quality of life postsurgical intervention (Peek et al., 2020). Moreover, Caragounis et al. (2016) argue that outcome measures following rib fixation should not be measured earlier than 12 months, as they observed improvement in the following parameters post 12 months, including quality of life, overall perceived health, and lung function testing.

There are also important implications for healthcare-related resource use and subsequent direct and indirect societal costs. Surgical intervention invariably increases costs due to equipment, personnel, and theater utilization. Increased length of stay following surgery, which may be secondary to postoperative complications, is also thought to increase costs when compared with conservative management (Sarode et al., 2021). However, studies have demonstrated broader cost-effectiveness with increased Quality Adjusted Life Years (QALY) for surgical intervention (Swart et al., 2017).

The incidence of complications after rib fixation is 10%−15% (Beks et al., 2019; Chen et al., 2022; Peek et al., 2020). The most common complications postsurgery can be broadly divided into pulmonary, surgical, implant-related, and bone healing (Peek et al., 2020). They can also be considered as early; empyema, wound infection, hematoma and effusion, and late postoperative complications; chest tightness, dyspnea, chronic pain, hardware dislodgement, hardware infection, and periprosthetic fracture.

Early procedure-related complications

Bleeding

The most common cause of postoperative bleeding is ongoing or recurrent bleeding secondary to the index traumatic injury. In cases of significant blood loss, this may present in the form of a hemothorax and hemodynamic instability. Clinically important bleeding as a direct result of surgical rib fracture fixation is uncommon, although it has been reported in approximately 1.5% of cases, secondary to iatrogenic trauma to the intercostal bundle during plate application and fracture manipulation (Peek et al., 2020).

Iatrogenic injury to the intercostal bundle can result in arterial or venous hemorrhage and if not identified swiftly may result in significant blood loss, need for blood transfusion, and reoperation. Plate systems designed for rib fixation sometimes require screw placement through the rib, whereby the screw is driven through the bone. While the direction of the screw should be perpendicular to the rib, malposition, and migration of screws can result in direct injury to the intercostal bundle. Likewise, the use of crimped plate systems, designed to attach to the ribs superiorly and inferiorly with small metal struts, also brings a risk for vascular injury (Nirula & Mayberry, 2010).

Formation of a wound hematoma is more common following thoracotomy for fracture fixation due to the increased manipulation of tissues (Bae et al., 2023). There is also an increased incidence of cases of flail chest (2.7%), likely secondary to the larger, more extensive incisions and tissue mobilization required for access (Beks et al., 2019).

Pulmonary complications

The incidence of postoperative pneumonia is reportedly as high as 64% (Chen et al., 2022) and is the most common complication postrib fixation (Peek et al., 2020). In a propensity-matched cohort of 147 patients, those who underwent rib fixation experienced an increased incidence of atelectasis, compared to those who received conservative management (12.2% vs 5.4%) (Shibahashi et al., 2019). Furthermore, there appears to be a significant association between delayed rib fracture fixation of more than 3 days postinjury, and an increased risk of developing atelectasis (Shibahashi et al., 2019) and pneumonia (Wang et al., 2023).

Pleural effusion

Postoperative pleural effusion has an incidence of 3%−6% (Beks et al., 2019; Hoepelman et al., 2023; Peek et al., 2020). Further progression to empyema occurs in approximately 5% of patients following rib fracture fixation, however, this is much lower than the 10% incidence in patients managed conservatively (Nirula, 2018; Peek et al., 2020). Opportunistic evacuation of hemothorax and washout of the pleural cavity during rib fixation is likely a contributing factor.

Pneumothorax

In a recent systematic review of almost 2000 patients postrib fixation, the incidence of pneumothorax postoperatively was low (3.1%). Most are easily managed with the insertion of a chest drain, and resolve without the need for further intervention (Peek et al., 2020). Prolonged air leak is uncommon, but largely related to the quality of underlying parenchyma.

Ventilatory failure

Acute respiratory distress syndrome (ARDS), the need for prolonged ventilation, and tracheostomy placement (up to 45% of patients in some cohorts) are all complications associated with operative rib fixation but also frequently encountered in patients managed nonoperatively (Peek et al., 2020, 2021; Pieracci et al., 2016; Uchida et al., 2020). This circles back to the discussion of how it is often difficult to untangle complications of injury from those related to surgical treatment. However, when compared to conservative management, rib fixation has been shown to reduce mechanical ventilation time, especially in those with flail chests and those operated on early (Pieracci et al., 2016; Sawyer et al., 2022; Wang et al., 2023).

Wound infection

Surgical site infection following rib fixation is variable, ranging from 2% to30% across the published literature, with increased risk in those who sustain flail injuries (Lodin et al., 2020; Peek et al., 2020; Thiels et al., 2016). Factors contributing to the development of wound infections include the mechanism of injury, the presence of a traumatic, contaminated wound, tissue hypoperfusion, altered immune response, the need for multiple procedures, placement of prehospital chest tube traversing fracture segments, and preoperative diagnosis of pneumonia (Junker et al., 2019; Lodin et al., 2020).

Except for open fractures, most surgical wounds secondary to rib fracture fixation are classified as Class I (clean) wounds. However, nearby open wounds and concomitant infection can result in Class II (clean-contaminated) wounds (Lodin et al., 2020). The most common organisms found in surgical site infections are gram-positive organisms (Bauman et al., 2023; Thiels et al., 2016). A study by Wang et al. (2023), also demonstrated that those patients who underwent delayed fixation of more than 3 days from the time of injury, are more likely to develop surgical site infections, although this trend was not statistically significant.

Surgical techniques that can be used to reduce the risk of surgical site infections include; the removal of preexisting chest drains before surgical incisions, and the placement of a new drain away from the surgical wound and any implanted hardware (Fokin et al., 2020; Lafferty et al., 2011). For open wounds, effective irrigation and debridement of de-vascularized tissue is of paramount importance, with liberal use of perioperative antibiotics (Lafferty et al., 2011).

Neuromuscular weakness

Surgical approaches to rib fixation are varied and are dependent on the location of fractures and traumatic wounds, as well as the access required for fixation. When performing a standard posterolateral thoracotomy, the rhomboid muscles should be handled with care, as excessive intraoperative trauma may result in winging of the scapula (Fokin et al., 2020). Transection of the rhomboids and trapezius muscles is generally not required unless posterior fractures above the 4th rib require fixation. Significant muscle transection for access to rib fractures may result in significant alterations in shoulder cosmesis and function.th

Multiple anterolateral rib fractures may necessitate an axillary approach and splitting of the serratus anterior. Here the long thoracic nerve is at risk as it lies anterior to the muscle. Damage to this nerve will also manifest as winging of the scapula (Taylor et al., 2013). As such the long thoracic nerve should be visualized and protected during the operation to prevent such complications.

Fixation of the lower ribs may require mobilization of the external oblique muscle, and care must be taken to avoid excessive manipulation which may result in abdominal wall hernia formation. Likewise, cryoablation or damage to the intercostal nerves below the 10th rib should be avoided, as this may result in temporary bulging of the upper abdominal wall (Fokin et al., 2020).

Hardware infection and failure

Hardware complications following fixation of rib fractures are reported in up to 3% of cases and include implant infection, migration and breakage, fracture malunion, and nonunion, sometimes necessitating surgical removal or revision (Nirula, 2018; Sarani et al., 2019). The majority of patients with hardware failure are asymptomatic and, in such cases, complications are only identified on routine chest x-ray or other chest imaging incidentally. Symptoms that are commonly reported with hardware complications are pain, clicking, and signs of infection (Sarani et al., 2019).

Hardware infection

Hardware infection is a significant concern for surgeons, often resulting in readmission and/or prolonged hospitalization, antibiotic therapy, and in severe cases further surgical intervention to remove the affected hardware sometimes with associated mechanical sequelae (Bauman et al., 2023; Nirula, 2018). Patients may present with classical features of infection including, erythema, wound breakdown, and discharge of purulent material. However, one should maintain a high index of suspicion for those patients who present with worsening pain or more subtle indicators of infection with no overt source (Thiels et al., 2016).

A recent large, multicenter study of 450 patients demonstrated rates of hardware infection as low as 0.4%, however in the wider literature rates of 0.5%–10% have been reported (Bauman et al., 2023; Thiels et al., 2016). More recent reductions in infection may in part be due to improved trauma care, but also improved surgical techniques including muscle-sparing approaches, smaller incisions, and minimally invasive access, resulting in less chest wall tissue damage (Bauman et al., 2023).

Gram-positive organisms are the most commonly identified organisms, with *Staphylococcus* accounting for approximately 40% of hardware infections (Bauman et al., 2023). However, it is important to obtain samples for microbiological culture and sensitivity assessment, and consultation with a microbiologist to ensure adequate antimicrobial cover and identification of atypical organisms (Thiels et al., 2016).

Management of hardware infections is often a lengthy process and depends on the severity, organism present, and presence of a nonunion or malunion of the underlying fracture. The key management principle is to reduce bacterial load with antibiotics, allowing time for fracture healing to take place.

In the presence of early or superficial infection only, it may be sufficient to manage with antibiotics and if required surgical debridement of the wound and subcutaneous tissues only. Where deep space infection, exposed hardware due to wound dehiscence, or abscess is present, full exploration and debridement are indicated.

The use of antibiotic beads may be used to reduce bacterial load and allow time for fracture union, before removal of infected hardware, and is widely used in other site orthopedic practice. (Junker et al., 2019; Thiels et al., 2016). Current guidelines recommend a double dose of vancomycin (2 g) and gentamicin (2.4 g) mixed in bone cement and remain in place for 2–7 days (Thiels et al., 2016).

The placement of antibiotic beads as a prophylactic measure for patients at risk of hardware infection such as those with prehospital tube thoracostomy, pneumonia, chest wall wounds, high BMI, and hemorrhagic shock has also been evaluated. However, studies suggest that prophylactic use of antibiotic beads may reduce the duration of intravenous and or oral antibiotics, but with no other consistent benefit (Junker et al., 2019).

Another treatment strategy for deep-space infections includes the use of negative pressure wound therapy. Its application applies continuous subatmospheric pressure to surgical wounds, thereby reducing wound edema, bacterial load, and seroma formation, and improving blood flow to promote accelerated wound healing (Lodin et al., 2020).

All infected hardware should be removed, however, timing is crucial, and it is advised, where possible, that hardware removal be delayed for at least 3 months to allow for fracture healing and stabilization to take place (Thiels et al., 2016). Assessment of fracture healing can be made using computer tomography (CT) scanning. Where infection is severe, but bone healing is not satisfactory, partial hardware removal can be undertaken with a plan to remove remaining hardware in the future (Thiels et al., 2016).

Mechanical hardware failure

Rib fixation hardware must be durable and able to withstand continuous movement with up to 25,000 breathing cycles per day (Lafferty et al., 2011). Increased rigidity of plate and screw systems is hypothesized to be one of the causes of implant breakage, migration, and chronic pain (Prins et al., 2023). Unlike cortical bone, ribs are membranous bones with a 1−2 mm thick cortex and a very soft marrow, therefore they do not have a high-stress tolerance to hold screws and plates, which poses several challenges (Nirula, 2018). As such hardware failure is reported between 3% and 7%, with mechanical failure (broken/displaced hardware) accounting for 38% of malfunctions (Choi et al., 2021).

Where titanium plates and locking screw mechanisms are used, the most common failures are plate fracture (47%), screw migration (45%), and plate migration (26%). The most frequent location for hardware failure is in the lateral and posterolateral positions (Sarani et al., 2019). While the currently widely available plate systems are precontoured, significant manipulation by the surgeon before implantation is often needed, thus increasing the risk of plate fracture and subsequent migration (Fokin et al., 2020; Lafferty et al., 2011). However, without manipulation of plates, appropriate apposition of the plate to the rib surface is unlikely, leading to ineffective fracture reduction and subsequent nonunion (Lafferty et al., 2011).

Available plate and screw systems are designed using bicortical locking screws, designed to allow secure attachment to the plate and provide good purchase of the cortex of the rib anteriorly and posteriorly (Fokin et al., 2020). Nevertheless, complications can arise secondary to screw malposition. Screws that are driven too deep may protrude beyond the posterior cortex of the rib and into the pleural cavity leading to damage to lung tissue. Conversely, screws that are too short to have good purchase of the posterior cortex of the rib will be loose and prone to migration (Fokin et al., 2020). This is more common in elderly patients with osteoporotic ribs where achieving a secure screw hold is more difficult. Iatrogenic rib fracture may also occur as a result of screw insertion, so it is imperative that holes are pre-drilled with the correct size drill bit and that self-tapping screws are used (Fokin et al., 2020).

For rib fractures situated more posteriorly, beneath the scapula, intramedullary systems can be utilized for fracture fixation. Such systems are fed into the intramedullary canal via a single drill hole in the outer cortex. These systems do not require fixation to the rib on either side of the fracture site, so distraction forces are still able to pull fracture ends apart during movement and respiration and may limit bone healing (Marasco et al., 2015). Intramedullary plates can also perforate through the cortex of the rib and cause damage to adjacent pleura, lungs, and muscles. However, advantages to their use are the avoidance of the neurovascular bundle which can be impinged or damaged by crimped plates and screws (Zhang et al., 2019).

Long-term complications

Fracture nonunion or malunion

Failure of fracture healing following fixation has been reported to occur in up to 10% of patients (McClure et al., 2019). However, a recent large systematic review of 24 studies suggests a much lower rate of 1.3% for fracture nonunion (Peek et al., 2020). A nonunion or malunion can occur for several reasons and may be patient or procedure-related. Patient factors affecting fracture healing include advanced age, poor nutrition, diabetes, obesity, alcoholism, steroid and NSAID use, smoking, and vasculopathy (McClure et al., 2019).

Rib fracture patterns increasing the risk for poor healing include the presence of excessive distraction forces, high degree of displacement, presence of comminution, infection, and soft tissue interposition between fracture ends (DeGenova et al., 2023). Operative fixation is adopted to overcome some of these risks by stabilizing fracture ends and ensuring good approximation for healing.

Where intramedullary plate systems are used Marasco et al. (2015) found in a cohort of 15 patients that only 9% of fractures had a complete bony union at 3 months, with 85% of fractures having undergone only partial healing. However, by 6 months all partially healed fractures had undergone complete bony union.

Where plates and screws are used the average time to complete union is reportedly 3.4 months (Taylor et al., 2013).

Chronic pain and irritation

The long-term effects of chest trauma may be related to damage sustained in the initial injury or as a result of surgical intervention such as rib fixation. A recent systematic review and meta-analysis of 22 studies using a variety of plates, screws, and k-wires, and splints demonstrated that irritation secondary to the application of plates and screws is a common indication for hardware removal, affecting up to 45.7% of patients (Peek et al., 2020). Irritation may present as overlying skin sensitivity, clicking of plates on bony structures, migration of hardware, or pain. Removal of hardware to manage chronic irritation occurs in a reported 3%–15% of cases (Bae et al., 2023; Beks et al., 2019; Hoepelman et al., 2023). Timing of removal should be carefully considered to ensure adequate bony healing has occurred.

Postsurgical pain is a challenge across the spectrum of thoracic surgery and is a common problem following rib fixation, occurring in 5%–22% of patients at 12 months (Khandelwal et al., 2011; Kim et al., 2021).

References

Bae, C. M., Son, S. A., Lee, Y. J., & Lee, S. C. (2023). Clinical outcomes of minimally invasive surgical stabilization of rib fractures using video-assisted thoracoscopic surgery. *Journal of Chest Surgery, 56*(2), 120–125. Available from https://doi.org/10.5090/jcs.22.119, http://www.jchestsurg.org/main.html.

Baker, S. P., O'Neill, B., Haddon, W., & Long, W. B. (1974). The injury severity score: A method for describing patients with multiple injuries and evaluating emergency care. *Journal of Trauma, 14*(3), 187–196. Available from https://doi.org/10.1097/00005373-197403000-00001.

Bauman, Z. M., Sutyak, K., Daubert, T. A., Khan, H., King, T., Cahoy, K., Kashyap, M., Cantrell, E., Evans, C., & Kaye, A. (2023). Hardware infection from surgical stabilization of rib fractures is lower than previously reported. *Cureus.* Available from https://doi.org/10.7759/cureus.35732.

Beks, R. B., de Jong, M. B., Houwert, R. M., Sweet, A. A. R., De Bruin, I. G. J. M., Govaert, G. A. M., Wessem, K. J. P., Simmermacher, R. K. J., Hietbrink, F., Groenwold, R. H. H., & Leenen, L. P. H. (2019). Long-term follow-up after rib fixation for flail chest and multiple rib fractures. *European Journal of Trauma and Emergency Surgery, 45*(4), 645–654. Available from https://doi.org/10.1007/s00068-018-1009-5, http://link.springer.com/journal/68.

Boyd, C. R., Tolson, M. A., & Copes, W. S. (1987). Evaluating trauma care: The TRISS method. *Journal of Trauma - Injury, Infection and Critical Care, 27*(4), 370–378. Available from https://doi.org/10.1097/00005373-198704000-00005.

Callisto, E., Costantino, G., Tabner, A., Kerslake, D., & Reed, M. J. (2022). The clinical effectiveness of the STUMBL score for the management of ED patients with blunt chest trauma compared to clinical evaluation alone. *Internal and Emergency Medicine, 17*(6), 1785–1793. Available from https://doi.org/10.1007/s11739-022-03001-0, http://www.springer.com/italy/home?SGWID = 6-102-70-173668106-0&changeHeader = true.

Caragounis, E. C., Fagevik Olsén, M., Pazooki, D., & Granhed, H. (2016). Surgical treatment of multiple rib fractures and flail chest in trauma: A one-year follow-up study. *World Journal of Emergency Surgery, 11*(1). Available from https://doi.org/10.1186/s13017-016-0085-2, http://www.wjes.org/.

Chapman, B. C., Herbert, B., Rodil, M., Salotto, J., Stovall, R. T., Biffl, W., Johnson, J., Burlew, C. C., Barnett, C., Fox, C., Moore, E. E., Jurkovich, G. J., & Pieracci, F. M. (2016). RibScore: A novel radiographic score based on fracture pattern that predicts pneumonia, respiratory failure, and tracheostomy. *Journal of Trauma and Acute Care Surgery*, *80*(1), 95−101. Available from https://doi.org/10.1097/TA.0000000000000867, http://journals.lww.com/jtrauma.

Chen, J., Jeremitsky, E., Philp, F., Fry, W., & Smith, R. S. (2014). A chest trauma scoring system to predict outcomes. *Surgery*, *156*(4), 988−994. Available from https://doi.org/10.1016/j.surg.2014.06.045.

Chen, S. A., Liao, C. A., Kuo, L. W., Hsu, C. P., Ouyang, C. H., & Cheng, C. T. (2022). The surgical timing and complications of rib fixation for rib fractures in geriatric patients. *Journal of Personalized Medicine*, *12*(10). Available from https://doi.org/10.3390/jpm12101567, http://www.mdpi.com/journal/jpm.

Choi, J., Kaghazchi, A., Sun, B., Woodward, A., & Forrester, J. D. (2021). Systematic review and meta-analysis of hardware failure in surgical stabilization of rib fractures: Who, what, when, where, and why? *Journal of Surgical Research*, *268*, 190−198. Available from https://doi.org/10.1016/j.jss.2021.06.054, http://www.elsevier.com/inca/publications/store/6/2/2/9/0/1/index.htt.

DeGenova, D. T., Miller, K. B., McClure, T. T., Schuette, H. B., French, B. G., & Taylor, B. C. (2023). Operative fixation of rib fracture nonunions. *Archives of Orthopaedic and Trauma Surgery*, *143*(6), 3047−3054. Available from https://doi.org/10.1007/s00402-022-04540-z, https://www.springer.com/journal/402.

Fokin, A. A., Hus, N., Wycech, J., Rodriguez, E., & Puente, I. (2020). Surgical stabilization of rib fractures: Indications, techniques, and pitfalls. *JBJS Essential Surgical Techniques*, *10*(2). Available from https://doi.org/10.2106/JBJS.ST.19.00032, http://surgicaltechniques.jbjs.org/.

Fokin, A., Wycech, J., Crawford, M., & Puente, I. (2018). Quantification of rib fractures by different scoring systems. *Journal of Surgical Research*, *229*, 1−8. Available from https://doi.org/10.1016/j.jss.2018.03.025.

Hoepelman, R. J., Minervini, F., Beeres, F. J. P., van Wageningen, B., IJpma, F. F., van Veelen, N. M., Lansink, K. W. W., Hoogendoorn, J. M., van Baal, M. C. P., Groenwold, R. H. H., & Houwert, R. M. (2023). Quality of life and clinical outcomes of operatively treated patients with flail chest injuries: A multicentre prospective cohort study. *Frontiers in Surgery*, *10*. Available from https://doi.org/10.3389/fsurg.2023.1156489, journal.frontiersin.org/journal/surgery.

Junker, M. S., Kurjatko, A., Hernandez, M. C., Heller, S. F., Kim, B. D., & Schiller, H. J. (2019). Salvage of rib stabilization hardware with antibiotic beads. *American Journal of Surgery*, *218*(5), 869−875. Available from https://doi.org/10.1016/j.amjsurg.2019.02.032, www.elsevier.com/locate/amjsurg.

Khandelwal, G., Mathur, R. K., Shukla, S., & Maheshwari, A. (2011). A prospective single center study to assess the impact of surgical stabilization in patients with rib fracture. *International Journal of Surgery*, *9*(6), 478−481. Available from https://doi.org/10.1016/j.ijsu.2011.06.003.

Kheirbek, T., Martin, T. J., Cao, J., Tillman, A. C., Spivak, H. A., Heffernan, D. S., & Lueckel, S. N. (2022). Comparison of infectious complications after surgical fixation versus epidural analgesia for acute rib fractures. *Surgical Infections*, *23*(6), 532−537. Available from https://doi.org/10.1089/sur.2022.002, https://www.liebertonline.com/sur.

Kim, K. H., Lee, C. K., Kim, S. H., Kim, Y., Kim, J. E., Shin, Y. K., Seok, J., & Cho, H. M. (2021). Prevalence of chronic post-thoracotomy pain in patients with traumatic multiple rib fractures in South Korea: A cross-sectional study. *Scientific Reports, 11*(1). Available from https://doi.org/10.1038/s41598-021-82273-6, https://www.nature.com/srep/index.html.

Lafferty, P. M., Anavian, J., Will, R. E., & Cole, P. A. (2011). Operative treatment of chest wall injuries: Indications, technique, and outcomes. *Journal of Bone and Joint Surgery, 93*(1), 97–110. Available from https://doi.org/10.2106/JBJS.I.00696, http://www.ejbjs.org/cgi/reprint/93/1/97.

Lodin, D., Florio, T., Genuit, T., & Hus, N. (2020). Negative pressure wound therapy can prevent surgical site infections following sternal and rib fixation in trauma patients: Experience from a single-institution cohort study. *Cureus*. Available from https://doi.org/10.7759/cureus.9389.

Marasco, S., Lee, G., Summerhayes, R., Fitzgerald, M., & Bailey, M. (2015). Quality of life after major trauma with multiple rib fractures. *Injury, 46*(1), 61–65. Available from https://doi.org/10.1016/j.injury.2014.06.014.

Martin, T. J., Cao, J. L., Tindal, E., Adams, C. A., Lueckel, S. N., & Kheirbek, T. (2023). Comparison of surgical stabilization of rib fractures vs epidural analgesia on in-hospital outcomes. *Injury, 54*(1), 32–38. Available from https://doi.org/10.1016/j.injury.2022.07.038, https://www.elsevier.com/locate/injury.

McClure., Myers., Triplet., Johnson, J., & Taylor, B. (2019). Surgical treatment of flail chest and rib fractures: A systematic review of the literature. *International Journal of Orthopaedics, 6*, 1039–1044. Available from https://doi.org/10.17554/j.issn.2311-5106.2019.06.284.

Meyer, D. E., Harvin, J. A., Vincent, L., Motley, K., Wandling, M. W., Puzio, T. J., Moore, L. J., Cotton, B. A., Wade, C. E., & Kao, L. S. (2023). Randomized controlled trial of surgical rib fixation to nonoperative management in severe chest wall injury. *Annals of Surgery, 278*(3), 357–365. Available from https://doi.org/10.1097/SLA.0000000000005950, http://journals.lww.com/annalsofsurgery/pages/default.aspx.

Nirula, R., & Mayberry, J. C. (2010). Rib fracture fixation: Controversies and technical challenges. *American Surgeon, 76*(8), 793–802. Available from http://docserver.ingentaconnect.com/deliver/connect/sesc/00031348/v76n8/s20.pdf?expires = 1294763521&id = 60632121&titleid = 11737&accname = Elsevier + Science&checksum = C8CFD3ADA32A5D2078E1E4F2 DAE158D0.

Nirula, R. (2018). *Postoperative complications after rib fracture repair* (pp. 159–163). Springer Science and Business Media LLC. Available from https://doi.org/10.1007/978-3-319-91644-6_14.

Osler, T., Baker, S. P., & Long, W. (1997). A modification of the injury severity score that both improves accuracy and simplifies scoring. *The Journal of Trauma: Injury, Infection, and Critical Care, 43*(6), 922–926. Available from https://doi.org/10.1097/00005373-199712000-00009.

Peek, J., Beks, R. B., Hietbrink, F., Heng, M., De Jong, M. B., Beeres, F. J. P., Leenen, L. P. H., Groenwold, R. H. H., & Houwert, R. M. (2020). Complications and outcome after rib fracture fixation: A systematic review. *Journal of Trauma and Acute Care Surgery, 89*(2), 411–418. Available from https://doi.org/10.1097/ta.0000000000002716.

Peek, J., Beks, R. B., Kremo, V., van Veelen, N., Leiser, A., Houwert, R. M., Link, B.-C., Knobe, M., Babst, R. H., & Beeres, F. J. P. (2021). The evaluation of pulmonary function after rib fixation for multiple rib fractures and flail chest: a retrospective study and systematic review of the current evidence. *European Journal of Trauma and Emergency Surgery, 47*(4), 1105–1114. Available from https://doi.org/10.1007/s00068-019-01274-3.

Pieracci, F. M., Lin, Y., Rodil, M., Synder, M., Herbert, B., Tran, D. K., Stoval, R. T., Johnson, J. L., Biffl, W. L., Barnett, C. C., Cothren-Burlew, C., Fox, C., Jurkovich, G. J., & Moore, E. E.

(2016). A prospective, controlled clinical evaluation of surgical stabilization of severe rib fractures. *Journal of Trauma and Acute Care Surgery*, *80*(2), 187−194. Available from https://doi.org/10.1097/TA.0000000000000925, http://journals.lww.com/jtrauma.

Prins, J. T. H., Van Wijck, S. F. M., Leeflang, S. A., Kleinrensink, G. J., Lottenberg, L., de la Santa Barajas, P. M., Van Huijstee, P. J., Vermeulen, J., Verhofstad, M. H. J., Zadpoor, A. A., Wijffels, M. M. E., & Van Lieshout, E. M. M. (2023). Biomechanical characteristics of rib fracture fixation systems. *Clinical Biomechanics*, *102*. Available from https://doi.org/10.1016/j.clinbiomech.2023.105870, http://www.elsevier.com/locate/clinbiomech.

Sarani, B., Allen, R., Pieracci, F. M., Doben, A. R., Eriksson, E., Bauman, Z. M., Gupta, P., Semon, G., Greiffenstein, P., Chapman, A. J., Kim, B. D., Lottenberg, L., Gardner, S., Marasco, S., & White, T. (2019). Characteristics of hardware failure in patients undergoing surgical stabilization of rib fractures: A Chest Wall Injury Society multicenter study. *Journal of Trauma and Acute Care Surgery*, *87*(6), 1277−1281. Available from https://doi.org/10.1097/TA.0000000000002373, http://journals.lww.com/jtrauma.

Sarode, A. L., Ho, V. P., Pieracci, F. M., Moorman, M. L., & Towe, C. W. (2021). The financial burden of rib fractures: National estimates 2007 to 2016. *Injury*, *52*(8), 2180−2187. Available from https://doi.org/10.1016/j.injury.2021.05.027, http://www.elsevier.com/locate/injury.

Sawyer, E., Wullschleger, M., Muller, N., & Muller, M. (2022). Surgical rib fixation of multiple rib fractures and flail chest: A systematic review and meta-analysis. *Journal of Surgical Research*, *276*, 221−234. Available from https://doi.org/10.1016/j.jss.2022.02.055, http://www.elsevier.com/inca/publications/store/6/2/2/9/0/1/index.htt.

Shibahashi, K., Sugiyama, K., Okura, Y., & Hamabe, Y. (2019). Effect of surgical rib fixation for rib fracture on mortality: A multicenter, propensity score matching analysis. *Journal of Trauma and Acute Care Surgery*, *87*(3), 599−605. Available from https://doi.org/10.1097/ta.0000000000002358.

Swart, E., Laratta, J., Slobogean, G., & Mehta, S. (2017). Operative treatment of rib fractures in flail chest injuries: A meta-analysis and cost-effectiveness analysis. *Journal of Orthopaedic Trauma*, *31*(2), 64−70. Available from https://doi.org/10.1097/bot.0000000000000750.

Taylor, B. C., French, B. G., & Fowler, T. T. (2013). Surgical approaches for rib fracture fixation. *Journal of Orthopaedic Trauma*, *27*(7), e168. Available from https://doi.org/10.1097/BOT.0b013e318283fa2d.

Thiels, C. A., Aho, J. M., Naik, N. D., Zielinski, M. D., Schiller, H. J., Morris, D. S., & Kim, B. D. (2016). Infected hardware after surgical stabilization of rib fractures: Outcomes and management experience. *Journal of Trauma and Acute Care Surgery*, *80*(5), 819−823. Available from https://doi.org/10.1097/TA.0000000000001005, http://journals.lww.com/jtrauma.

Uchida, K., Miyashita, M., Kaga, S., Noda, T., Nishimura, T., Yamamoto, H., & Mizobata, Y. (2020). Long-term outcomes of surgical rib fixation in patients with flail chest and multiple rib fractures. *Trauma Surgery & Acute Care Open*, *5*(1), e000546. Available from https://doi.org/10.1136/tsaco-2020-000546.

Walters, S. T., Craxford, S., Russell, R., Khan, T., Nightingale, J., Moran, C. G., Taylor, A. M., Forward, D. P., & Ollivere, B. J. (2019). Surgical stabilization improves 30-day mortality in patients with traumatic flail chest: A comparative case series at a major trauma center. *Journal of Orthopaedic Trauma*, *33*(1), 15−22. Available from https://doi.org/10.1097/bot.0000000000001344.

Wang, Z., Jia, Y., & Li, M. (2023). The effectiveness of early surgical stabilization for multiple rib fractures: A multicenter randomized controlled trial. *Journal of Cardiothoracic Surgery*, *18*(1). Available from https://doi.org/10.1186/s13019-023-02203-7.

Williams, A., Bigham, C., & Marchbank, A. (2020). Anaesthetic and surgical management of rib fractures. *BJA Education*, *20*(10), 332−340. Available from https://doi.org/10.1016/j.bjae.2020.06.001, https://www.journals.elsevier.com/bja-education.

Wycech, J., Fokin, A. A., & Puente, I. (2020). Evaluation of patients with surgically stabilized rib fractures by different scoring systems. *European Journal of Trauma and Emergency Surgery*, *46*(2), 441−445. Available from https://doi.org/10.1007/s00068-018-0999-3, http://link.springer.com/journal/68.

Zhang, Q., Song, L., Ning, S., Xie, H., Li, N., & Wang, Y. (2019). Recent advances in rib fracture fixation. *Journal of Thoracic Disease*, *11*(S8), S1070. Available from https://doi.org/10.21037/jtd.2019.04.99.

Chapter 9

Rib malunion

Athanasios Kleontas[1] and Kostas Papagiannopoulos[2]
[1]Department of Thoracic Surgery, European Interbalkan Medical Centre, Thessaloniki, Northern Prefecture, Greece, [2]Department of Thoracic Surgery, St. James' University Hospital, Leeds, West Yorkshire, United Kingdom

Rib fractures are prevalent injuries, occurring in approximately 10%–20% of trauma patients and affecting roughly 60%–80% of individuals who experience blunt chest trauma. These fractures vary in severity, ranging from single nondisplaced fractures to multiple segmental fractures, sometimes resulting in flail chest, with mortality rates reaching up to 33%. Patients with reduced pulmonary function are at even greater risk of adverse outcomes. Despite the significant pain and disability associated with these injuries, the majority are managed conservatively, and most heal without surgical intervention. Nonoperative treatment typically involves a combination of multimodal pain management, bronchodilator inhalers, pulmonary physical therapy, oxygen support, and mechanical ventilation if necessary. The majority of fractures heal without significant morbidity. However, in rare cases, nonoperative management is unsuccessful, resulting in symptomatic nonunion, with chronic pain being the most common symptom (DeGenova et al., 2023).

Definition

Malunion, in the context of bone fractures, refers to the improper healing of a broken bone that results in a misalignment of the bone fragments. This condition arises due to inadequate formation of the fracture callus between adjacent ribs, leading to noticeable rib misalignment. This deviation from the normal alignment during the healing process can lead to functional impairment, altered joint mechanics, and other complications (Van Wijck & Wijffels, 2023) (Fig. 9.1).

138 Chest Blunt Trauma

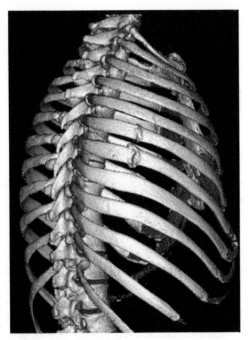

FIGURE 9.1 3D reconstruction of ribs reveals malunion of 6th and 7th rib with significant, ongoing pain. 3D reconstruction of ribs following a computed tomography scan. The study shows the previously broken ribs and the lack of full healing with malunion. This has resulted in significant ongoing pain with movement, deep breathing, and coughing.

Types

Malunion can manifest in various forms based on the nature of the misalignment during the healing of a bone fracture. Some common types include (Schowalter et al., 2022):

- Angular Malunion: The bone fragments heal at an angle, resulting in a deviation from the normal alignment.
- Rotational Malunion: The improper healing causes a twist or rotation of the bone fragments, affecting the natural orientation.
- Shortening Malunion: The bone heals with a length discrepancy compared to its original state, often leading to functional issues.
- Translation Malunion: The bone fragments shift horizontally, causing a lateral displacement during the healing process.
- Heterotopic ossifications (HO) in rib malunion: Rib malunion can also be associated with the formation of abnormal bone growths (HO) in the vicinity of the healed fracture site, further complicating the condition.

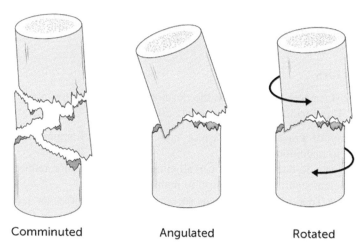

FIGURE 9.2 Different types of rib fractures which can lead to malunion without surgical correction and splinting. This figure shows the different rib fractures developed depending on the mechanism of injury.

- Complex rib malunion: In some cases, rib malunion may involve a combination of angulation, displacement, and heterotopic ossifications, requiring comprehensive treatment strategies.

These types of malunion can occur alone or in combination, and the specific form depends on factors such as the location and severity of the fracture (Fig. 9.2).

Incidence

Fractures of the ribs are more frequently observed in adults compared to children, with the occurrence and prevalence of such fractures being contingent upon the nature and severity of the injury (Schowalter et al., 2022). Elderly individuals are also at higher risk of rib fractures and associated complications (Van Wijck & Wijffels, 2023).

Although specific data regarding the prevalence of rib malunion among various age groups is unavailable, it's noteworthy that children possess more flexible ribs and are less prone to experiencing fractures after blunt chest trauma compared to adults. As individuals age, the probability of sustaining rib fractures and encountering complications such as malunion rises (Peters et al., 2008).

In essence, while detailed statistics on the occurrence of rib malunion across different age demographics are lacking, it is evident that children exhibit lower vulnerability to rib fractures and subsequent malunion, whereas older individuals face increased risks. Adults generally experience rib fractures and malunion more frequently, although precise data for distinct age brackets remains elusive (Peters et al., 2008).

Risk factors

Several risk factors contribute to the development of malunion after a bone fracture. These include (Wuermser et al., 2011):

- Severity of Fracture: Complex or highly displaced fractures are more prone to malunion.
- Inadequate Reduction: Improper alignment during fracture reduction or insufficient stabilization can lead to malunion.
- Delayed Treatment: A delay in seeking medical attention or receiving appropriate treatment can affect the healing process.
- Age: Older individuals may experience slower healing and are more susceptible to malunion.
- Poor Blood Supply: Inadequate blood flow to the fracture site hinders proper healing and increases the risk of malunion.
- Nutritional Deficiencies: Poor nutrition, particularly insufficient intake of essential nutrients like calcium and vitamin D, can impact bone healing.
- Infection: Infections at the fracture site can disrupt the healing process and contribute to malunion.
- Comorbidities: Certain medical conditions, such as diabetes or osteoporosis, can affect bone health and increase the risk of malunion.
- Smoking: Tobacco use can impede blood flow and delay healing, increasing the risk of complications.
- Steroids and nonsteroidal antiinflammatory drug (NSAID) use: These medications can also increase the risk of non union of rib fractures.
- Alcohol abuse: Heavy alcohol use has been associated with an increased risk of rib fractures.
- Diabetes: Diabetes is a risk factor for nonunion of rib fractures.

Identifying and addressing these risk factors is crucial in preventing malunion and promoting optimal bone healing.

Main symptom

The main sign of malunion is typically a visible or palpable deformity or misalignment of the affected rib or joint. In severe cases with a multiplicity of rib malunion the chest wall loses its original shape, the chest cavity becomes flat, and shoulder position can change, especially in upper rib fractures. This can result in functional impairment, pain, and limitations in range of motion. The degree of the deformity and its impact on the individual's ability to perform daily activities often determine the prominence of these symptoms. The main focus in diagnosing malunion is often on assessing the structural misalignment and its consequences on the affected rib or joint (Van Wijck & Wijffels, 2023).

Chronic pain or difficulty in breathing (dyspnea) are other primary symptoms associated with rib malunion, stemming from a previously fractured rib that has healed with displacement and angulation. Such a condition can disrupt regular breathing patterns and physical movement, significantly affecting the overall quality of life over time (Van Wijck & Wijffels, 2023).

Diagnostic approach

Diagnosing rib malunion relies on clinical symptoms, findings from physical examination, and imaging studies. The typical diagnostic process for rib malunion includes the following steps (Holzmacher & Sarani, 2017):

- Clinical Assessment:
- History: Gather information about the initial fracture, treatment received, and the progression of symptoms.
- Physical Examination: Conduct a thorough examination, including visual inspection, range of motion testing, palpation, and functional assessment.
- Pain is a cardinal symptom of rib malunion and an experienced physician must probe the description of pain in great detail.
- What is important is to characterize pain, record intensity with visual analogue score (VAS) score, and connect such with daily activities and restrictions.
- It is also imperative to differentiate organic or mechanical pain from neuropathic pain as such will be important information to discuss treatment options with the patient.
- Imaging Studies:
- X-rays: These are often the initial imaging modality to visualize bone alignment and assess the degree of misalignment in the fracture site.
- Plain chest films might be enough to offer an initial assessment but the information revealed depends on the position of the malunion.
- In a nutshell and for plain rib fractures the anteroposterior (AP) chest film might offer useful information.
- In the rare cases of 1st rib malunion and associated pain, specific lordotic views might give an initial answer.
- Such rib malunions are unusual and rare but should always be suspected in professional athletes as they are related to stress fractures (Fig. 9.3).
- Computed tomography (CT) Scan: In most cases, a CT scan should be used to provide more detailed imaging, especially for complex fractures or when a three-dimensional assessment is needed. The 3D reconstruction of the chest wall is important as it provides a detailed configuration of the malunion, the displacement of the fracture, and possible rib overlapping, all important points for considering the correct treatment modality required to alleviate symptoms Fig. 9.4.

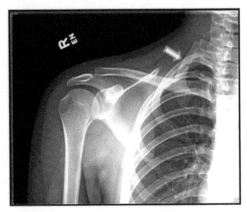

FIGURE 9.3 Right first rib malunion. The patient presented with ongoing pain 6 months following the injury.

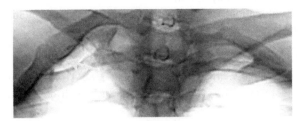

FIGURE 9.4 Right first rib malunion in a tennis player with a stress fracture. This is a chest film of a tennis player who had a spontaneous stress fracture of the first rib.

- Other Diagnostic Tests:
- MRI (magnetic resonance imaging): While less commonly used for malunion, an MRI may be employed to assess soft tissue involvement or complications.
- Bone Scans: In certain situations, a bone scan might be utilized to evaluate blood flow and bone metabolism.

The combination of clinical assessment and imaging studies helps establish a comprehensive understanding of the malunion, guiding healthcare professionals in determining the severity and planning appropriate interventions for correction and management.

Differential diagnosis

When evaluating a patient with symptoms suggestive of malunion, one should consider several conditions in the differential diagnosis. These may include (Van Wijck & Wijffels, 2023):

- Nonunion: A situation where the fracture fails to heal completely, resulting in persistent symptoms similar to malunion.

- Delayed Union: Slower-than-expected healing of a fracture, which may mimic some features of malunion.
- Osteoarthritis: Joint degeneration over time, leading to pain, stiffness, and functional limitations.
- Soft Tissue Injuries: Damage to muscles, tendons, or ligaments around the fracture site, causing pain and functional impairment.
- Infection: Bone infections (osteomyelitis) can present with similar symptoms, including pain and swelling.
- Neurological Disorders: Conditions affecting nerves can result in pain, weakness, or altered sensation.
- Post-Traumatic Stress Disorder: Emotional distress and psychological factors may contribute to perceived symptoms

Treatment options

Treatment modalities for rib malunion may encompass initial conservative management and/or surgical interventions targeted at stabilizing the site of malunion, removing all fibrotic tissue to promote bone migration to the old fracture site, and protecting and releasing, if necessary, the intercostal nerve:

- Conservative Management: Not all patients may be eligible candidates for surgery, and some might prefer nonoperative management approaches, particularly if they have multiple severely displaced rib fractures but do not meet the criteria for surgical intervention (Lafferty et al., 2011). Common sense dictates that displaced malunited fractures need reconstruction but timing becomes an increasingly important factor.

There is mounting evidence and experience that severely displaced fractures which are allowed to heal conservatively can lead to ongoing chronic and debilitating pain.

Unfortunately, late reconstruction, over 6 months from the initial injury, reduces significantly the success of reconstructive surgery as neuropathic pain establishes and surgery becomes much less effective Fig. 9.5.

- Physical Therapy: Targeted exercises to improve range of motion, strength, and functional abilities.
- Corrective Surgery: The decision to pursue surgical intervention should be made following a thorough assessment of symptoms, imaging results, and the potential advantages of surgery in enhancing the quality of life and relieving pain associated with rib malunion (Van Wijck & Wijffels, 2023). In simple terms, corrective surgery should encompass all 3 elements:
- Surgical stabilization of rib fractures (SSRF);
- Remodeling and Excision of malunited part and;
- Rib Reconstruction.

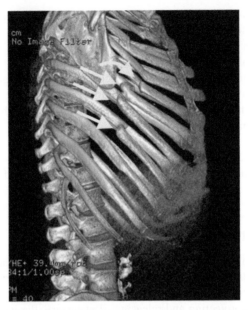

FIGURE 9.5 3D reconstruction of the chest reveals multiple malunited ribs following a bike accident. The patient experienced significant band-type pain and chest wall "stiffness." This is a computed tomography scan with a 3D reconstruction of the chest wall. it shows clearly the healing process of rib fractures and areas of malunion.

- Pain Management: Irrespective of the chosen treatment method, effective pain management is crucial in the management of rib malunion to prevent respiratory insufficiency resulting from pain (Holzmacher & Sarani, 2017). Pain management for rib malunion aims to alleviate ongoing discomfort and prevent complications. Approaches include:
- Pharmacological treatments: Begin pain management with NSAIDs and escalate to opioids if necessary. Avoid medications that could compromise respiratory function, such as benzodiazepines and certain anticholinergics.
- Physical therapies: Advocate for deep breathing exercises and the utilization of incentive spirometry to mitigate the risk of complications like pneumonia.
- Interventions: Contemplate regional blocks or continuous infusions of local anesthetic agents for targeted pain relief (Malekpour et al., 2017).

Surgical treatment

Indications for surgical intervention

Indications for surgical intervention in rib malunion include (Holzmacher & Sarani, 2017; Lafferty et al., 2011; Malekpour et al., 2017; Van Wijck & Wijffels, 2023):

- Restrictive defects leading to compromised respiration or chronic pain.

- Ongoing symptoms persist despite nonoperative management.
- Severe malunion resulting in difficulty breathing, pain, or dependence on narcotic analgesia.
- Evidence of paradoxical breathing in patients being weaned from mechanical ventilation.
- HO cause local compression or pain.
- Complex malunion and nonunion requiring surgical resection of rib synostosis, reconstruction of rib nonunion, and contouring of rib malunion.

Surgical options

Surgical treatment for malunion aims to correct the misalignment of fractured bones and restore optimal function. Depending on the specific circumstances, various surgical procedures may be considered:

- **SSRF**: Surgical intervention is typically recommended for rib malunion cases characterized by a constrictive defect that hampers regular breathing or leads to persistent pain or breathing difficulties. Subsequent segmental rib resection and fixation (SSRF) might be contemplated for individuals experiencing pain due to malunion or nonunion, with limited data indicating potential enhancements in long-term pain scores and evidence of healing on radiographic assessment post-surgery (Holzmacher & Sarani, 2017).
- **Open technique** (Fokin et al., 2020): The choice of incision is dictated by the malunion location, while the surgical technique is tailored to the type of malunion. Single-lung intubation is preferred when respiratory reserves are sufficient. Otherwise, two lung ventilation might be employed. Surgery on malunited ribs usually takes place weeks after the initial injury and lung contusions have already resolved.

The lateral approach serves as the primary surgical approach due to its accessibility to the majority of rib fractures. A curvilinear skin incision is made overlying the malunited ribs. Posterior rib malunions are accessed through a vertical incision within the triangle of auscultation, while anterior malunions are approached through a transverse inframammary incision. It is not unusual to design small and multiple incisions over malunited ribs, especially in areas where the scapula body hinders exposure. Multiple small incisions allow exposure of all targets with muscle fiber sparing and minimal soft tissue trauma compared with single large mutilating incisions (Figs. 9.1 and 9.6).

Whenever possible, the muscle-sparing technique, which involves splitting alongside fibers without transection, is employed and supplemented by muscle retraction.

Stabilization of ribs is usually performed with precontoured rib-specific plates featuring threaded holes and self-tapping locking screws.

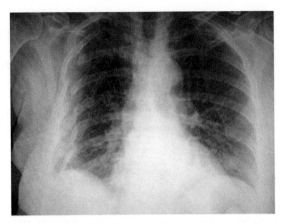

FIGURE 9.6 Severe chest injury 8 months ago, with a combination of nonunion and malunion of rib fractures. Pain was the cardinal symptom. Despite reconstructive surgery patient complained of ongoing, significant pain. This is a chest film taken several months after an injury of the chest. the chest film has been added to show a lack of rib fracture union and significant pain.

There are several such materials available in the market. Sometimes reinforcement with Polymer cables is utilized to reinforce the plating of osteoporotic bones and when bone splinting might happen while threading the screws.

What is most important is to debride generously the area of the malunion and expose cancellous bone, prevent any injury to the intercostal bundle with sparse use of diathermy, respect the periosteum, release any rib overlapping and recontour the rib to the initial shape to prevent impairment of chest wall mechanics.

In case bone gaps present after debridement, these can be bridged with bone grafts. They can be stabilized with one or 2 screws, depending on the length underneath the plate, guaranteeing bone opposition and growth. An alternative and cheaper solution in small defects is to utilize part of the extracted malunited "joints" as long as good quality bone is still available in the resected specimens (Figs. 9.7, 9.8 and Fig. 9.9).

- **Thoracoscopic technique** (Pieracci, 2019): SSRF has become a routine procedure in many high-volume trauma centers. Increased experience with this surgery has led to various technical refinements aimed at reducing incision length, muscle disruption, scapular retraction, and overall tissue trauma. One notable advancement is the adoption of completely thoracoscopic SSRF, utilizing video-assisted thoracoscopic surgery (VATS) techniques to both reduce and fix rib fractures or malunion within the thoracic cavity.

  The benefits of thoracoscopic SSRF can be categorized into those related to rib fracture repair itself and those pertaining to additional procedures

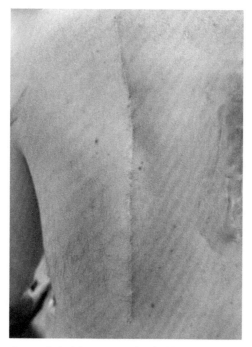

FIGURE 9.7 Posterior vertical right parascapular incision to allow reconstructive surgery on ribs 2–8. This is a picture showing a type of incision utilized for accessing posterior rib fractures.

during rib fracture repair. Regarding the former, potential advantages include enhanced visualization of rib fractures, particularly those located posteriorly or subscapularly, minimized trauma to surrounding muscles and nerves, reduced trauma to intrathoracic structures, and avoidance of palpable plates. In terms of the latter, theoretical benefits encompass drainage of retained haemothorax, precise placement of local anesthesia and chest tubes, and identification and treatment of associated intrathoracic injuries. Additionally, VATS may facilitate trainee education.

However, despite these theoretical advantages, early attempts at thoracoscopic SSRF have been hindered by user inexperience and inadequate instrumentation. Furthermore, there is currently no data comparing the efficacy of completely thoracoscopic SSRF to contemporary minimally invasive extrathoracic SSRF or nonoperative management strategies.

Furthermore, for late reconstructions and diagnosed malunions, the intrapleural space might not be anymore a friendly and clean operating field. Adhesions will certainly hinder dissection close to the malunited ribs and visualization of the target areas is difficult following bleeding, tissue damage, and subsequent recovery.

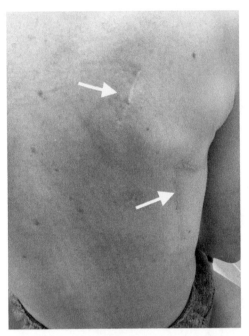

FIGURE 9.8 Reconstruction of rib malunions (ribs 4, 7,8,9) through small individual incisions in order to circumvent the body of the scapula. This is a snapshot from a patient's chest that shows various small skin incisions utilized to access and stabilize rib fractures.

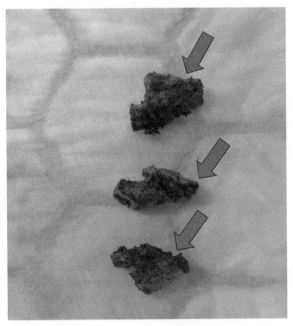

FIGURE 9.9 Resected malunited rib fragments. A picture that shows rib segments removed which contain the pseudoarthroses and areas of malunited bony tissue.

The choice of surgical approach depends on factors such as the location and severity of malunion, the type of fracture, and the overall health of the patient. Rehabilitation and postoperative care are crucial components to optimize outcomes and facilitate recovery.

Associated complications

Potential complications associated with nonsurgical treatment options for rib malunion include (Holzmacher & Sarani, 2017; Van Wijck & Wijffels, 2023):

- Persistent pain: Ongoing discomfort may impact daily activities and overall quality of life.
- Pulmonary complications: Prolonged pain can lead to compromised respiratory function, heightening the risk of pneumonia and impairing lung capacity.
- Dependence on opioids: Extended pain management may necessitate continued use of potent pain relievers, potentially resulting in adverse effects and addiction concerns.
- Secondary osteoarthritis: Malunion-related joint abnormalities may increase the susceptibility to developing arthritis.
- Progression of deformity: Left untreated, malunion may advance over time with worsening symptoms and even affecting the esthetic appearance of the chest wall.

Complications arising from surgical treatment for rib malunion may include (Holzmacher & Sarani, 2017):

- Hardware Infection: This complication may necessitate reoperation and removal of the hardware. Reported infection rates typically range from 0% to 10%.
- Hardware Failure: Though less common, hardware failure, such as screw displacement in anterior plating systems or plate failure, may require additional intervention. It is important to respect all aspects of appropriate plating procedures, according to the manufacturer's recommendations.

It is also imperative to allow bone regeneration and healing by bridging bone defects with grafts and providing an elastic reconstruction of the chest wall without undue stiffness.

For anterior malunion and lower rib malunion, the neurovascular bundle needs to be left intact and respected at the surgery to avoid injury with subsequent abdominal upper quadrant muscle weakness.

Malunions of cartilages can be treated with plating systems but the repair needs to be softer and the perichondrial and periosteal beds left intact.

Chronic pain or increased pain intensity: one of the main purposes of surgical treatment of rib malunion is the anticipated pain from the instability of the rib. Common sense dictates that stabilization of the rib with trimming of the pseudoarthrosis will prevent further irritation of the neurovascular bundle and improve or completely alleviate pain.

It is though important, as mentioned earlier, to differentiate mechanical pain from neuropathic pain or apportion these 2 entities in order to offer a meaningful prognosis to the patient.

Such a statement is of paramount importance as patients with neuropathic pain will not improve following surgical treatment and in fact, further surgical trauma to the affected area will exacerbate symptoms.

In summary, postoperative complications of surgical treatment for rib malunion may encompass hardware infection, plating system failure, the need for reoperation, and potential long-term sequelae associated with these issues.

Prognosis

The prognosis of rib malunion is contingent upon the severity of the malunion and the patient's symptoms. While most cases of rib fractures heal without complications and do not necessitate surgical intervention, a small subset of patients may require surgical repair to address rib malunions and nonunions. The decision for surgery is influenced by the extent of the damage and the patient's symptoms. Conservative management is often favored to mitigate the risks associated with surgery, such as infection and damage to surrounding tissues. Untreated rib malunion can result in chronic pain, impaired lung function, and other complications. Surgical intervention may be warranted for severe cases, particularly when symptoms include paradoxical breathing, respiratory distress, or reliance on pain medication. The long-term consequences of rib malunion are not extensively documented, emphasizing the need for further prospective data collection to inform treatment decisions regarding symptomatic rib fracture malunion and nonunion (Van Wijck & Wijffels, 2023).

The reader should be aware that appropriate patient selection remains the single main crucial factor for surgical stabilization of rib malunion.

For patients with significant neuropathic pain the statement "less is more" should be carefully considered.

Prevention strategies

To reduce the likelihood of rib malunion, it is advisable to implement the following preventive measures (Holzmacher & Sarani, 2017; Van Wijck & Wijffels, 2023):

- Early detection and timely treatment of rib fractures, especially in vulnerable populations like older adults and individuals with underlying medical conditions.

- Effective pain management encourages early mobility and engagement in physical therapy to foster proper chest wall movement and preserve optimal lung function.
- Consideration of surgical stabilization for rib fractures in specific cases, such as flail chest injuries, multiple rib fractures, displaced fractures, and in patients where pain management continues to be an issue within the first 10−15 days from injury.
- Close monitoring of patients for indications of progressive deformity or misalignment, with prompt surgical correction as needed.

These strategies are aimed at optimizing recovery and reducing the incidence of rib malunion, ultimately leading to improved long-term outcomes for affected individuals.

References

DeGenova, D. T., Miller, K. B., McClure, T. T., Schuette, H. B., French, B. G., & Taylor, B. C. (2023). Operative fixation of rib fracture nonunions. *Archives of Orthopaedic and Trauma Surgery*, *143*(6), 3047−3054. Available from https://doi.org/10.1007/s00402-022-04540-z, https://www.springer.com/journal/402.

Fokin, A. A., Hus, N., Wycech, J., Rodriguez, E., & Puente, I. (2020). Surgical stabilization of rib fractures: Indications, techniques, and pitfalls. *JBJS Essential Surgical Techniques*, *10*(2). Available from https://doi.org/10.2106/JBJS.ST.19.00032, http://surgicaltechniques.jbjs.org/.

Holzmacher, J. L., & Sarani, B. (2017). Surgical stabilization of rib fractures. *Current Surgery Reports*, *5*(9). Available from https://doi.org/10.1007/s40137-017-0185-2, springer.com/journal/40137.

Lafferty, P. M., Anavian, J., Will, R. E., & Cole, P. A. (2011). Operative treatment of chest wall injuries: Indications, technique, and outcomes. *Journal of Bone and Joint Surgery*, *93*(1), 97−110. Available from https://doi.org/10.2106/JBJS.I.00696, http://www.ejbjs.org/cgi/reprint/93/1/97.

Malekpour, M., Hashmi, A., Dove, J., Torres, D., & Wild, J. (2017). Analgesic choice in management of rib fractures: Paravertebral block or epidural analgesia? *Anesthesia and Analgesia*, *124*(6), 1906−1911. Available from https://doi.org/10.1213/ANE.0000000000002113, http://journals.lww.com/anesthesia-analgesia/toc/publishahead.

Peters, M. L., Starling, S. P., Barnes-Eley, M. L., & Heisler, K. W. (2008). The presence of bruising associated with fractures. *Archives of Pediatrics and Adolescent Medicine*, *162*(9), 877−881. Available from https://doi.org/10.1001/archpedi.162.9.877UnitedStates, http://archpedi.ama-assn.org/cgi/reprint/162/9/877.

Pieracci, F. M. (2019). Completely thoracoscopic surgical stabilization of rib fractures: Can it be done and is it worth it. *Journal of Thoracic Disease*, *11*(S8), S1061. Available from https://doi.org/10.21037/jtd.2019.01.70.

Schowalter, S., Le, B., Creps, J., & McInnis, K. C. (2022). Rib fractures in professional baseball pitchers: Mechanics, epidemiology, and management. *Open Access Journal of Sports Medicine*, *13*, 89−105. Available from https://doi.org/10.2147/OAJSM.S288882, www.dovepress.com/open-access-journal-of-sports-medicine-journal.

Van Wijck, S. F. M., & Wijffels, M. M. E. (2023). Surgical strategy for treating multiple symptomatic rib fracture malunions with bridging heterotopic ossifications: A case report. *Trauma Case Reports*, 45. Available from https://doi.org/10.1016/j.tcr.2023.100825, http://www.journals.elsevier.com/trauma-case-reports/.

Wuermser, L.-A., Achenbach, S. J., Amin, S., Khosla, S., & Melton, L. J. (2011). What accounts for rib fractures in older adults. *Journal of Osteoporosis*, *2011*, 1−6. Available from https://doi.org/10.4061/2011/457591.

Chapter 10

Slipping rib syndrome

Athanasios Kleontas[1] and Kostas Papagiannopoulos[2]

[1]Department of Thoracic Surgery, European Interbalkan Medical Centre, Thessaloniki, Northern Prefecture, Greece, [2]Department of Thoracic Surgery, St. James' University Hospital, Leeds, West Yorkshire, United Kingdom

Definition

Slipping Rib Syndrome (SRS), also known as rib tip syndrome or lower rib subluxation, is a musculoskeletal condition characterized by hypermobility of the false ribs (ribs 8–12) where they attach to the spine or the cartilage of the upper ribs (the costal cartilage slipping out of its normal anatomical position). This hypermobility can lead to symptoms, primarily localized pain and discomfort in the upper abdomen or lower chest. The condition is often marked by tenderness and can be exacerbated by certain movements or activities, such as twisting or bending. Individuals with SRS may also experience a clicking or popping sensation in the affected area (Gress et al., 2020).

SRS may be confused with Tietze Syndrome, but they are not the same, although they can share similar symptoms. Tietze Syndrome is characterized mainly by inflammation of the cartilage where the upper ribs attach to the sternum (de Carvalho, 2022).

The literature is rich in various definitions for SRS, including (McMahon, 2018; Patel et al., 2021):

- Clicking rib
- Displaced ribs
- Rib tip syndrome
- Costal margin syndrome
- Floating rib syndrome
- Nerve nipping
- Painful rib syndrome
- Slipping-rib-cartilage syndrome
- Gliding ribs
- Traumatic intercostal neuritis
- Twelfth rib syndrome

- Cyriax syndrome
- Interchondral subluxation, among others

History

The condition was first detailed in 1919 by Edgar Ferdinand Cyriax, an English-Swedish orthopedic physician and physiotherapist. He stated that disrupted ribs were responsible for aggravating downstream branches of the nerve. Cyriax also assigned the lesser-known name "Cyriax syndrome" to this condition (Cyriax, 1919). The initial recorded treatment of the condition is attributed to Eleanor Davies-Colley, the inaugural female fellow in the history of the Royal College of Surgeons. In 1922, she documented the cases of two female patients whom she treated by surgically removing the mobile cartilage of the tenth rib, resulting in immediate symptomatic relief (Davies-Colley, 1922).

Types

There are two different types of SRS:

- Primary SRS: Occurs due to inherent ligament laxity or congenital factors, contributing to the hypermobility of the false ribs.
- Secondary SRS: Results from external factors such as trauma, repetitive strain, or sports-related injuries leading to the displacement of the cartilage.

Forms

The disease can be categorized as follows:

- Acute: Characterized by sudden onset of symptoms, often following a specific triggering event like a trauma or strenuous activity.
- Chronic: Involves persistent or recurrent symptoms over an extended period, potentially indicating long-term ligamentous or structural issues.
- Intermittent: Symptoms may come and go, with periods of relief followed by exacerbation, making diagnosis challenging.
- Unilateral: Hypermobility and pain are localized to one side of the ribcage, affecting a specific set of false ribs.
- Bilateral: Involves hypermobility and symptoms on both sides of the ribcage, impacting a broader area.

Understanding the various types and forms helps clinicians tailor treatment approaches to the specific characteristics and underlying causes of SRS in individual cases.

Demographics

SRS can affect individuals of any age, but it is often seen in adolescents and adults. The condition may be more prevalent in certain demographics due to factors such as physical activity levels and underlying health conditions (Gress et al., 2020).

Incidence

It is regarded as an uncommon syndrome, constituting approximately five percent of all musculoskeletal chest pain cases in primary care. Exact incidence rates may vary, and the syndrome may be underdiagnosed due to its diverse and sometimes subtle symptomatology. Incidence can be influenced by factors such as age, lifestyle, and participation in activities that put strain on the ribcage (Gress et al., 2020).

Etiology

The etiology of SRS involves several factors (Gress et al., 2020):

- Ligamentous Laxity: Weakened or laxed ligaments in the costovertebral or costotransverse joints contribute to hypermobility of the false ribs.
- Trauma: Direct impact or trauma to the ribcage can disrupt the normal alignment of the ribs and contribute to the syndrome.
- Repetitive Movements: Activities involving repeated twisting or bending of the torso can strain the ribcage, leading to subluxation.
- Congenital Factors: Inherent ligament laxity or structural abnormalities from birth may predispose individuals to SRS.

Pathogenesis

Understanding both the etiology and pathogenesis of SRS is crucial for developing targeted interventions and treatments that address the underlying cause and mechanisms contributing to the condition. Irrespective of the mechanism, the inadequate interchondral attachments become loose and may even rupture, permitting the tips of the cartilage to curl up and override the superior rib. The ensuing pain is a consequence of the impingement of branches of the intercostal nerve due to the described subluxation (Ayloo et al., 2013).

The basic conditions of pathogenesis are analyzed below:

- Rib Displacement: The primary characteristic is the abnormal movement or displacement of the false ribs, particularly ribs 8–12. The anterior false ribs slide out of orientation and become pinned underneath their adjacent superior ribs

- Costovertebral and Costotransverse Joint Involvement: Instability in the joints where the ribs articulate with the spine (costovertebral) or the transverse processes (costotransverse) is a key aspect of pathogenesis.
- Cartilage and Soft Tissue Involvement: The cartilage connecting the ribs to the sternum can also be affected, causing inflammation and contributing to symptoms.
- Pain Sensation: Displacement and irritation of surrounding structures lead to pain, manifesting as localized discomfort, tenderness, or referred pain.
- Inflammatory Response: Inflammation in the affected joints and surrounding soft tissues may exacerbate symptoms, contributing to the overall pathogenesis.

Main symptom

The primary and most prominent symptom of SRS is localized pain. The pain typically occurs in the upper abdomen or lower chest, often on one side of the ribcage. It can be sharp, aching, or both, and individuals may describe it as a stabbing sensation. Pain intensity may vary but is often exacerbated by specific movements, such as twisting, bending, or deep breathing. The pain may be reproducible during a physical examination, confirming the involvement of the ribs. Recognition and understanding of this main symptom guide the diagnostic process, helping healthcare providers differentiate SRS from other conditions presenting with chest or abdominal pain. A comprehensive history of the present disease is crucial for healthcare professionals to establish the timeline, characteristics, and impact of SRS. This information aids in accurate diagnosis and the development of an effective treatment plan (Lum-Hee & Abdulla, 1997).

Diagnostic approach flow chart

Clinical assessment (Madeka et al., 2023)

- Onset and Duration: Inquire about when the pain started and its duration. Determine if the onset was sudden or gradual.
- Nature of Pain: Understand the character of pain (sharp, aching, stabbing) and its specific location. Ask about factors that worsen or alleviate the pain.
- Activities and Triggers: Explore activities or movements that seem to trigger or exacerbate the pain. Inquire about any recent trauma or repetitive actions involving the torso.
- Previous Episodes: Ask if the individual has experienced similar symptoms in the past. Explore any patterns of recurrence or specific triggers.

- Associated Symptoms: Inquire about any associated symptoms such as swelling, tenderness, or clicking sensations. Explore the presence of referred pain in other areas like the back or shoulder.
- Impact on Daily Life: Assess how the pain affects daily activities, work, and sleep. Explore any modifications in posture or behavior to alleviate symptoms.
- Medical History: Gather information on any previous rib injuries, surgeries, or relevant medical conditions. Inquire about underlying factors like hypermobility syndromes or congenital conditions.
- Treatments Trialed: Ask about any self-management strategies or treatments attempted for pain relief. Inquire about the effectiveness of these interventions.
- Psychosocial Factors: Explore the impact of pain on the individual's mood, stress levels, and overall well-being. Inquire about any psychosocial factors that may contribute to the symptomatology.

Physical examination

A thorough clinical examination helps confirm the diagnosis of SRS, ruling out other potential causes of chest or abdominal pain. It also provides valuable information for tailoring an effective treatment plan (Mazzella et al., 2020).

- Inspection: Observe for any asymmetry, swelling, or deformities in the ribcage. Note any visible signs of discomfort or guarding.
- Palpation: Palpate the ribcage, specifically focusing on the costovertebral and costotransverse joints.
- Identify tenderness or localized pain during palpation.
- Rib Maneuvers: Perform specific maneuvers to reproduce or exacerbate symptoms, such as the Hooking Maneuver or Rib Mobilization Test.
- The Hooking Maneuver was initially described by Heinz and Zaval in 1977. In this technique, the examiner palpates under the costal margin and pulls the entire ribcage superiorly and anteriorly. During the maneuver, both the examiner and the patient may sometimes perceive a click, disconnected cartilage, or hypermobility, all of which can result in significant pain (Fam & Smythe, 1985; Turcios, 2017) (Fig. 10.1).
- The Rib Mobilization Test is a diagnostic examination used to assess the mobility and potential abnormalities in the ribs. During this test, a healthcare professional typically evaluates the movement, flexibility, and integrity of the ribs to identify any issues such as subluxations, dislocations, or restrictions in the ribcage. One common technique involves applying gentle pressure or manipulation to the ribs to assess their response and detect any abnormalities in movement. This test is often employed in the evaluation of conditions related to the thoracic spine, costovertebral

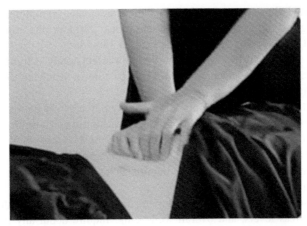

FIGURE 10.1 This picture shows the position of the patient and the "hook maneuver" to examine for possible slipping rib syndrome. The operator stands on the side of the patient and cephalad.

joints, and intercostal muscles, helping clinicians diagnose and address issues affecting the ribcage. It can be part of a broader physical examination for individuals experiencing chest or ribcage pain, discomfort, or related symptoms (Heiderscheit & Boissonnault, 2008).

- Cough Test: Instruct the individual to cough and observe for increased pain, which may indicate rib involvement.
- Respiratory Assessment: Assess respiratory patterns and note any discomfort during deep breathing. Evaluate for potential diaphragmatic involvement.
- Range of Motion: Assess the range of motion of the thoracic spine and observe for pain or restriction during movements such as twisting and bending.
- Neurological Examination: Rule out neurological involvement by conducting a basic neurological assessment, focusing on sensory and motor functions.
- Muscle Strength: Evaluate the strength of the surrounding muscles, especially those supporting the ribcage.
- Reflexes: Check reflexes to rule out neurological issues contributing to pain.
- Provocative Tests: Perform tests like the Rib Springing Test or Rib Compression Test to elicit pain or reproduce symptoms.
- Breastbone Examination: Assess for tenderness or inflammation at the junction of the ribs and the sternum in case of Tietze Syndrome.
- Functional Assessment: Evaluate how the symptoms impact daily activities and function.

Differential diagnosis

Thorough evaluation, including history, physical examination, and appropriate diagnostic tests, is crucial for differentiating SRS from other potential causes of chest or abdominal pain. Collaboration with healthcare professionals helps ensure an accurate diagnosis and tailored treatment plan (Fu et al., 2012). The following list of differential diagnoses might not be exhaustive but offers an opportunity to understand why SRS is often misdiagnosed.

- Costochondritis: Inflammation of the cartilage where the upper ribs attach to the sternum, presenting with localized chest pain.
- Musculoskeletal Strain or Injury: Trauma or overuse injuries affecting muscles, ligaments, or joints in the chest or abdominal region.
- Rib Fractures: Fractures of the ribs may result in localized pain, often exacerbated by movement.
- Intercostal Neuralgia: Irritation or inflammation of the intercostal nerves, leading to sharp or burning chest pain.
- Tietze Syndrome: Inflammation of the costal cartilage, causing localized chest pain and swelling.
- Scoliosis: Abnormal curvature of the spine may contribute to chest or back pain.
- Herniated Disks: Disc issues in the spine may cause radiating pain into the chest or abdomen.
- Pleuritis or Pleurisy: Inflammation of the parietal pleura, leading to sharp chest pain exacerbated by breathing.
- Pneumonia: lower respiratory infections can lead to chest pain, cough, and respiratory symptoms.
- Referred Pain from Abdominal Organs: Conditions affecting abdominal organs, such as kidney stones or liver disease, may cause referred pain to the chest.
- Gastrointestinal Issues: Conditions such as gastritis, peptic ulcers, or gallbladder disease may cause abdominal pain.
- Psychogenic Factors: Anxiety, stress, or psychosomatic factors can manifest as chest or abdominal pain.

Diagnostic examinations

Diagnostic examinations are selected based on clinical judgment and may be performed sequentially to rule out other potential causes and confirm the diagnosis of SRS (Hussain, 2020).

- Plain chest films: To visualize the ribcage and assess the alignment of the ribs. Helpful in ruling out fractures or identifying any bony abnormalities.

160 Chest Blunt Trauma

- Magnetic Resonance Imaging (MRI): Provides detailed images of soft tissues, including cartilage and ligaments. Helps evaluate the condition of the costovertebral and costotransverse joints.
- Dynamic Ultrasound: This can be used to assess the movement and alignment of the ribs in real time. Useful for visualizing soft tissues and detecting inflammation (Scholbach, Ribinjuryclinic.com).
- CT Scan: Offers detailed cross-sectional images of the chest and ribcage. May be used to identify structural abnormalities or assess fractures.
- Nerve Block or Local Anesthetic Injection: Diagnostic injections near the affected ribs can help confirm if the pain is originating from the ribs.
- Laboratory Tests: Blood tests to rule out inflammatory or autoimmune conditions contributing to symptoms.
- Electromyography and Nerve Conduction Studies: Assess nerve function and rule out neurological involvement.
- Pulmonary Function Tests: Evaluate lung function and assess for any impact on respiratory mechanics.
- Cardiac Evaluation: Electrocardiogram (ECG) and other cardiac assessments to rule out heart-related issues contributing to chest pain.
- Gastrointestinal Studies: If abdominal symptoms are prominent, consider endoscopy or imaging to evaluate the gastrointestinal tract.
- Psychological Assessment: Consideration of psychological factors through interviews or assessments, especially if stress or anxiety may contribute to symptomatology.
- Provocative Maneuvers and Clinical Tests: Specific clinical tests, such as the Hooking Maneuver or Rib Mobilization Test, reproduce or exacerbate symptoms.

Treatment options flow chart

The choice of treatment depends on the severity of symptoms, individual factors, and response to initial interventions. A multidisciplinary approach involving healthcare professionals, physical therapists, and, if necessary, surgeons can provide comprehensive care for individuals with SRS (McMahon, 2018).

- Rest and Activity Modification: Avoidance of activities that exacerbate symptoms, allowing the ribcage to heal. Modification of daily activities to minimize strain on the affected area.
- Pain Medications: Nonsteroidal anti-inflammatory drugs for pain and inflammation relief. Analgesics for pain management.
- Physical Therapy: Targeted exercises to strengthen the muscles supporting the ribcage. Stretching exercises to improve flexibility and mobility.
- Manual therapy, involving the manipulation of the costovertebral joint and electric stimulation, may assist in pain management, though long-term relief is uncertain (Udermann et al., 2005).

- Taping of the ribs may offer temporary relief. To determine the optimal location and direction for taping, administer a manual superior compression force through the postero-lateral aspect of the rib cage. Instruct the patient to take a deep breath or rotate. If the patient experiences a notable improvement in symptoms, apply the tape at that level (Bahram, 2015).
- Rib mobilization with movement (MWM), as introduced by Brian Mulligan, involves evaluating the range of motion and pain level. A cranial glide is applied over the lateral aspect of the rib above the painful region. While sustaining this rib elevation (unloading), the patient is instructed to rotate, and both range of motion and pain are reevaluated. If there is no change, the technique is repeated on a rib above or below. If MWM on a rib at a specific level is found to reduce or eliminate the pain, the process is repeated 10 times (Bahram, 2015).
- A home program of MWM may be recommended. Instructions include using the web space of one hand to lift the rib up and actively rotating towards the painful direction; repeat as often as necessary. The objective is to move the irritated costovertebral joint without pain as frequently as possible, aiming to diminish both protective muscle spasms and local inflammation (Bahram, 2015; Hansen et al., 2020).
- Rib Maneuvers and Manipulation: Manual manipulation by a qualified healthcare professional to reposition displaced ribs. Specific maneuvers, such as the Hooking Maneuver, to alleviate symptoms.
- Breathing Exercises: Diaphragmatic breathing techniques to improve respiratory mechanics and reduce strain on the ribs.
- Supportive Measures: Application of heat or cold packs to alleviate pain and reduce inflammation. Use of supportive braces or wraps for added stability.
- Injections: Local anesthetic or corticosteroid injections around the affected ribs for pain relief and inflammation reduction.
- Surgical Intervention: Reserved for severe cases where conservative measures have failed. Surgical stabilization of the ribcage may be considered to prevent recurrent subluxations.
- Psychological Support: Counseling or psychological support if stress or anxiety contributes to symptoms. Mindfulness or relaxation techniques.
- Education and Lifestyle Modification: Patient education on proper body mechanics and posture to prevent recurrence. Lifestyle modifications to address contributing factors.

Surgical treatment

Surgical treatment is usually considered for SRS if all other measures have failed and decisions are made on a case-by-case basis. It is essential for individuals to discuss the potential benefits, risks, and alternatives with their healthcare team to make informed decisions about surgical intervention (Hansen et al., 2020).

Indications for surgical intervention

- Persistent Symptoms: Severe and persistent pain despite conservative measures.
- Recurrent Subluxations: Frequent recurrence of rib subluxations despite attempts at manual reduction.
- Functional Impairment: Impaired daily activities and reduced quality of life.

Surgical options

Several techniques have been described, either as an isolated procedure or hybrid combinations.

Resection of affected cartilage (Madeka et al., 2023; Mazzella et al., 2020): Often surgeons, considering that the hypermobile cartilage impinges on the intercostal nerve, opt to resect the cartilage of the involved false rib. Such cases have been successfully reported either in the adult or pediatric population.

We would exercise caution when employing such a technique for the following reasons:

- Complete resection of the affected cartilage with the perichondrial bed could produce a defect on the chest wall with potential chest wall weakness and herniation;
- Using significant energy in the lower intercostal nerves could damage the nerves, induce pain, and also an upper abdominal quadrant weakness presenting as a pseudo hernia of the abdominal wall.
- Often the cartilage resection in itself can produce a true floating rib and such could be responsible for symptom recurrence.

We would therefore recommend removal of the cartilage by filing it off from the cartilaginous "bed" (similar resection performed for pectus deformities) and reconstructing the defect with reefing sutures of the perichondrium and re-attachment to adjacent cartilage. In this manner impingement of the cartilage to the nerve is alleviated, the neurovascular bundle is protected and preserved, the cartilage is allowed to re-grow as a good fibrous start, and the stability of the false ribs is secured with the re-establishment of a stable costal arch.

Stabilization of costal arch and cartilages with absorbable or non-absorbable implants (Madeka et al., 2023): There are now reports in the literature of stabilization of the cartilages with vertical rib plating. Most authors have utilized bio-absorbable plates which dissolve after months leaving a strong fibrous band supporting the cartilage. Most cases seem to experience good overall pain relief, although long-term results are not yet available. Additionally, plating has been employed in combination with cartilage excision and therefore these two techniques might complement each other but not necessarily have equal benefits when employed on their own.

Additionally, one should remain cautious in employing a complex and at times lengthy procedure with unnecessary tissue dissection when treating a pathology with pain being the cardinal symptom.

Rib Stabilization Surgery: Involves stabilizing the affected cartilages using simple suturing. This technique seems to be by far the simplest and has been described well by Hansen et al. (2020).

The most important points are identification of the laxed cartilage/s, a small incision on the top of the pathology, careful dissection of overlying muscle fibers, and stabilization of cartilages with the simple figure of eight sutures avoiding a breach of the pleural cavity.

The technique can be employed with simple sedation and local anesthetic as a day case. Nonabsorbable heavy orthopedic sutures or ethibond sutures can be applied to stabilize the cartilage as shown in the following pictures Figs. 10.2 and 10.3.

Picture showing the floating cartilage of the 10th rib with a sizeable gap between the ribs at the anterior aspect of the costal arch.

Two figures of eight stitches have been applied and the cartilages are now approximated and ribs stabilized.

We would recommend to try and place the suture around the upper rib or cartilage and protect the intercostal bundle of the upper rib from injury.

1. Local ablative treatments:
a. Intercostal Nerve Blocks: these could alleviate pain, if successfully applied, in patients with mild symptoms and those who do not wish to

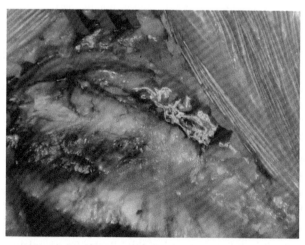

FIGURE 10.2 The floating cartilage of the 10th rib with a sizeable gap between the ribs at the anterior aspect of the costal arch.

FIGURE 10.3 Eight stitches have been applied and the cartilages are now approximated and ribs stabilized.

consider any form of surgery. There are reports of pain relief with the combination of local anesthetic agents and steroids. The long-term effect of these agents is limited considering the pathophysiology of the SRS and the constant impingement of the intercostal nerve with activity.

b. Cryoablation (Migliore et al., 2014): Cryoprobes have been employed recently in the management of SRS. It should be noted that such were not used as a single therapeutic agent but as part of multimodality treatment including stabilization of the cartilages. It has been noted that cryoablation can reduce postoperative pain and length of hospital stay. There is no real evidence that long-term results should be expected with isolated use of cryoprobes for SRS. The reasons are obvious as such have to breach the pleural cavity if they were to be employed underneath the rib and those used outside the chest cavity would require a large incision and multiple tunnels to reach the neurovascular bundle.

Additionally, the manufacturers of the cryoprobes advise against their use below the 9th rib as the lower intercostal nerves, innervate the upper abdominal wall muscles with the risk of denervation and gradual redundancy with weakness of the upper abdominal wall.

2. Use of botox injections (Pirali et al., 2013). Such experience is limited to case reports. It is only referenced in this chapter for completion purposes and it is not recommended with the current evidence.

Patient monitoring

Patient monitoring is essential for optimizing the management of SRS. Regular follow-up and open communication between the patient and healthcare team contribute to effective treatment and improved quality of life.

- Regular Follow-up Appointments: Schedule periodic follow-up appointments with healthcare professionals to assess progress and adjust the treatment plan as needed.
- Pain Assessment: Continuously monitor pain levels and inquire about any changes or fluctuations in symptoms during follow-up visits.
- Functional Evaluation: Assess the impact of SRS on daily activities and overall functional status.
- Imaging Studies: Consider repeat imaging studies (X-rays, MRI) if necessary to evaluate the alignment of the ribs and assess the success of interventions.
- Physical Examination: Conduct regular physical examinations, including palpation and specific maneuvers, to check for tenderness, swelling, or signs of recurrence.
- Response to Treatment: Evaluate the response to ongoing treatments, including physical therapy, medications, or injections.
- Psychological Support: Monitor psychological well-being and provide additional support if stress, anxiety, or mood changes are noted.
- Educational Reinforcement: Reinforce proper body mechanics, posture, and lifestyle modifications to prevent recurrence.
- Rehabilitation Progress: Track progress in rehabilitation exercises and adjust the exercise regimen as needed.
- Patient Education: Provide ongoing education about SRS, its management, and potential triggers for symptoms.
- Long-Term Management Strategies: Develop and implement long-term strategies to prevent recurrence, including lifestyle modifications and ongoing exercises.

Prevention strategies

Prevention of SRS involves a combination of individual awareness, lifestyle modifications, and population-level strategies to minimize risk factors and promote musculoskeletal health (Gress et al., 2020).

- Proper Body Mechanics: Educate individuals on maintaining good posture and body mechanics to reduce strain on the ribcage during daily activities.
- Core Strengthening Exercises: Promote exercises that strengthen the core muscles, which play a role in supporting the ribcage.

- Avoidance of Trauma: Encourage caution and safety measures to prevent direct trauma to the chest or ribcage.
- Posture Awareness: Raise awareness about the importance of maintaining proper posture, especially during activities involving repetitive or prolonged torso movements.
- Condition-Specific Education: Provide education to individuals with hypermobility syndromes or congenital factors that may predispose them to SRS.
- Periodic Health Check-ups: Include routine health check-ups where healthcare professionals assess musculoskeletal health and address any predisposing factors.
- Physical Activity Guidance: Offer guidance on appropriate physical activities and exercises to ensure a balanced and supportive musculoskeletal system.
- Early Intervention for Trauma: Encourage seeking medical attention promptly for any chest or ribcage trauma to address potential issues early.
- Psychological Support: Address stress management and provide psychological support, as stress can contribute to musculoskeletal issues.

References

Ayloo, A., Cvengros, T., & Marella, S. (2013). Evaluation and treatment of musculoskeletal chest pain. *Primary Care: Clinics in Office Practice*, *40*(4), 863–887. Available from https://doi.org/10.1016/j.pop.2013.08.007.

Bahram, J. (2015) *Ribs don't sublux, ribs don't "go out" ... so what's going on?* Advanced Physical Therapy Education Institute.

de Carvalho, J. F. (2022). Tietze's syndrome. *Mediterranean Journal of Rheumatology*, *33*(4), 467–468.

Cyriax, E. (1919). On various conditions that may simulate the referred pains of visceral disease, and a consideration of these from the point of view of cause and effect. *The Practitioner*, *102*, 314–322.

Davies-Colley, R. (1922). Slipping rib. *British Medical Journal*, *1*(3194), 432. Available from https://doi.org/10.1136/bmj.1.3194.432.

Fam, A. G., & Smythe, H. A. (1985). Musculoskeletal chest wall pain. *Canadian Medical Association Journal*, *133*(5), 379–389.

Fu, R., Iqbal, C. W., Jaroszewski, D. E., & St. Peter, S. D. (2012). Costal cartilage excision for the treatment of pediatric slipping rib syndrome. *Journal of Pediatric Surgery*, *47*(10), 1825–1827. Available from https://doi.org/10.1016/j.jpedsurg.2012.06.003.

Gress, K., Charipova, K., Kassem, H., Berger, A. A., Cornett, E. M., Hasoon, J., Schwartz, R., Kaye, A. D., Viswanath, O., & Urits, I. (2020). A comprehensive review of slipping rib syndrome: Treatment and management. *Psychopharmacology Bulletin*, *50*(4), 189–196.

Hansen, A. J., Toker, A., Hayanga, J., Buenaventura, P., Spear, C., & Abbas, G. (2020). Minimally invasive repair of adult slipped rib syndrome without costal cartilage excision. *Annals of Thoracic Surgery*, *110*(3), 1030–1035. Available from https://doi.org/10.1016/j.athoracsur.2020.02.081www.elsevier.com/locate/athoracsur.

Heiderscheit, B., & Boissonnault, W. (2008). Reliability of joint mobility and pain assessment of the thoracic spine and rib cage in asymptomatic individuals. *Journal of Manual and Manipulative Therapy*, *16*(4), 210−216. Available from https://doi.org/10.1179/106698108790818369, http://www.tandfonline.com/loi/yjmt20#.VwHawE1f1Qs.

Hussain, A. (2020). Diagnosing and treating Slipping Rib Syndrome: An unusual case of undiagnosed pain for 5 years. *COPD*.

Lum-Hee, N., & Abdulla, A. J. J. (1997). Slipping rib syndrome: An overlooked cause of chest and abdominal pain. *International Journal of Clinical Practice*, *51*(4), 252−253.

Madeka, I., Alaparthi, S., Moreta, M., Peterson, S., Mojica, J. J., Roedl, J., & Okusanya, O. (2023). A review of Slipping Rib Syndrome: Diagnostic and treatment updates to a rare and challenging problem. *Journal of Clinical Medicine*, *12*(24), 7671. Available from https://doi.org/10.3390/jcm12247671.

Mazzella, A., Fournel, L., Bobbio, A., Janet-Vendroux, A., Lococo, F., Hamelin, E. C., Icard, P., & Alifano, M. (2020). Costal cartilage resection for the treatment of slipping rib syndrome (Cyriax syndrome) in adults. *Journal of Thoracic Disease*, *12*(1), 10−16. Available from. Available from https://doi.org/10.21037/jtd.2019.07.83, http://www.jthoracdis.com/.

McMahon, L. E. (2018). Slipping Rib Syndrome: A review of evaluation, diagnosis and treatment. *Seminars in Pediatric Surgery*, *27*(3), 183−188. Available from. Available from https://doi.org/10.1053/j.sempedsurg.2018.05.009, http://www.elsevier.com/inca/publications/store/6/2/3/1/8/7/index.htt.

Migliore, M., Signorelli, M., Caltabiano, R., & Aguglia, E. (2014). Flank pain caused by slipping rib syndrome. *The Lancet*, *383*(9919), 844. Available from https://doi.org/10.1016/s0140-6736(14)60156-2.

Patel, N., John, J. K., Pakeerappa, P., Aiyer, R., & Zador, L. N. (2021). Slipping rib syndrome: Case report of an iatrogenic result following video-assisted thoracic surgery and chest tube placement. *Pain Management*, *11*(5), 555−559. Available from https://doi.org/10.2217/pmt-2020-0028, http://www.futuremedicine.com/loi/pmt.

Pirali, C., Santus, G., Faletti, S., & De Grandis, D. (2013). Botulinum toxin treatment for slipping rib syndrome: A case report. *Clinical Journal of Pain*, *29*(10), e1. Available from https://doi.org/10.1097/AJP.0b013e318278d497.

Scholbach, T. *Abdominal paindue to liver compression in slipping rib syndromein an EDS patient*. Available from https://scholbach.de/abdominal-pain-due-to-liver-compression-in-slipping-rib-syndrome-in-an-eds-patient#gsc.tab = 0, https://www.ribinjuryclinic.com/conditions/slipped-rib-syndrome/.

Turcios, N. L. (2017). Slipping Rib Syndrome: An elusive diagnosis. *Paediatric Respiratory Reviews*, *22*, 44−46. Available from https://doi.org/10.1016/j.prrv.2016.05.003, http://www.elsevier.com/inca/publications/store/6/2/3/0/6/6/index.htt.

Udermann, B. E., Cavanaugh, D. G., Gibson, M. H., Doberstein, S. T., Mayer, J. M., & Murray, S. R. (2005). Slipping rib syndrome in a collegiate swimmer: A case report. *Journal of Athletic Training*, *40*(2), 120−122.

Chapter 11

Role of VATS in nonpenetrating chest trauma

Jury Brandolini[1], Fabrizio Minervini[2] and Pietro Bertoglio[1]
[1]*Department of Thoracic Surgery, Azienda Ospedaliero-Universitaria di Bologna, Bologna, Italy,* [2]*Division of Thoracic Surgery, Cantonal Hospital Lucerne, Lucerne, Switzerland*

Introduction

In the last decade, minimally invasive surgery, and in particular video assisted thoracic surgery (VATS) has gained increasing importance in the treatment of thoracic diseases. Trials have been run to assess its feasibility for the management of lung malignancies showing significant benefit compared to thoracotomy (Bendixen et al., 2016; Lim et al., 2022). VATS is now considered a standard of care for early-stage lung cancer, but with the rapid improvement of skills and technology also complex procedures have been described to be managed by minimally invasive techniques with good results. Consequently, since its development in the early nineties, VATS has been increasingly used for the treatment of chest trauma patients. This chapter will give a general view of the role of VATS in chest trauma, while its specific use according to different clinical issues will be discussed more in detail in the next chapters.

The development of VATS surgery and its use in chest trauma

On the other hand, the use of minimally invasive techniques for chest trauma is less standardized and it depends on the surgeon's skill, available technologies, patients' conditions, and trauma characteristics (Bertoglio et al., 2019); Ludwig and Koryllos (2017) suggested that indications for minimally invasive approach in thoracic trauma should be the followings:

1. (Penetrating) injury with little blood loss in a stable patient.
2. Persistent hemothorax.
3. Empyema.

4. Persistent air-leakage.
5. Suspicion of diaphragmatic rupture.

These general indications encompass a large series of procedures that can be carried out with the use of VATS such as diagnostic exploration, pulmonary decortication, or lung resections. From the early reports that date back to the 1980s (Jones et al., 1981), VATS use has constantly increased as the first approach for trauma even though this increment has been slower than in the nontraumatic setting (Ahmed & Jones, 2004). Consistently, Alwatari et al. (2022) analyzed outcomes of the American College of Surgeons Trauma Quality Improvement Program database and 1% (2454 patients) received a nonurgent thoracic procedure (defined as a procedure performed at least after 24 h after admission). Among them, only 406 received a minimally invasive approach, with a conversion rate of 1.9% and they did not find a significant difference in the use of a minimally invasive approach according to the hospital's number of beds or level of trauma center. On the other hand, a study comparing the routine use of thoracoscopic exploration during chest wall fixation.

Several studies showed that the presence of diaphragm lesions was a risk factor for thoracotomy or conversion to thoracotomy (Alwatari et al., 2022; DuBose et al., 2012) even though the use of VATS has been showed to be able to avoid negative thoracotomies in case of false positive radiological findings (Ochsner et al., 1993).

Nevertheless, the use of minimally invasive techniques in blunt chest trauma requires a thoughtful and precise surgical and anesthesiological evaluation that should evaluate cardiovascular stability. In the case of a stable patient, there are no clear contraindications for VATS surgery which are more related to the surgeons and to the clinical conditions: in the case of tracheobronchial injuries which are technically challenging the use of a minimally invasive approach depends on the surgeon's skills; moreover, the presence of lung contusions might prevent single lung ventilation which is essential for a correct minimally invasive approach and might therefore contraindicate this technique (Lodhia et al., 2019).

Benefits, features, and timing of VATS surgery in blunt trauma patients

As already mentioned, the minimally invasive approach has been increasingly used in the management of thoracic surgery procedures; its benefit lies in its lower impact on patients and the possibility of improving outcomes in terms of reduced postoperative pain and complications and shorter postoperative in-hospital stay (Mowery et al., 2011).

Ben-Nun et al. (2007) compared the outcomes of thoracotomy and VATS trauma patients based on their institutional database. VATS group had a significantly shorted in-hospital length of stay, fluid output from chest drain, and shorter chest drain period; moreover, in-hospital opioid painkillers were required from 90% of thoracotomy patients and 54% of VATS patients with a significant difference. VATS also showed some significant benefits on the long-term outcomes. After 3 months 68% of VATS patients returned to their daily life versus only 15% of the thoracotomy group. At 3-year follow-up, VATS patients had significantly less pain and were more likely to have come back to their daily life.

DuBose et al. (2012) highlighted the effectiveness of the VATS approach. In their prospective comparison between VATS and thoracotomy, the need for further surgery was required in 30% and 21% of cases respectively.

The effectiveness of VATS was also evaluated compared to intrapleural thrombolytics in trauma patients. In the thrombolytic group failure rate was 29%, significantly higher than the VATS group (6%) which showed a significantly lower in-hospital length of stay.

VATS approach could be divided into early or late according to the time that lasts since the trauma and surgery. Alwatari suggested defining a late approach when it occurs later than 5 days after trauma (Alwatari et al., 2022), while the Eastern Association for the Surgery of Trauma guidelines suggest performing surgery in the first 72 h (Patel et al., 2021). Regardless of the definition, all the results consistently demonstrated that an early VATS approach allows significantly better outcomes compared to delayed VATS in terms of postoperative unplanned intubation, surgical site infection, and pneumonia; moreover, earlier VATS is associated with a significantly lower incidence of conversion to thoracotomy (Ahmad et al., 2013; Alwatari et al., 2022; Lin et al., 2014; Smith et al., 2011; Vassiliu et al., 2001). Nevertheless, some authors suggest that a VATS approach should be at least attempted even after 7 days from the trauma, despite the increased technical difficulties, risk for intraoperative complications, and conversion to thoracotomy (Lang-Lazdunski & Pilling, 2007; Navsaria et al., 2004). A significantly shorter length of stay for patients treated in the first 72 h was confirmed by the results of a meta-analysis (Ziapour et al., 2020) that revealed a higher success rate of an early VATS approach (compared to patients with a 7-day delay); nonetheless, no difference in mortality was found according to different VATS timing. Concurrently, Kazempoor et al. (2023) based on the Trauma Quality Improvement Program (TQIP) database built a scoring tool to predict the need for an early surgical exploration in pediatric trauma patients taking into account demographic data, features of trauma, and the number and size of involved organs.

From a technical point of view, multiportal (bi-, tri-, or four-port) VATS has been described as the preferred access in the majority of reported experiences. Nevertheless, uniportal approach has also been described as safe and feasible in trauma patients (Sanna et al., 2017); with adequate surgical experience and a trained anesthesiologist team, this technique can be safely performed in awake patients under epidural or paravertebral nerve block (Tacconi et al., 2010).

VATS for posttraumatic haemothorax

Hemothorax is a severe medical condition characterized by the accumulation of blood in the pleural cavity, the space between the lungs and the chest wall. This accumulation can compress the lung, impairing its ability to function effectively and potentially leading to life-threatening complications.

Thoracic trauma is the most common cause of hemothorax, occurring in approximately 60% of all polytrauma cases (Zeiler et al., 2020). A significant portion (up to 15%) of patients with thoracic trauma require emergent thoracotomy; indications include cardiac tamponade, massive hemothorax, large thoracic wounds, major thoracic vascular injuries, tracheobronchial injuries, and suspected esophageal perforation (Cetindag et al., 2007). Concurrently, delayed hemothorax is a rare complication of thoracic trauma that occurs after initial imaging shows no evidence of bleeding. Delayed hemothorax has been reported to occur anywhere from 2 h to 44 days after initial presentation. Rib fractures are the most common cause of delayed hemothorax, occurring in 30%−80% of patients with thoracic trauma (Lin et al., 2016). Despite an initial treatment with chest tube placement, up to 30%, may develop persistent hemothorax.[7] If left untreated, persistent hemothorax can lead to the formation of fibrothorax and pulmonary trapping, which can impair lung function (Coselli et al., 1984).

According to Helling et al. (1989), VATS can play a role in the acute phase of management, particularly when the blood loss exceeds 100 mL/h and clots have not been entirely evacuated by a chest tube. The objectives of this upfront approach are manifold:

1. Rapid and early clot evacuation lowers the risk of fibrinolysis activation, maintaining hemostasis.
2. Enhanced chest tube placement.
3. Direct treatment of bleeding-causing injuries.
4. Avoidance of additional and multiple unnecessary transfusions.
5. Shorter hospital stay and consequently reduced healthcare expenses.

The etiology of the hemothorax plays a crucial role in guiding the surgical strategy. In most cases of post-traumatic retained hemothorax, clot removal, thorough irrigation of the cavity, and partial lung decortication are sufficient for a successful outcome (Fig. 11.1). In contrast, in case further lung or diaphragmatic lesions are found, VATS exploration necessitates a meticulous examination of the lung parenchyma and chest wall to identify the bleeding source and achieve definitive hemostasis.

FIGURE 11.1 Intraoperative view of a post-traumatic retained haemothorax, treated by VATS approach.

VATS for posttraumatic empyema

Chest trauma is a significant cause of empyema, following parenchymal infection and prior surgical procedures, with an incidence estimated to be between 2% and 10% (Fallon, 1994).

In trauma patients, several factors could contribute to the development of empyema: residual hemothorax; repeated chest tube insertions or prolonged presence of chest tubes or other foreign bodies; soft tissue damage; and development of ventilator-associated pneumonia. Additionally, the cumulative severity of injuries, in particular the presence of complicated intra-abdominal injuries, can further compromise host defenses and increase susceptibility to infectious complications, including empyema (Heniford et al., 1997).

VATS can be most effectively performed in the early exudative or purulent phase of empyema when the empyema is characterized by infected clots or fibrinous deposits without significant organized fibrin deposits on the visceral pleural surface. VATS can effectively remove the infected material and allow lung re-expansion. However, if the empyema has progressed to the fibrinopurulent or a later stage, VATS may be less effective and thoracotomy could be therefore necessary to achieve a proper pleural debridement and a complete lung decortication. VATS targets in empyema are lysing adhesions, debriding inflammatory and organizing reactions, and effectively draining the empyema cavity. The surgical approach mirrors that employed for retained hemothorax management.

VATS decortication has been reported to be successful in up to 54% of chronic empyema of all causes. However, the success rate may vary on the etiology and stage of the empyema. For example, empyema caused by trauma tends to be more successfully managed with VATS than empyema caused by other causes (Morales et al., 1997). In the study by Striffeler et al. 28% of patients with stage II empyema required conversion to an open procedure due to advanced disease; all the cases of post-traumatic pleural empyema (7%) were successfully operated via the VATS approach (Striffeler et al., 1998).

Patient selection is a fundamental step for the use of VATS in the treatment of empyema. Effective management of empyema necessitates prompt identification and control of the infection source and administration of culture-specific antibiotics. As for hemothorax, also for empyema timing for the surgical approach is crucial. Empyema collections less than 10 days old are optimally managed thoracoscopically. A preoperative CT scan is invaluable in planning port placement, and it is advisable to aspirate fluid in the operating room before port insertion.

A consensus among most authors indicates that beyond 10–14 days, surgical complexity and complication rates significantly increase.

VATS for posttraumatic hernias

Diaphragmatic injuries constitute approximately 2%–5% of all trauma cases. These injuries are frequently caused by high-impact events that result in a sudden surge in intra-abdominal pressure, leading to rupture. Due to the liver's protective role over the right diaphragm, 80% of ruptures occur on the left side. These tears, typically large and located in the center of the diaphragm, can result in the herniation of intra-abdominal contents into the chest (Karmy-Jones & Jurkovich, 2008).

For individuals with penetrating thoracoabdominal trauma who do not require laparotomy or thoracotomy, thoracoscopy serves as an invaluable tool for evaluating diaphragmatic injuries. This technique allows thorough examination of the entire diaphragmatic surface without the need for insufflation. While some authors favor laparoscopy, both thoracoscopy and laparoscopy offer excellent diagnostic and potentially therapeutic options in this context (Murray et al., 1998). Thoracoscopy offers a more comprehensive and detailed view of the diaphragm, particularly on the right side and in the posterior aspect, compared to laparoscopy.

Lung herniation is an uncommon medical condition characterized by the protrusion of a portion of the lung parenchyma beyond the confines of the rib cage (Wani et al., 2015). In case of blunt chest trauma, damage to the chest wall, such as rib fractures or chondral-costal or clavicle-sternal dislocations, can create interruption through which the lung tissue can herniate; pulmonary herniation is most frequently seen in the anterior part of the chest wall (Sulaiman et al., 2006): more in details, the two most recurring types of thoracic lung hernias are anterior intercostal and parasternal hernias, with anterior intercostal hernias accounting for over 98% of all lung hernias

(Lang-Lazdunski et al., 2002). More rarely, also cervical and diaphragmatic lung hernias have been anecdotically reported.

The development of pulmonary hernia is a complex process that involves a combination of anatomical and physiological factors. For lung tissue to migrate outside its normal boundaries, there must be a compromise in the structural integrity of the chest wall, coupled with either a sudden or sustained elevation in intrathoracic pressure (Bikhchandani et al., 2012; Kaliyadan et al., 2011; Sulaiman et al., 2006). Brandolini et al. (2023) was the first to describe a case of posttraumatic trans-mediastinal pulmonary hernia in a young woman, following blunt force trauma. The herniated portion of the right lower lobe was successfully reduced through a biportal VATS approach and the mediastinal defect between the aorta and esophagus was securely repaired using continuous PDS 3/0 sutures (Fig. 11.2 and 11.3).

Due to the rarity of LH, there is currently no standardized approach to its management, and treatment decisions are often made on a case-by-case basis, taking into account the individual patient's circumstances and the severity of the herniation. While conservative management is often employed in cases of pulmonary hernias, surgery was deemed necessary in this instance to prevent potential strangulation of the herniated lung parenchyma.

Minimally invasive surgical approaches may be considered, particularly if performed within the first few hours following the incident. Surgical intervention typically involves closure of the herniary breach, with the potential use of prostheses depending on the size of the herniation defect. In our case, a direct suture of the mediastinal pleura between the esophagus and aorta was performed without the need for prostheses.

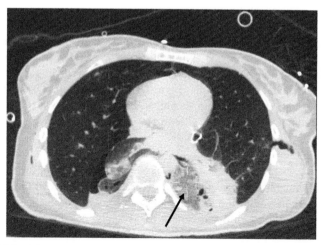

FIGURE 11.2 CT scan reveals a portion of the right lower lung lobe herniating into the contralateral mediastinum, displacing the aorta and esophagus.

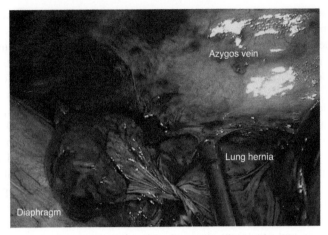

FIGURE 11.3 Intraoperative view of the right lower lobe after surgical reduction using a by-portal VATS technique.

VATS for posttraumatic chest wall injuries

As discussed in another chapter, a flail chest is a potentially life-threatening traumatic injury.

Some authors Bui et al. (2020), Lin et al. (2019), and Wu et al. (2020) analyzed the outcomes of patients who received VATS pleural exploration In the case of open chest wall fixation; results suggested that VATS exploration is related to a faster ventilator weaning and reduce postoperative complication rate. The same conclusions were shared by a large retrospective study (van Gool et al., 2022), which also one-year outcomes of patients treated with VATS-assisted rib fixation.

Despite open surgery is still the gold standard for rib fixation, some authors described a VATS approach for internal rib fixation (Castater et al., 2022; Zhang et al., 2022); this approach could potentially be offered for all kinds of rib fracture, but it could be particularly convenient for those fractures located under the scapula. In a series of 35 consecutive patients, 24-month follow-up showed no technical complications related to internal device detachment (Zhang et al., 2022).

Conclusions

VATS is a valuable and effective tool for thoracic surgeons in the treatment of trauma patients. With the development of surgeons' skills, it may expand its indications with a significant benefit for patients' postoperative outcomes. Nevertheless, the timing and strategies of the surgical approach and patients' clinical conditions should be carefully evaluated before surgery in order to improve the success of the surgical procedure.

References

Ahmad, T., Ahmed, S. W., Soomro, N. H., & Sheikh, K. A. (2013). Thoracoscopic evacuation of retained post-traumatic hemothorax. *Journal of the College of Physicians and Surgeons Pakistan*, 23(3), 234–236. Available from http://www.jcpsp.pk/archive/2013/Mar2013/20.pdf.

Ahmed, N., & Jones, D. (2004). Video-assisted thoracic surgery: State of the art in trauma care. *Injury*, 35(5), 479–489. Available from https://doi.org/10.1016/s0020-1383(03)00289-4.

Alwatari, Y., Simmonds, A., Ayalew, D., Khoraki, J., Wolfe, L., Leichtle, S. W., Aboutanos, M. B., & Rodas, E. B. (2022). Early video-assisted thoracoscopic surgery (VATS) for non-emergent thoracic trauma remains underutilized in trauma accredited centers despite evidence of improved patient outcomes. *European Journal of Trauma and Emergency Surgery*, 48(4), 3211–3219. Available from https://doi.org/10.1007/s00068-022-01881-7.

Bendixen, M., Jørgensen, O. D., Kronborg, C., Andersen, C., & Licht, P. B. (2016). Postoperative pain and quality of life after lobectomy via video-assisted thoracoscopic surgery or anterolateral thoracotomy for early stage lung cancer: A randomised controlled trial. *The Lancet Oncology*, 17(6), 836–844. Available from. Available from https://doi.org/10.1016/S1470-2045(16)00173-X, http://www.journals.elsevier.com/the-lancet-oncology/.

Ben-Nun, A., Orlovsky, M., & Best, L. A. (2007). Video-assisted thoracoscopic surgery in the treatment of chest trauma: Long-term benefit. *Annals of Thoracic Surgery*, 83(2), 383–387. Available from https://doi.org/10.1016/j.athoracsur.2006.09.082.

Bertoglio, P., Guerrera, F., Viti, A., Terzi, A. C., Ruffini, E., Lyberis, P., & Filosso, P. L. (2019). Chest drain and thoracotomy for chest trauma. *Journal of Thoracic Disease*, 11(S2), S186. Available from https://doi.org/10.21037/jtd.2019.01.53.

Bikhchandani, J., Balters, M. W., & Sugimoto, J. T. (2012). Conservative management of traumatic lung hernia. *Annals of Thoracic Surgery*, 93(3), 992–994. Available from https://doi.org/10.1016/j.athoracsur.2011.08.023.

Brandolini, J., Bertoglio, P., Kawamukai, K., Bonfanti, B., Parri, S. N. F., Garelli, E., & Solli, P. (2023). Posttraumatic transmediastinal pulmonary hernia: An extremely rare clinical entity. *JTCVS Techniques*, 18, 168–170. Available from https://doi.org/10.1016/j.xjtc.2023.01.014.

Bui, J. T., Browder, S. E., Wilson, H. K., Kindell, D. G., Ra, J. H., Haithcock, B. E., & Long, J. M. (2020). Does routine uniportal thoracoscopy during rib fixation identify more injuries and impact outcomes? *Journal of Thoracic Disease*, 12, 5281–5288, 10. Available from https://doi.org/10.21037/jtd-20-2087, http://jtd.amegroups.com/article/view/44963/html.

Castater, C., Hazen, B., Davis, C., Hoppe, S., Butler, C., Grant, A., Archer-Arroyo, K., Maceroli, M., Todd, S. R., & Nguyen, J. (2022). Video-assisted thoracoscopic internal rib fixation. *The American Surgeon*, 88(5), 994–996. Available from https://doi.org/10.1177/00031348211060450.

Cetindag, I. B., Neideen, T., & Hazelrigg, S. R. (2007). Video-assisted thoracic surgical applications in thoracic trauma. *Thoracic Surgery Clinics*, 17(1), 73–79. Available from https://doi.org/10.1016/j.thorsurg.2007.02.007.

Coselli, J. S., Mattox, K. L., & Beall, A. C. (1984). Reevaluation of early evacuation of clotted hemothorax. *The American Journal of Surgery*, 148(6), 786–790. Available from https://doi.org/10.1016/0002-9610(84)90438-0.

DuBose, J., Inaba, K., Demetriades, D., Scalea, T. M., O'Connor, J., Menaker, J., Morales, C., Konstantinidis, A., Shiflett, A., & Copwood, B. (2012). Management of post-traumatic retained hemothorax: A prospective, observational, multicenter AAST study. *Journal of Trauma and Acute Care Surgery*, 72(1), 11–24. Available from https://doi.org/10.1097/TA.0b013e318242e368.

Fallon, W. F. (1994). Post-traumatic empyema. *Journal of the American College of Surgeons*, *179*(4), 483−492.

van Gool, M. H., van Roozendaal, L. M., Vissers, Y. L. J., van den Broek, R., van Vugt, R., Meesters, B., Pijnenburg, A. M., Hulsewé, K. W. E., & de Loos, E. R. (2022). VATS-assisted surgical stabilization of rib fractures in flail chest: 1-year follow-up of 105 cases. *General Thoracic and Cardiovascular Surgery*, *70*(11), 985−992. Available from. Available from https://doi.org/10.1007/s11748-022-01830-6, https://www.springer.com/journal/11748.

Helling, T. S., Gyles, N. R., Eisenstein, C. L., & Soracco, C. A. (1989). Complications following blunt and penetrating injuries in 216 victims of chest trauma requiring tube thoracostomy. *Journal of Trauma - Injury, Infection and Critical Care*, *29*(10), 1367−1370. Available from https://doi.org/10.1097/00005373-198910000-00013.

Heniford, B. T., Carrillo, E. H., Spain, D. A., Sosa, J. L., Fulton, R. L., & Richardson, J. D. (1997). The role of thoracoscopy in the management of retained thoracic collections after trauma. *Annals of Thoracic Surgery*, *63*(4), 940−943. Available from https://doi.org/10.1016/S0003-4975(97)00173-2.

Jones, J. W., Kitahama, A., Webb, W. R., & McSwain, N. (1981). Emergency thoracoscopy: A logical approach to chest trauma management. *Journal of Trauma - Injury, Infection and Critical Care*, *21*(4), 280−284. Available from https://doi.org/10.1097/00005373-198104000-00004.

Kaliyadan, A., Kebede, A., Ali, T., Karchevsky, M., Vasseur, B., & Patel, N. (2011). Spontaneous transient lateral thoracic lung herniation resulting in systemic inflammatory response syndrome (SIRS) and subsequent contralateral lung injury. *Clinical Medicine Insights: Case Reports*, *4*, 39−42. Available from. Available from https://doi.org/10.4137/CCRep.S7002, http://www.la-press.com/redirect_file.php?fileId = 3673&filename = 2749-CCRep-Spontaneous-Transient-Lateral-Thoracic-Lung-Herniation-Resulting-in-Sy.pdf&fileType = pdf.

Karmy-Jones, R., & Jurkovich, G. (2008). *Management of blunt chest and diaphragmatic injuries* (pp. 1768−1776). Elsevier BV. Available from 10.1016/b978-0-443-06861-4.50149-0.

Kazempoor, B., Nahmias, J., Clark, I., Schubl, S., Lekawa, M., Swentek, L., Keshava, H. B., & Grigorian, A. (2023). Scoring tool to predict need for early video-assisted thoracoscopic surgery (VATS) after pediatric trauma. *World Journal of Surgery*, *47*(11), 2925−2931. Available from https://doi.org/10.1007/s00268-023-07141-y.

Lang-Lazdunski, L., Bonnet, P. M., Pons, F., Brinquin, L., & Jancovici, R. (2002). Traumatic extrathoracic lung herniation. *Annals of Thoracic Surgery*, *74*(3), 927−929. Available from https://doi.org/10.1016/S0003-4975(02)03669-X.

Lang-Lazdunski, L., & Pilling, J. E. (2007). Video assisted thoracoscopic surgery is still the standard [2]. *British Medical Journal*, *334*(7588), 273.

Lim, E., Harris, R. A., McKeon, H. E., Batchelor, T. J. P., Dunning, J., Shackcloth, M., Anikin, V., Naidu, B., Belcher, E., Loubani, M., Zamvar, V., Dabner, L., Brush, T., Stokes, E. A., Wordsworth, S., Paramasivan, S., Realpe, A., Elliott, D., Blazeby, J., & Rogers, C. A. (2022). Impact of video-assisted thoracoscopic lobectomy versus open lobectomy for lung cancer on recovery assessed using self-reported physical function: VIOLET RCT. *Health Technology Assessment*, *26*(48), 1−162. Available from https://doi.org/10.3310/thbq1793.

Lin, F. C. F., Li, R. Y., Tung, Y. W., Jeng, K. C., & Tsai, S. C. S. (2016). Morbidity, mortality, associated injuries, and management of traumatic rib fractures. *Journal of the Chinese Medical Association*, *79*(6), 329−334. Available from https://doi.org/10.1016/j.jcma.2016.01.006, http://www.sciencedirect.com/science/journal/17264901.

Lin, H. L., Huang, W. Y., Yang, C., Chou, S. M., Chiang, H. I., Kuo, L. C., Lin, T. Y., & Chou, Y. P. (2014). How early should VATS be performed for retained haemothorax in blunt chest trauma. *Injury, 45*(9), 1359−1364. Available from. Available from https://doi.org/10.1016/j.injury.2014.05.036, www.elsevier.com/locate/injury.

Lin, H. L., Tarng, Y. W., Wu, T. H., Huang, F. D., Huang, W. Y., & Chou, Y. P. (2019). The advantages of adding rib fixations during VATS for retained hemothorax in serious blunt chest trauma − A prospective cohort study. *International Journal of Surgery, 65*, 13−18. Available from https://doi.org/10.1016/j.ijsu.2019.02.022, http://www.elsevier.com/wps/find/journaldescription.cws_home/705107/description#description.

Lodhia, J. V., Konstantinidis, K., & Papagiannopoulos, K. (2019). Video-assisted thoracoscopic surgery in trauma: Pros and cons. *Journal of Thoracic Disease, 11*(4), 1662−1667. Available from https://doi.org/10.21037/jtd.2019.03.55, http://www.jthoracdis.com/.

Ludwig, C., & Koryllos, A. (2017). Management of chest trauma. *Journal of Thoracic Disease, 9*(S3), S172. Available from https://doi.org/10.21037/jtd.2017.03.52.

Morales, C. H., Salinas, C. M., Henao, C. A., Patino, P. A., & Munoz, C. M. (1997). Thoracoscopic pericardial window and penetrating cardiac trauma. *The Journal of Trauma: Injury, Infection, and Critical Care, 42*(2), 273−275. Available from https://doi.org/10.1097/00005373-199702000-00015.

Mowery, N. T., Gunter, O. L., Collier, B. R., Diaz, J. J., Haut, E., Hildreth, A., Holevar, M., Mayberry, J., & Streib, E. (2011). Practice management guidelines for management of hemothorax and occult pneumothorax. *Journal of Trauma − Injury, Infection and Critical Care, 70*(2), 510−518. Available from https://doi.org/10.1097/TA.0b013e31820b5c31.

Murray, J. A., Demetriades, D., Asensio, J. A., Cornwell, E. E., Velmahos, G. C., Belzberg, H., & Berne, T. V. (1998). Occult injuries to the diaphragm: Prospective evaluation of laparoscopy in penetrating injuries to the left lower chest. *Journal of the American College of Surgeons, 187*(6), 626−630. Available from https://doi.org/10.1016/S1072-7515(98)00246-4.

Navsaria, P. H., Vogel, R. J., & Nicol, A. J. (2004). Thoracoscopic evacuation of retained post-traumatic hemothorax. *Annals of Thoracic Surgery, 78*(1), 282−285. Available from. Available from https://doi.org/10.1016/j.athoracsur.2003.11.029, www.elsevier.com/locate/athoracsur.

Ochsner, M. G., Rozycki, G. S., Lucente, F., Wherry, D. C., & Champion, H. R. (1993). Prospective evaluation of thoracoscopy for diagnosing diaphragmatic injury in thoracoabdominal trauma: A preliminary report. *Journal of Trauma − Injury, Infection and Critical Care, 34*(5), 704−710. Available from https://doi.org/10.1097/00005373-199305000-00013.

Patel, N. J., Dultz, L., Ladhani, H. A., Cullinane, D. C., Klein, E., McNickle, A. G., Bugaev, N., Fraser, D. R., Kartiko, S., Dodgion, C., Pappas, P. A., Kim, D., Cantrell, S., Como, J. J., & Kasotakis, G. (2021). Management of simple and retained hemothorax: A practice management guideline from the Eastern Association for the Surgery of Trauma. *The American Journal of Surgery, 221*(5), 873−884. Available from https://doi.org/10.1016/j.amjsurg.2020.11.032.

Sanna, S., Bertolaccini, L., Brandolini, J., Argnani, D., Mengozzi, M., Pardolesi, A., & Solli, P. (2017). Uniportal video-assisted thoracoscopic surgery in hemothorax. *Journal of Visualized Surgery, 3*, 126. Available from https://doi.org/10.21037/jovs.2017.08.06, 126.

Smith, J. W., Franklin, G. A., Harbrecht, B. G., & Richardson, J. D. (2011). Early VATS for blunt chest trauma: A management technique underutilized by acute care surgeons. *Journal of Trauma − Injury, Infection and Critical Care, 71*(1), 102−107. Available from https://doi.org/10.1097/TA.0b013e3182223080.

Striffeler, H., Gugger, M., Im Hof, V., Cerny, A., Furrer, M., & Ris, H. B. (1998). Video-assisted thoracoscopic surgery for fibrinopurulent pleural empyema in 67 patients. *Annals of Thoracic Surgery, 65*(2), 319−323. Available from https://doi.org/10.1016/S0003-4975(97)01188-0.

Sulaiman, A., Cottin, V., De Souza Neto, E. P., Orsini, A., Cordier, J. F., Gamondes, J. P., & Tronc, F. (2006). Cough-induced intercostal lung herniation requiring surgery: Report of a case. *Surgery Today, 36*(11), 978–980. Available from https://doi.org/10.1007/s00595-006-3284-8.

Tacconi, F., Pompeo, E., Fabbi, E., & Mineo, T. C. (2010). Awake video-assisted pleural decortication for empyema thoracis. *European Journal of Cardio-thoracic Surgery, 37*(3), 594–601. Available from https://doi.org/10.1016/j.ejcts.2009.08.003.

Vassiliu, P., Velmahos, G. C., & Toutouzas, K. G. (2001). Timing, safety, and efficacy of thoracoscopic evacuation of undrained post-traumatic hemothorax. *The American Surgeon, 67* (12), 1165–1169. Available from https://doi.org/10.1177/000313480106701210.

Wani, A. S., Kalamkar, P., Alhassan, S., & Farrell, M. J. (2015). Spontaneous intercostal lung herniation complicated by rib fractures: A therapeutic dilemma. *Oxford Medical Case Reports, 2015*(12), 378–381. Available from. Available from https://doi.org/10.1093/omcr/omv069, https://academic.oup.com/omcr.

Wu, T. H., Lin, H. L., Chou, Y. P., Huang, F. D., Huang, W. Y., & Tarng, Y. W. (2020). Facilitating ventilator weaning through rib fixation combined with video-assisted thoracoscopic surgery in severe blunt chest injury with acute respiratory failure. *Critical Care, 24* (1). Available from. Available from https://doi.org/10.1186/s13054-020-2755-4, http://ccforum.com/content/17.

Zeiler, J., Idell, S., Norwood, S., & Cook, A. (2020). Hemothorax: A review of the literature. *Clinical Pulmonary Medicine, 27*(1), 1–12. Available from https://doi.org/10.1097/cpm.0000000000000343.

Zhang, J., Hong, Q., Mo, X., & Ma, C. (2022). Complete video-assisted thoracoscopic surgery for rib fractures: Series of 35 cases. *The Annals of Thoracic Surgery, 113*(2), 452–458. Available from https://doi.org/10.1016/j.athoracsur.2021.01.065.

Ziapour, B., Mostafidi, E., Sadeghi-Bazargani, H., Kabir, A., & Okereke, I. (2020). Timing to perform VATS for traumatic-retained hemothorax (a systematic review and meta-analysis). *European Journal of Trauma and Emergency Surgery, 46*(2), 337–346. Available from https://doi.org/10.1007/s00068-019-01275-2.

Chapter 12

Hydropneumothorax: why, when, how

Savvas Lampridis[1], Fabrizio Minervini[2] and Marco Scarci[3,4]

[1]*Department of Cardiothoracic Surgery, Hammersmith Hospital, Imperial College Healthcare NHS Trust, London, United Kingdom,* [2]*Division of Thoracic Surgery, Cantonal Hospital Lucerne, Lucerne, Switzerland,* [3]*Department of Thoracic Surgery, Imperial College NHS Healthcare Trust, London, United Kingdom,* [4]*National Heart and Lung Institute, Imperial College, London, United Kingdom*

Introduction

Blunt chest injuries frequently result in hemothorax, pneumothorax, or hemopneumothorax, presenting significant challenges in the management of thoracic trauma patients. The presence of air, blood, or both within the pleural cavity can critically impair respiratory function, leading to substantial morbidity and mortality if not promptly and effectively addressed. The advent of modern imaging techniques and advanced therapeutic interventions has dramatically enhanced the prognosis of patients with traumatic hemothorax and/or pneumothorax. However, these advancements have also introduced new intricacies into management strategies. This chapter aims to provide an in-depth understanding of traumatic hemothorax, pneumothorax, and hemopneumothorax, with a particular focus on blunt thoracic injuries.

Historical background

The management of thoracic trauma, with its inherent complexities and high mortality rates, has been a formidable challenge in the medical field since antiquity. The concept of draining air and blood from the pleural cavity using a tube, a technique that originated in ancient Greece, underwent millennia of evolution before becoming the standard of care for hemothorax and pneumothorax in the latter half of the 20th century.

The use of tube thoracostomy as a treatment for hemothorax began to gain traction in Europe during the Late Medieval Ages, primarily as a response to penetrating chest wounds sustained on the battlefield. However,

182 Chest Blunt Trauma

over the ensuing centuries, the medical community remained divided over the optimal treatment for hemothorax resulting from such injuries. Some physicians advocated for wound closure without drainage, while others championed an initial open approach to evacuate the hemothorax. The 18th century saw the development of various tubes and cannulas to facilitate hemothorax drainage, and the 19th century witnessed the creation of the first closed chest drainage systems. Despite these advances, consensus on the best initial management of traumatic hemothorax remained elusive.

Although battlefield experiences are frequently credited with driving major advancements in trauma care, the journey of thoracic trauma management has charted a unique trajectory. The military medical practices during World War I, World War II, and the Korean War paradoxically delayed the widespread acceptance of emergency tube thoracostomy as the primary management approach for combat casualties with chest injuries. Instead, the management strategy during these conflicts leaned towards evacuating hemothorax via thoracentesis and reserving thoracotomy for addressing uncontrolled hemorrhage and tension pneumothorax.

The paradigm shift in this approach was not ignited by warfare, but rather by civilian encounters with penetrating chest trauma in the United States during the 1950s, and it was cemented through combat casualty care during the Vietnam War. As the number of motor vehicle accidents surged in the 1960s, the principles of tube thoracostomy and underwater seal chest drainage were eventually adopted for managing hemothorax and pneumothorax in patients suffering from blunt thoracic trauma (Felton, 1963; Maloney & McDonald, 1963). This change in consensus among surgeons in urban trauma centers led to the widespread adoption of tube thoracostomy, which eventually became the mainstay of treatment for traumatic hemothorax and pneumothorax. Subsequent significant advancements in surgical techniques, chest tubes, and drainage systems have further optimized patient outcomes.

Terminology

Before delving deeper into the chapter, it is essential to clarify certain key terms.

Hemothorax refers to the accumulation of blood within the pleural cavity. The term also encompasses pleural fluid collections with a hematocrit level exceeding 50% of the peripheral blood hematocrit.

Massive hemothorax typically denotes a hemothorax resulting in blood loss greater than 1500 mL (approximately 30% of the total blood volume of around 5 L in a typical adult) or blood loss greater than 200 mL per hour (approximately 3 mL per kg of body weight per hour) for a duration of two to four hours.

Retained hemothorax is usually defined as a residual bloody pleural effusion or blood clots larger than 500 mL persisting after 72 hours of chest tube placement.

Pneumothorax denotes the presence of air in the pleural space. The term occult pneumothorax describes a pneumothorax that is not immediately identifiable through physical examination or plain chest radiography but is later revealed through more sensitive imaging modalities, commonly computed tomography (CT).

Hydropneumothorax, in the context of trauma, refers to hemopneumothorax, that is, the simultaneous presence of hemothorax and pneumothorax.

Epidemiology

Hemothorax and pneumothorax rank among the most frequent sequelae of blunt thoracic trauma, surpassed only by rib fractures. Studies have reported the incidence of pneumothorax in patients with blunt trauma to range from 6.1% to 81.6% (Ball et al., 2005, 2009; Blaivas et al., 2005; Charbit et al., 2015; de Moya et al., 2007; Kim et al., 2023; Lamb et al., 2007; Lee et al., 2010; Shorr et al., 1987; Soldati et al., 2006, 2008; Wilson et al., 2009; Zhang et al., 2006). The corresponding incidence of hemothorax, while less well-documented, is estimated to fall between 2.5% and 37.5% (Bilello et al., 2005; Kulshrestha et al., 2004; Shorr et al., 1987; Trupka et al., 1997). This considerable variation in reported incidences can be attributed to a multitude of factors, including the characteristics of the study population, the nature of the trauma under investigation, and the capabilities of the trauma center.

A comprehensive study conducted at a Level I US trauma center, which offers a representative snapshot of urban trauma experiences, retrospectively analyzed 1359 consecutive patients over five years from 1995 to 2000 (Kulshrestha et al., 2004). The vast majority (90.1%) of these patients had sustained blunt trauma, while the remaining had suffered penetrating injuries. Within this cohort, 20.4% of patients were diagnosed with either hemothorax, pneumothorax, or both. More specifically, pneumothorax was identified in 16.7% of patients, hemothorax in 2.5%, and hemopneumothorax in 1.2%.

Similar findings have been reported in studies focusing exclusively on blunt thoracic trauma. In one such study conducted at a hospital specializing in thoracic diseases, 1490 patients with blunt chest injuries were retrospectively reviewed over two years (Liman et al., 2003). Hemothorax, pneumothorax, or hemopneumothorax were observed in 22.8% of these patients. Notably, a significant correlation was identified between the presence of hemothorax and/or pneumothorax and the number of rib fractures. Specifically, the incidence of hemothorax and/or pneumothorax was 6.7% in patients without a rib fracture, 24.9% in patients with one or two rib fractures, and 81.4% in patients with more than two rib fractures.

Turning our attention to an earlier study, we find more valuable insights into the patterns of blunt thoracic trauma. This investigation involved a retrospective analysis of 515 patients at a US trauma referral center (Shorr et al., 1987). In this study, fractures of the bony thorax emerged as the most prevalent injury, present in 75.3% of cases. Interestingly, in the absence of a bony fracture, the majority (62.2%) of patients presented with hemothorax, pneumothorax, or a combination of the two. Hemothorax was more frequent, affecting 37.5% of the study cohort, while pneumothorax was observed in 18.4% of the patients.

The epidemiology of occult pneumothorax in the context of blunt trauma has become a focal point of attention due to its potentially severe outcomes. The reported incidence of occult pneumothorax exhibits a wide range, from a modest 1.8% to an alarming 56.3% in patients with blunt trauma (Ball et al., 2005, 2009; Blaivas et al., 2005; Charbit et al., 2015; de Moya et al., 2007; Kim et al., 2023; Lamb et al., 2007; Lee et al., 2010; Mahmood et al., 2020; Misthos et al., 2004; Soldati et al., 2006, 2008; Wilson et al., 2009; Zhang et al., 2006; Zhang et al., 2016). Furthermore, the proportion of occult pneumothoraces among all diagnosed pneumothoraces resulting from blunt trauma spans from 22.1% to 75.7% (Ball et al., 2005, 2009; Blaivas et al., 2005; Charbit et al., 2015; de Moya et al., 2007; Kim et al., 2023; Lamb et al., 2007; Lee et al., 2010; Soldati et al., 2006, 2008; Wilson et al., 2009; Zhang et al., 2006). This substantial variation can be ascribed to the inherent heterogeneity in the reporting studies, which includes differences in hospital environments, radiological assessments (e.g., supine vs erect chest radiograph, thoracic vs. abdominal CT), and the qualifications of the interpreting personnel (e.g., board-certified radiologists vs emergency medicine physicians). It is also worth noting that up to 20% of traumatic pneumothoraces that are classified as occult may be missed diagnoses (Brar et al., 2010).

Pathophysiology

Hemothorax

Hemothorax can result from a disruption of tissues within the chest wall or intrathoracic structures. The most prevalent cause is a laceration of the lung or an injury to intercostal vessels, often as a consequence of rib fractures. On less frequent occasions, the source of bleeding may be the internal thoracic arteries, the great vessels, or even the heart. In cases where a diaphragmatic rupture is present, the hemothorax may be caused by bleeding from an intrabdominal source.

The physiological response to hemothorax is primarily contingent on the volume and rate of blood loss. For example, a loss of less than 15% of the circulating blood volume[1] generally does not trigger significant hemodynamic changes. However, a blood volume loss ranging from 15% to 30% can precipitate the early signs of hypovolemic shock. More severe shock symptoms, indicative of end-organ hypoperfusion, typically manifest when blood volume loss exceeds 30%. It is noteworthy that the pleural cavity can

---

1. The total blood volume in adults is estimated to be approximately 70 ml per kg of body weight.

accommodate nearly the entire circulating blood volume, thus enabling a potentially life-threatening hemorrhage to occur without visible external blood loss. A substantial accumulation of blood within the pleural cavity can also hinder respiratory function due to its space-occupying effect, potentially leading to ventilation and oxygenation abnormalities. These effects can be exacerbated when combined with lung and chest wall injuries or the simultaneous presence of a pneumothorax.

The blood entering the pleural cavity undergoes a degree of defibrination due to the motion of the diaphragm, lungs, and mediastinum, leading to incomplete clotting. In the hours succeeding bleeding cessation, enzymatic activity within the pleural cavity initiates the degradation of the formed clots. The lysis of erythrocytes prompts a significant increase in the protein concentration of the pleural fluid, which results in an elevation in the oncotic pressure within the pleural space. This increased intrapleural oncotic pressure contributes to the production of a net pressure gradient that drives fluid from the surrounding tissues into the pleural cavity. Consequently, a small, initially asymptomatic hemothorax can evolve into a large, symptomatic bloody pleural effusion.

The natural progression of a large hemothorax is a complex and dynamic process. In the initial phase, which can occur as early as the seventh day following the traumatic event, there is a proliferation of fibroblasts at the periphery of the clotted blood. This marks the onset of a transformative period in which the hemothorax begins to evolve from a liquid state into a more organized form. As the days turn into weeks, this fibrous tissue matures, gradually encapsulating the blood clot and forming a peel that adheres to the pleura. This fibrinous peel, a testament to the body's attempt to wall off and isolate the hemothorax, continues to thicken and harden over time. This process may ultimately culminate in the development of a fibrothorax, which significantly compromises lung function and poses a considerable challenge to subsequent therapeutic interventions.

Pneumothorax

In instances of pneumothorax, the pleural cavity becomes infiltrated with air typically due to an inwardly displaced rib fracture that disrupts the pleura and injures the pulmonary parenchyma. Alternatively, a deceleration injury may precipitate an abrupt increase in intra-alveolar pressure, leading to alveolar rupture. Less frequently, air may enter the pleural space via a tear in the tracheobronchial tree or esophagus, potentially resulting in concomitant pneumomediastinum. A tension pneumothorax can develop when the pleural injury creates a one-way valve mechanism, permitting air entry but obstructing its escape. Importantly, tension pneumothorax can manifest even in the absence of associated rib fractures.

The clinical implications of pneumothorax are largely determined by the volume of air within the pleural cavity. This volume dictates the extent of ipsilateral lung collapse, which can compromise oxygenation and ventilation. In the case of tension pneumothorax, the continuous inflow of air leads to complete collapse of the ipsilateral lung. As the intrapleural pressure continues to rise, it induces a mediastinal shift towards the contralateral side. This shift compresses the contralateral lung and vena cavae, thereby impeding the return of blood to the right atrium and precipitating cardiovascular compromise. If left untreated, this condition can rapidly progress to respiratory insufficiency, cardiovascular collapse, and ultimately death.

Primary evaluation

The initial assessment of a trauma patient should be conducted in tandem with resuscitation efforts, adhering to the guidelines set forth by the Advanced Trauma Life Support (ATLS) program (American College of Surgeons, Committee on Trauma, 2018). It is crucial to consider the potential presence of pneumothorax and hemothorax in all patients presenting to the emergency department following blunt thoracic trauma. A high degree of clinical suspicion, coupled with a comprehensive physical examination, can often guide appropriate interventions even before imaging studies are obtained.

Hemothorax can manifest with decreased breath sounds, dullness to percussion, respiratory distress, as well as signs and symptoms of hypovolemic shock, depending on the severity of the bleeding. In clinically unstable patients requiring tube thoracostomy, the use of a chest tube of at least 24 or 28 French is recommended. Immediate drainage of an amount of blood exceeding 20 mL per kg of body weight (approximately 1500 mL in a typical adult) is generally considered an indication for thoracotomy in the operating room. In the setting of blunt thoracic trauma, emergency department thoracotomy has narrow indications and rarely results in successful resuscitation (Slessor & Hunter, 2015). The management of massive hemothorax is discussed in more detail in the Management section.

Patients with pneumothorax may exhibit a variety of signs and symptoms, including tachypnea, hypoxia, dyspnea, and pleuritic chest pain. These manifestations largely depend on the size of the pneumothorax. Physical examination may reveal diminished or absent breath sounds and hyperresonance to percussion. The presence of crepitus, indicative of subcutaneous emphysema, may also be noted, although this is a less common finding. In situations where a patient presents with respiratory distress and hypotension, in conjunction with any of the aforementioned physical signs, a tension pneumothorax should be suspected. Additional signs of tension pneumothorax may include tracheal deviation to the contralateral side and neck vein distension; however, the neck veins may appear flat in cases of systemic hypovolemia. In such critical scenarios, needle thoracostomy decompression with a 14-gauge angiocatheter should be performed immediately. Tube thoracostomy should be carried out promptly afterwards, before obtaining a chest radiograph.

Imaging investigations
Plain radiography

A plain chest radiograph is generally recommended for all hemodynamically stable patients who have experienced blunt chest trauma. This imaging modality can reveal evidence of hemothorax and/or pneumothorax, among other abnormalities, such as rib fractures, pulmonary contusions, and signs of blunt aortic injury.

The radiographic appearance of a hemothorax shares similarities with that of a pleural effusion and depends largely on patient positioning. On an erect chest radiograph, features suggesting hemothorax may include blunting of the costophrenic and cardiogenic angles, fluid in the interlobar fissures, and a meniscus-like density; however, this meniscus may be absent with concurrent pneumothorax. Typically, at least 300 to 400 mL of fluid are required before obvious changes appear on an erect frontal view. An erect lateral projection can detect smaller volumes since the costophrenic angles are deepest posteriorly. Even better sensitivity is provided by a lateral decubitus film obtained with the patient's affected side down, allowing very small amounts of blood to be visualized. Meanwhile, large hemothorax in supine patients may show only subtle increased opacity, as fluid collects posteriorly. Bilateral hemothorax of similar size can also be challenging to discern on supine films due to comparable lung densities. Lastly, it is worth mentioning that preexisting adhesions can alter the typical distribution of blood and thus the usual radiographic appearance of a hemothorax.

Pneumothorax is readily apparent on erect chest radiography, showing the visceral pleura as a thin curvilinear opacity with the absence of peripheral pulmonary vascular markings. In tension pneumothorax, the mediastinal shift towards the contralateral side is evident. If the pneumothorax is suspected but not visualized on a standard frontal film, an expiratory chest radiograph or a lateral decubitus view with the affected side up can provide further investigation. In contrast, detection can be challenging on supine chest radiography. Common signs in this view include relative lucency of the ipsilateral hemithorax, sharpened adjacent mediastinal and diaphragmatic margins, and deep costophrenic sulcus.

Ultrasonography

Ultrasonography, specifically the Focused Assessment with Sonography for Trauma (FAST), is a critical component of the initial trauma evaluation and resuscitation. The extended FAST incorporates additional views to assess for pneumothorax. FAST should be performed during the secondary survey if not done in the primary evaluation, unless CT is planned. A low-frequency curvilinear or phased array probe is typically used to optimize viewing depth.

The most dependent portion of the chest when supine is the costophrenic area, where free intrathoracic fluid accumulates. In hemothorax, an anechoic stripe is seen above the diaphragm, with the lower lung margin sometimes visible. However, differentiating hemothorax from chronic effusion can be challenging, as both may exhibit fluid, fibrin strands, and clots. If the clinical scenario is ambiguous, further evaluation with CT or tube thoracostomy may be warranted.

When assessing for pneumothorax with ultrasonography, certain considerations apply. In the supine patient, air rises towards nondependent areas within the thoracic cavity. The ultrasound probe should be oriented perpendicular to the skin, with the scanning plane adjusted to survey intercostal spaces anteriorly. The lung point sign is diagnostic of pneumothorax but may not always be visible. Absent lung sliding or lung pulse could indicate pneumothorax, although these findings may also result from underlying lung or pleural diseases. Conversely, the detection of lung sliding or lung pulse effectively rules out the presence of pneumothorax at that location. To reliably exclude pneumothorax quickly, multiple interspaces should be scanned over a brief period.

Ultrasonography versus radiography

The diagnostic accuracy of ultrasonography for hemothorax detection in trauma patients has been consistently demonstrated across multiple studies (Brooks et al., 2004; Ma & Mateer, 1997; Sisley et al., 1998). In one prospective study, 54 blunt trauma patients were assessed with FAST by emergency physicians or general surgeons, each with experience exceeding 50 scans (Brooks et al., 2004). For detecting hemothorax, FAST showed 90% sensitivity, 100% specificity and positive predictive value, and 97.8% negative predictive value. Comparable results were reported in studies of both blunt and penetrating trauma, whether ultrasonography was performed by surgeons (Sisley et al., 1998) or emergency physicians (Ma & Mateer, 1997). Collectively, these studies revealed that ultrasonography had equivalent accuracy to portable chest radiography for diagnosing hemothorax.

Ultrasonography has been shown to provide superior detection of traumatic pneumothorax when compared to supine chest radiography. A systematic review and meta-analysis included 13 prospective studies comparing these imaging modalities for identifying pneumothorax due to blunt or penetrating chest trauma (Chan et al., 2020). In a focused analysis of five included studies encompassing 754 blunt trauma patients, 223 of whom had pneumothorax, ultrasonography exhibited an overall sensitivity of 94% (95% confidence interval [CI], 86% to 98%) and specificity of 99% (95% CI, 97% to 100%) for detecting pneumothorax. In comparison, supine chest radiography had a markedly lower sensitivity of just 38% (95% CI, 19% to 62%), although specificity remained high at 100% (95% CI, 92% to 100%).

However, these results should be interpreted cautiously due to the limitations of the included studies. Indeed, all studies were found to have a high or unclear risk of bias in their patient selection methodology. Additionally, many of the included studies involved clinicians with extensive experience in ultrasonography, while others were conducted in clinical settings where point-of-care ultrasound was already widely adopted into standard protocols.

Ultrasonography offers several advantages over plain radiography for the initial trauma evaluation when hemothorax or pneumothorax is suspected. In addition to providing accurate and actionable information at the bedside, ultrasonography avoids ionizing radiation exposure, is cost-effective, and can be performed rapidly and repeatedly. In contrast, plain radiography demands a greater number of resources and is more time-consuming, potentially delaying diagnosis and treatment.

Despite its advantages, ultrasonography is not without its limitations. Most notably, its effectiveness depends on the experience and skill level of the operator. Furthermore, ultrasonography is patient-dependent, with challenges in image quality for certain individuals, such as obese patients. Moreover, subcutaneous emphysema can interfere with image acquisition. Therefore, while ultrasonography has many benefits, its operator- and patient-dependence introduces potential pitfalls.

Computed tomography

Thoracic CT provides a detailed evaluation of intrathoracic structures and is typically reserved for trauma patients with concerning clinical findings, abnormal chest radiographs, or high-energy mechanisms of injury. Compared to chest radiography, CT offers superior sensitivity for diagnosing hemothorax and pneumothorax and enables their precise quantification (Fig. 12.1).

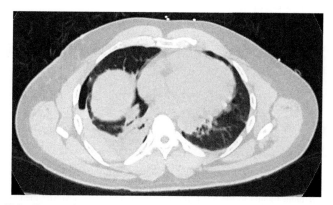

FIGURE 12.1 Computed tomography image of hemopneumothorax. Axial thoracic computed tomography image of a 51-year-old male demonstrating right hemopneumothorax following blunt chest trauma sustained in a motor vehicle accident.

Additional findings that are more readily evident on CT may include pneumomediastinum and subcutaneous emphysema (Fig. 12.2).

CT can identify hemothorax or pneumothorax unapparent on plain radiographs (Fig. 12.3), often prompting significant changes in patient management. In a prospective trial of 169 blunt trauma patients, chest radiography missed pneumothorax in 11.8% of cases and hemothorax in 8.3% (Omert et al., 2001). While none of the overlooked hemothoraces required chest tube drainage, 30% of the missed pneumothoraces were eventually managed with tube thoracostomy. Even higher discrepancy compared to CT was observed in another prospective study of 104 patients with severe chest injuries, where chest radiography missed 46.2% of hemothoraces and 40.3% of pneumothoraces (Guerrero-López et al., 2000). Drainage was subsequently performed for the missed hemothoraces and pneumothoraces in 9 and 19 patients, respectively. Despite this added diagnostic value, however, CT incurs substantially increased cost and radiation exposure, especially when performed after a normal chest radiograph (Rodriguez et al., 2014).

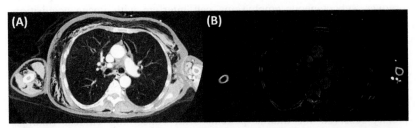

FIGURE 12.2 Computed tomography images of pneumothorax, pneumomediastinum, subcutaneous emphysema, and rib fracture. Axial thoracic computed tomography images of a 62-year-old male after sustaining blunt chest trauma from a fall. (A) Lung window revealing shallow bilateral pneumothoraces, pneumomediastinum, and diffuse subcutaneous emphysema. (B) Bone window of the same slice displaying a right rib fracture.

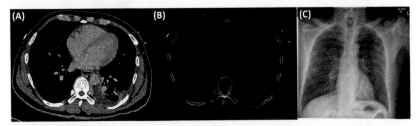

FIGURE 12.3 Computed tomography and plain radiography images of pleural effusion and rib fracture. Axial thoracic computed tomography (CT) and chest radiography images of a 56-year-old female after blunt trauma to the left hemithorax from an assault. (A) The mediastinal window of the CT scan shows a small left pleural effusion with adjacent lung atelectasis. (B) Bone window of the same CT slice demonstrating a left rib fracture. (C) Posteroanterior erect chest radiograph displaying subtle blunting of the left costophrenic angle.

CT can also be valuable in the later stages of blunt chest trauma management, particularly in assessing persistent radiographic abnormalities or lack of clinical progress. Several experts advocate for the early use of CT to localize and quantify retained hemothorax and empyema that can develop after blunt thoracic trauma and may impact recovery (Velmahos et al., 1999; Watkins et al., 2000). Timely CT assessment of residual hemothorax and empyema may enable interventions that reduce morbidity and mortality.

Management

Most patients with hemothorax, pneumothorax, or hemopneumothorax resulting from blunt thoracic trauma can be effectively managed with supportive measures and potential drainage of the pleural cavity. Consequently, the initial care for these patients can be competently administered by emergency physicians and trauma surgeons. However, some patients may not respond adequately to initial management measures or can later develop pleural complications. These complex cases may require tertiary care and a multidisciplinary approach involving a thoracic surgeon to improve outcomes.

Massive hemothorax

When massive hemothorax is present, surgical exploration through thoracotomy is the procedure of choice. Traditional indications for urgent thoracotomy include immediate bloody drainage exceeding 20 mL per kg of body weight (approximately 1500 mL in an average adult) or continuing bleeding over 3 mL per kg of body weight per hour (around 200 mL per hour) for two to four hours. However, the decision for urgent thoracotomy should not merely rely on chest tube output but be guided by physiological parameters, fluid requirements to preserve hemodynamic stability, as well as the presence and type of concomitant injuries.

Most guidelines for urgent thoracotomy draw largely from studies that focused predominantly on penetrating chest injuries (Kish et al., 1976; Mattox, 1983; McNamara et al., 1970; Oparah & Mandal, 1979; Reul et al., 1973; Siemens et al., 1977). One study attempted to address this knowledge gap by analyzing 2316 cases and comparing indications for urgent thoracotomy between blunt and penetrating injuries (Mansour et al., 1992). Among the 83 patients who required urgent thoracotomy in the operating room, 32.5% had sustained blunt trauma, while the rest (67.5%) had incurred penetrating wounds. The investigators concluded that chest tube output alone seldom justified urgent thoracotomy in cases of blunt trauma. Instead, refractory shock due to ongoing intrathoracic bleeding was found to be a more reliable indicator.

In a more recent endeavor to shed light on the subject, researchers retrospectively analyzed the cases of 157 patients who had undergone thoracotomy to manage hemorrhage within 48 hours of trauma (R Karmy-Jones et al., 2001). In this patient cohort, 22.9% had suffered a blunt injury, while the rest (77.1%) had sustained penetrating trauma. The study found a clear correlation between the volume of blood lost and mortality, with the risk of death being three times higher at a blood loss of 1500 mL compared to 500 mL. This supports the proposed threshold of 1500 mL blood loss as an indicator for thoracotomy. However, the study lacked detailed data regarding physiological parameters at different chest tube outputs.

Hemothorax

Tube thoracostomy drainage is the primary initial treatment for hemothorax. However, the optimal chest tube size for trauma patients remains unclear. Traditionally, tube sizes of 24 French or larger have been used in adult patients to ensure adequate drainage. Some studies suggest that smaller tubes (14–22 French) may be equally effective, with similar complication and failure rates (Bauman et al., 2018, 2021; Choi et al., 2021; Kulvatunyou et al., 2021; Maezawa et al., 2020; Tanizaki et al., 2017). Nevertheless, these studies largely excluded hemodynamically unstable patients requiring emergency tube thoracostomy. Thus large-bore tubes may be preferable for significant trauma, while smaller tubes can be a reasonable option for blunt trauma patients who are hemodynamically stable.

The size threshold at which a hemothorax can be managed conservatively is debated. This is relevant considering that tube thoracostomy, while a common intervention, can be associated with complications in up to 21% of trauma patients (Etoch et al., 1995). Hemothoraces that are clinically insignificant, measuring less than 300 mL or presenting as a lamellar fluid stripe less than 1.5 cm on axial thoracic CT, may be managed conservatively. A systematic review and meta-analysis of six studies involving 1405 patients with occult hemothorax, mostly (96.3%) due to blunt trauma, found that 56.3% were initially managed expectantly, while the rest (43.7%) were managed with tube thoracostomy (Gilbert et al., 2020). Of the patients managed expectantly, 23.1% (95% CI, 17.1%–29.1%) eventually required tube thoracostomy drainage due to failure of conservative management. Predictive factors of failure included hemothorax size larger than 300 mL and requirement for mechanical ventilation. Crucially, the mortality rates between the groups were comparable.

Retained hemothorax

Retained hemothorax has been reported to occur in 17.6%–31.1% of trauma patients managed with tube thoracostomy (Kumar et al., 2015; Prakash et al., 2020; Scott et al., 2017). The clinical significance of retained hemothorax cannot be understated, as it can contribute to a 33% rate of empyema (Karmy-Jones et al., 2008). To mitigate the risk of empyema and avoid potential thoracotomy, early video-assisted thoracoscopic surgery (VATS) is suggested as more appropriate than the insertion of a second chest tube. This was illustrated in a prospective, randomized trial examining early VATS for evacuation of retained hemothorax (Meyer et al., 1997). Of 39 patients who developed retained hemothorax within 72 hours following initial tube thoracostomy, 15 underwent subsequent VATS, while 24 had a second chest tube inserted. The patients treated primarily with VATS had significantly shorter duration of tube drainage (2.5 ± 1.4 days vs 4.5 ± 2.8 days; $P < 0.02$), shorter hospitalization (5.4 ± 2.2 days vs 8.1 ± 4.6 days; $P < 0.02$), and lower hospital costs ($P < 0.02$). Importantly, 41.7% of the patients managed with a second chest tube eventually required VATS or thoracotomy.

The optimal timing for surgical intervention in cases of retained hemothorax remains a topic of considerable discussion. Some studies suggest that performing VATS between the third and fifth-day post-trauma can reduce the risk of empyema, need for conversion to open thoracotomy, and requirements for subsequent interventions, as compared to later surgery (Heniford et al., 1997; Morales Uribe et al., 2008). Other research advocates for VATS within the first three days of admission, citing benefits that include reduced hospital stay, decreased likelihood of positive cultures in residual blood, and less operative difficulty compared to VATS performed after this three-day window (Vassiliu et al., 2001). Despite these findings, however, there are no absolute contraindications to performing VATS in a delayed manner.

Prompted by a shift towards nonoperative management of retained hemothorax, researchers have explored the use of fibrinolytics. For instance, a prospective observational study examined intrapleural thrombolysis in managing undrained traumatic hemothorax (Kimbrell et al., 2007). Of 25 patients treated with intrapleural streptokinase or urokinase, 92% achieved successful hemothorax resolution within an average of 3.4 ± 1.4 days. Notably, no significant bleeding or other complications related to the procedure were reported. Moreover, intrapleural streptokinase demonstrated comparable results to VATS in a randomized controlled trial for post-traumatic residual hemothorax (Kumar et al., 2015). Hemothorax resolution was equivalent between VATS and intrapleural streptokinase groups (72% vs 71%, respectively), with no significant differences in morbidity. However, the role of intrapleural thrombolysis requires further research. Currently, fibrinolytic should be considered secondary to surgery, particularly when surgical risks could outweigh the benefits.

Pneumothorax

Traumatic pneumothorax is typically managed with tube thoracostomy (Fig. 12.4), as recommended in the ATLS guidelines (American College of Surgeons, Committee on Trauma, 2018). This approach is suggested for any radiographically apparent pneumothorax. However, optimal management is less established for small, asymptomatic pneumothoraces. Considering the known complications associated with chest tubes in a trauma setting (Etoch

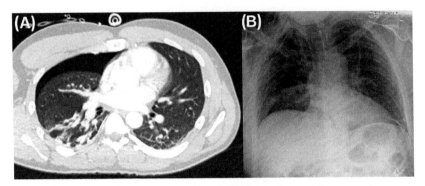

FIGURE 12.4 Computed tomography and chest radiography images of a pneumothorax managed with tube thoracostomy. Thoracic computed tomography (CT) and chest radiography images of a 37-year-old male after sustaining blunt chest trauma in a motor vehicle accident. (A) Axial CT scan displaying right pneumothorax. (B) Anteroposterior chest radiograph following drainage of the pneumothorax with tube thoracostomy.

et al., 1995), alternative strategies, such as percutaneous catheter insertion, needle aspiration, or careful observation, could be reasonable options for these minor pneumothoraces.

The optimal management of occult pneumothorax is often debated. For blunt trauma patients with small occult pneumothoraces (measuring less than 1 cm on CT), observation alone may be an appropriate course of action (Ball et al., 2005; Moore et al., 2011; Wilson et al., 2009). However, occult pneumothoraces can enlarge, occasionally evolving into tension pneumothorax. Regular clinical and radiographic monitoring can help identify such cases. If pneumothorax enlargement occurs or concerning symptoms develop, chest tube insertion is warranted.

The risk of an occult pneumothorax expansion is higher during mechanical ventilation, although optimal management remains uncertain in this patient population (Brasel et al., 1999; Clements et al., 2021; Enderson et al., 1993). Aiming to shed light on the safety and effectiveness of conservative management versus prophylactic intercostal catheter insertion for occult pneumothorax, a recent systematic review and meta-analysis included 12 studies with a total of 354 mechanically ventilated patients (Smith et al., 2021). The conservative management group demonstrated a relatively low rate of tension pneumothorax (2.8%), as well as a noticeably lower rate of complications (5.8% vs 19.5%). Overall, patients expected to require short-term (less than four days) mechanical ventilation and who could tolerate a brief episode of hypotension or hypoxia, can be potentially managed with observation alone. However, if tube thoracostomy is not performed in mechanically ventilated patients with occult pneumothorax, careful monitoring for signs of an expanding pneumothorax and immediate availability of an experienced proceduralist become imperative.

Prognosis

Hemothoraces less than 300—400 mL generally resolve spontaneously within a few weeks following trauma. However, larger undrained hemothorax may fail to reabsorb, potentially causing empyema (Fig. 12.5). Retained hemothorax also risk empyema. Multiple studies have sought to identify risk factors for the development of retained hemothorax after initial management with tube thoracostomy (Prakash et al., 2020; Rossmann et al., 2022; Scott et al., 2017). Single-center, retrospective studies found a range of independent prognostic factors, including bilateral injuries, sternal fracture, preadmission hypoxemia, need for mechanical ventilation upon admission, and prolonged chest tube duration (Rossmann et al., 2022; Scott et al., 2017). A multiinstitutional, prospective study involving 369 patients (28.7% of whom developed retained hemothorax) discovered that larger volumes of initial hemothorax were independently associated with the development of retained hemothorax (Prakash et al., 2020). Specifically, every additional 100 mL of initial hemothorax increased the risk of retained hemothorax by 15% (odds ratio [OR], 1.15; 95% CI, 1.08—1.21; $P < 0.001$).

Post-traumatic empyema has been reported in 2%—26% of patients, with retained hemothorax recognized as a distinctive risk factor (Arom et al., 1977; Caplan et al., 1984; Coselli et al., 1984; Eddy et al., 1989). A multiinstitutional, prospective observational study including 328 patients with

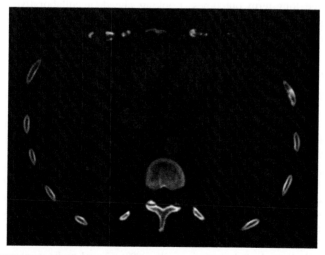

FIGURE 12.5 Computed tomography image of post-traumatic empyema. Axial thoracic computed tomography image of a 42-year-old female with a history of blunt trauma to the left hemithorax sustained 24 days prior. The scan demonstrates a large, hyperdense, biconvex pleural fluid collection with adjacent lung atelectasis. The contiguous soft tissue collection in the left anterolateral chest wall is consistent with empyema necessitans. A healing left rib fracture is also visible.

retained hemothorax sought to identify risk factors for the development of post-traumatic empyema (DuBose et al., 2012). Independent predictors included rib fractures (OR, 2.3; 95% CI, 1.3−4.1; $P = 0.006$), Injury Severity Score of 25 or higher (OR, 2.4; 95% CI, 1.3−4.4; $P = 0.005$), and additional interventions to evacuate retained blood (OR, 28.8; 95% CI, 6.6−125.5; $P < 0.001$). Patients developing empyema had significantly longer intensive care unit stay (mean difference, 4.1 days; 95% CI, 1.3−6.9; $P = 0.008$) and hospitalization (mean difference, 7.9 days; 95% CI, 12.7−3.2; $P = 0.01$).

Conclusions

The management of hemothorax and pneumothorax from blunt chest trauma has progressed through pivotal advancements. The latter half of the 20th century witnessed the widespread adoption of tube thoracostomy as the standard of care. Subsequently, imaging modalities like ultrasonography and CT enabled more accurate and timely diagnosis. However, determining optimal treatment still requires carefully weighing the risks and benefits of invasive versus conservative approaches, using clinical judgment tailored to each patient. While mortality has markedly improved, complications such as retained hemothorax and empyema remain concerning. Ongoing research is still needed to better understand trauma-induced pathophysiological changes and enhance long-term outcomes. Looking ahead, a collaborative, patient-centered approach provides optimism that progress in both the science and practice of trauma care will continue improving patient outcomes.

References

American College of Surgeons, Committee on Trauma. (2018). *ATLS: Advanced trauma life support student course manual. 10th*. Chicago, IL: American College of Surgeons.

Arom, K. V., Grover, F. L., Richardson, J. D., & Trinkle, J. K. (1977). Posttraumatic empyema. *The Annals of Thoracic Surgery*, 23(3), 254−258. Available from https://doi.org/10.1016/s0003-4975(10)64119-7.

Ball, C. G., Kirkpatrick, A. W., Laupland, K. B., Fox, D. I., Nicolaou, S., Anderson, I. B., Hameed, S. M., Kortbeek, J. B., Mulloy, R. R., Litvinchuk, S., & Boulanger, B. R. (2005). Incidence, risk factors, and outcomes for occult pneumothoraces in victims of major trauma. *The Journal of Trauma*, 59(4), 917−924. Available from https://doi.org/10.1097/01.ta.0000174663.46453.86.

Ball, C. G., Ranson, K., Dente, C. J., Feliciano, D. V., Laupland, K. B., Dyer, D., Inaba, K., Trottier, V., Datta, I., & Kirkpatrick, A. W. (2009). Clinical predictors of occult pneumothoraces in severely injured blunt polytrauma patients: A prospective observational study. *Injury*, 40(1), 44−47. Available from https://doi.org/10.1016/j.injury.2008.07.015.

Bauman, Z. M., Kulvatunyou, N., Joseph, B., Gries, L., O'Keeffe, T., Tang, A. L., & Rhee, P. (2021). Randomized Clinical Trial of 14-French (14F) Pigtail Catheters versus 28-32F Chest Tubes in the Management of Patients with Traumatic Hemothorax and Hemopneumothorax.

World Journal of Surgery, *45*(3), 880−886. Available from https://doi.org/10.1007/s00268-020-05852-0.

Bauman, Z. M., Kulvatunyou, N., Joseph, B., Jain, A., Friese, R. S., Gries, L., O'Keeffe, T., Tang, A. L., Vercruysse, G., & Rhee, P. (2018). A Prospective Study of 7-Year Experience Using Percutaneous 14-French Pigtail Catheters for Traumatic Hemothorax/Hemopneumothorax at a Level-1 Trauma Center: Size Still Does Not Matter. *World Journal of Surgery*, *42*(1), 107−113. Available from https://doi.org/10.1007/s00268-017-4168-3.

Bilello, J. F., Davis, J. W., & Lemaster, D. M. (2005). Occult traumatic hemothorax: when can sleeping dogs lie? *American Journal of Surgery*, *190*(6), 841−844. Available from https://doi.org/10.1016/j.amjsurg.2005.05.053.

Blaivas, M., Lyon, M., & Duggal, S. (2005). A prospective comparison of supine chest radiography and bedside ultrasound for the diagnosis of traumatic pneumothorax. *Academic emergency medicine: official journal of the Society for Academic Emergency Medicine*, *12*(9), 844−849. Available from https://doi.org/10.1197/j.aem.2005.05.005.

Brar, M. S., Bains, I., Brunet, G., Nicolaou, S., Ball, C. G., & Kirkpatrick, A. W. (2010). Occult pneumothoraces truly occult or simply missed: redux. *The Journal of Trauma*, *69*(6), 1335−1337. Available from https://doi.org/10.1097/TA.0b013e3181f6f525.

Brasel, K. J., Stafford, R. E., Weigelt, J. A., Tenquist, J. E., & Borgstrom, D. C. (1999). Treatment of occult pneumothoraces from blunt trauma. *The Journal of Trauma*, *46*(6), 987−990. Available from https://doi.org/10.1097/00005373-199906000-00001.

Brooks, A., Davies, B., & Smethhurst, M. (2004). J Connolly, Emergency ultrasound in the acute assessment of haemothorax. *Emergency medicine journal: EMJ.*, *21*(1), 44−46. Available from https://doi.org/10.1136/emj.2003.005438.

Caplan, E. S., Hoyt, N. J., Rodriguez, A., & Cowley, R. A. (1984). Empyema occurring in the multiply traumatized patient. *The Journal of Trauma*, *24*(9), 785−789. Available from https://doi.org/10.1097/00005373-198409000-00002.

Chan, K. K., Joo, D. A., McRae, A. D., Takwoingi, Y., Premji, Z. A., Lang, E., & Wakai, A. (2020). Chest ultrasonography versus supine chest radiography for diagnosis of pneumothorax in trauma patients in the emergency department. *The Cochrane Database of Systematic Reviews*, *7*(7). Available from https://doi.org/10.1002/14651858.CD013031.pub2.

Charbit, J., Millet, I., Maury, C., Conte, B., Roustan, J.-P., Taourel, P., & Capdevila, X. (2015). Prevalence of large and occult pneumothoraces in patients with severe blunt trauma upon hospital admission: experience of 526 cases in a French level 1 trauma center. *The American Journal of Emergency Medicine*, *33*(6), 796−801. Available from https://doi.org/10.1016/j.ajem.2015.03.057.

Choi, J., Villarreal, J., Andersen, W., Min, J. G., Touponse, G., Wong, C., Spain, D. A., & Forrester, J. D. (2021). Scoping review of traumatic hemothorax: Evidence and knowledge gaps, from diagnosis to chest tube removal. *Surgery*, *170*(4), 1260−1267. Available from https://doi.org/10.1016/j.surg.2021.03.030.

Clements, T. W., Sirois, M., Parry, N., Roberts, D. J., Trottier, V., Rizoli, S., Ball, C. G., Xiao, Z. J., & Kirkpatrick, A. W. (2021). OPTICC: A multicentre trial of Occult Pneumothoraces subjected to mechanical ventilation: The final report. *American Journal of Surgery*, *221*(6), 1252−1258. Available from https://doi.org/10.1016/j.amjsurg.2021.02.012.

Coselli, J. S., Mattox, K. L., & Beall, A. C. (1984). Reevaluation of early evacuation of clotted hemothorax. *American Journal of Surgery*, *148*(6), 786−790. Available from https://doi.org/10.1016/0002-9610(84)90438-0.

DuBose, J., Inaba, K., Okoye, O., Demetriades, D., Scalea, T., O'Connor, J., Menaker, J., Morales, C., Shiflett, T., Brown, C., & Copwood, B. (2012). Development of posttraumatic empyema in patients with retained hemothorax: results of a prospective, observational AAST study. *The Journal of Trauma and Acute Care Surgery*, *73*(3), 752−757. Available from https://doi.org/10.1097/TA.0b013e31825c1616.

Eddy, A. C., Luna, G. K., & Copass, M. (1989). Empyema thoracis in patients undergoing emergent closed tube thoracostomy for thoracic trauma. *American Journal of Surgery*, *157*(5), 494−497. Available from https://doi.org/10.1016/0002-9610(89)90643-0.

Enderson, B. L., Abdalla, R., Frame, S. B., Casey, M. T., Gould, H., & Maull, K. I. (1993). Tube thoracostomy for occult pneumothorax: a prospective randomized study of its use. *The Journal of Trauma*, *35*(5), 726−729. Available from https://doi.org/10.1097/00005373-199311000-00013.

Etoch, S. W., Bar-Natan, M. F., & Miller, F. B. (1995). J D Richardson, Tube thoracostomy. Factors related to complications. *Archives of Surgery (Chicago, Ill.: 1960).*, *130*(5), 521−525. Available from https://doi.org/10.1001/archsurg.1995.01430050071012.

Felton, W. L. (1963). Initial evaluation and management of the patient with a chest injury. *American Journal of Surgery*, *105*, 445−453. Available from https://doi.org/10.1016/0002-9610(63)90314-3.

Gilbert, R. W., Fontebasso, A. M., Park, L., Tran, A., & Lampron, J. (2020). The management of occult hemothorax in adults with thoracic trauma: A systematic review and meta-analysis. *The Journal of Trauma and Acute Care Surgery*, *89*(6), 1225−1232. Available from https://doi.org/10.1097/TA.0000000000002936.

Guerrero-López, F., Vázquez-Mata, G., Alcázar-Romero, P. P., Fernández-Mondéjar, E., Aguayo-Hoyos, E., & Linde-Valverde, C. M. (2000). Evaluation of the utility of computed tomography in the initial assessment of the critical care patient with chest trauma. *Critical Care Medicine*, *28*(5), 1370−1375. Available from https://doi.org/10.1097/00003246-200005000-00018.

Heniford, B. T., Carrillo, E. H., Spain, D. A., Sosa, J. L., Fulton, R. L., & Richardson, J. D. (1997). The role of thoracoscopy in the management of retained thoracic collections after trauma. *The Annals of Thoracic Surgery*, *63*(4), 940−943. Available from https://doi.org/10.1016/s0003-4975(97)00173-2.

Karmy-Jones, R., Jurkovich, G. J., Nathens, A. B., Shatz, D. V., Brundage, S., Wall, M. J., Engelhardt, S., Hoyt, D. B., Holcroft, J., & Knudson, M. M. (2001). Timing of urgent thoracotomy for hemorrhage after trauma: a multicenter study. *Archives of Surgery (Chicago, Ill.: 1960).*, *136*(5), 513−518. Available from https://doi.org/10.1001/archsurg.136.5.513.

Karmy-Jones, R., Holevar, M., Sullivan, R. J., Fleisig, A., & Jurkovich, G. J. (2008). Residual hemothorax after chest tube placement correlates with increased risk of empyema following traumatic injury. *Canadian Respiratory Journal*, *15*(5), 255−258. Available from https://doi.org/10.1155/2008/918951.

Kim, C.-W., Park, I.-H., Youn, Y.-J., & Byun, C.-S. (2023). Occult Pneumothorax in Blunt Thoracic Trauma: Clinical Characteristics and Results of Delayed Tube Thoracostomy in a Level 1 Trauma Center. *Journal of Clinical Medicine*, *12*(13). Available from https://doi.org/10.3390/jcm12134333.

Kimbrell, B. J., Yamzon, J., Petrone, P., Asensio, J. A., & Velmahos, G. C. (2007). Intrapleural thrombolysis for the management of undrained traumatic hemothorax: a prospective observational study. *The Journal of Trauma*, *62*(5), 1175−1178. Available from https://doi.org/10.1097/TA.0b013e3180500654.

Kish, G., Kozloff, L., Joseph, W. L., & Adkins, P. C. (1976). Indications for early thoracotomy in the management of chest trauma. *The Annals of Thoracic Surgery*, *22*(1), 23−28. Available from https://doi.org/10.1016/s0003-4975(10)63946-x.

Kulshrestha, P., Munshi, I., & Wait, R. (2004). Profile of chest trauma in a level I trauma center. *The Journal of Trauma*, *57*(3), 576−581. Available from https://doi.org/10.1097/01.ta.0000091107.00699.c7.

Kulvatunyou, N., Bauman, Z. M., Zein Edine, S. B., de Moya, M., Krause, C., Mukherjee, K., Gries, L., Tang, A. L., Joseph, B., & Rhee, P. (2021). The small (14 Fr) percutaneous catheter (P-CAT) versus large (28-32 Fr) open chest tube for traumatic hemothorax: A multicenter randomized clinical trial. *The Journal of Trauma and Acute Care Surgery*, *91*(5), 809−813. Available from https://doi.org/10.1097/TA.0000000000003180.

Kumar, S., Rathi, V., Rattan, A., Chaudhary, S., & Agarwal, N. (2015). VATS versus intrapleural streptokinase: A prospective, randomized, controlled clinical trial for optimum treatment of post-traumatic Residual Hemothorax. *Injury*, *46*(9), 1749−1752. Available from https://doi.org/10.1016/j.injury.2015.02.028.

Lamb, A. D. G., Qadan, M., & Gray, A. J. (2007). Detection of occult pneumothoraces in the significantly injured adult with blunt trauma. *European Journal of Emergency Medicine: Official Journal of the European Society for Emergency Medicine*, *14*(2), 65−67. Available from https://doi.org/10.1097/01.mej.0000228439.87286.ed.

Lee, K. L., Graham, C. A., Yeung, J. H. H., Ahuja, A. T., & Rainer, T. H. (2010). Occult pneumothorax in Chinese patients with significant blunt chest trauma: incidence and management. *Injury*, *41*(5), 492−494. Available from https://doi.org/10.1016/j.injury.2009.12.017.

Liman, S. T., Kuzucu, A., Tastepe, A. I., Ulasan, G. N., & Topcu, S. (2003). Chest injury due to blunt trauma. *European Journal of Cardio-thoracic Surgery: Official Journal of the European Association for Cardio-thoracic Surgery*, *23*(3), 374−378. Available from https://doi.org/10.1016/s1010-7940(02)00813-8.

Ma, O. J., & Mateer, J. R. (1997). Trauma ultrasound examination versus chest radiography in the detection of hemothorax. *Annals of Emergency Medicine*, *29*(3), 312−315. Available from https://doi.org/10.1016/s0196-0644(97)70341-x.

Maezawa, T., Yanai, M., Huh, J. Y., & Ariyoshi, K. (2020). Effectiveness and safety of small-bore tube thoracostomy (≤20 Fr) for chest trauma patients: A retrospective observational study. *The American Journal of Emergency Medicine*, *38*(12), 2658−2660. Available from https://doi.org/10.1016/j.ajem.2020.09.028.

Mahmood, I., Younis, B., Ahmed, K., Mustafa, F., El-Menyar, A., Alabdallat, M., Parchani, A., Peralta, R., Nabir, S., Ahmed, N., & Al-Thani, H. (2020). Occult Pneumothorax in Patients Presenting with Blunt Chest Trauma: An Observational Analysis. *Qatar Medical Journal*, *2020*(1). Available from https://doi.org/10.5339/qmj.2020.10.

Maloney, J. V., & McDonald, L. (1963). Treatment of blunt trauma to the thorax. *American Journal of Surgery*, *105*, 484−489. Available from https://doi.org/10.1016/0002-9610(63)90319-2.

Mansour, M. A., Moore, E. E., Moore, F. A., & Read, R. R. (1992). Exigent postinjury thoracotomy analysis of blunt versus penetrating trauma. *Surgery, Gynecology & Obstetrics*, *175*(2), 97−101.

Mattox, K. L. (1983). Thoracic injury requiring surgery. *World Journal of Surgery*, *7*(1), 49−55. Available from https://doi.org/10.1007/BF01655912.

McNamara, J. J., Messersmith, J. K., Dunn, R. A., Molot, M. D., & Stremple, J. F. (1970). Thoracic injuries in combat casualties in Vietnam. *The Annals of Thoracic Surgery*, *10*(5), 389−401. Available from https://doi.org/10.1016/s0003-4975(10)65367-2.

Meyer, D. M., Jessen, M. E., Wait, M. A., & Estrera, A. S. (1997). Early evacuation of traumatic retained hemothoraces using thoracoscopy: a prospective, randomized trial. *The Annals of Thoracic Surgery*, *64*(5), 1396−1400. Available from https://doi.org/10.1016/S0003-4975(97)00899-0.

Misthos, P., Kakaris, S., Sepsas, E., Athanassiadi, K., & Skottis, I. (2004). A prospective analysis of occult pneumothorax, delayed pneumothorax and delayed hemothorax after minor blunt thoracic trauma. *European Journal of Cardio-thoracic Surgery: Official Journal of the European Association for Cardio-Thoracic Surgery*, 25(5), 859−864. Available from https://doi.org/10.1016/j.ejcts.2004.01.044.

Moore, F. O., Goslar, P. W., Coimbra, R., Velmahos, G., Brown, C. V. R., Coopwood, T. B., Lottenberg, L., Phelan, H. A., Bruns, B. R., Sherck, J. P., Norwood, S. H., Barnes, S. L., Matthews, M. R., Hoff, W. S., de Moya, M. A., Bansal, V., Hu, C. K. C., Karmy-Jones, R. C., Vinces, F., . . . Haan, J. M. (2011). Blunt traumatic occult pneumothorax: is observation safe?--results of a prospective, AAST multicenter study. *The Journal of Trauma*, 70(5), 1019−1023. Available from https://doi.org/10.1097/TA.0b013e318213f727.

Morales Uribe, C. H., Villegas Lanau, M. I., & Petro Sánchez, R. D. (2008). Best timing for thoracoscopic evacuation of retained post-traumatic hemothorax. *Surgical Endoscopy*, 22(1), 91−95. Available from https://doi.org/10.1007/s00464-007-9378-6.

de Moya, M. A., Seaver, C., Spaniolas, K., Inaba, K., Nguyen, M., Veltman, Y., Shatz, D., Alam, H. B., & Pizano, L. (2007). Occult pneumothorax in trauma patients: development of an objective scoring system. *The Journal of Trauma*, 63(1), 13−17. Available from https://doi.org/10.1097/TA.0b013e31806864fc.

Omert, L., Yeaney, W. W., & Protetch, J. (2001). Efficacy of thoracic computerized tomography in blunt chest trauma. *The American Surgeon*, 67(7), 660, 4.

Oparah, S. S., & Mandal, A. K. (1979). Operative management of penetrating wounds of the chest in civilian practice. Review of indications in 125 consecutive patients. *The Journal of Thoracic and Cardiovascular Surgery*, 77(2), 162, 8.

Prakash, P. S., Moore, S. A., Rezende-Neto, J. B., Trpcic, S., Dunn, J. A., Smoot, B., Jenkins, D. H., Cardenas, T., Mukherjee, K., Farnsworth, J., Wild, J., Young, K., Schroeppel, T. J., Coimbra, R., Lee, J., Skarupa, D. J., Sabra, M. J., Carrick, M. M., Moore, F. O., . . . Cannon, J. W. (2020). Predictors of retained hemothorax in trauma: Results of an Eastern Association for the Surgery of Trauma multi-institutional trial. *The Journal of Trauma and Acute Care Surgery*, 89(4), 679−685. Available from https://doi.org/10.1097/TA.0000000000002881.

Reul, G. J., Beall, A. C., Jordan, G. L., & Mattox, K. L. (1973). The early operative management of injuries to great vessels. *Surgery*, 74(6), 862−873.

Rodriguez, R. M., Baumann, B. M., Raja, A. S., Langdorf, M. I., Anglin, D., Bradley, R. N., Medak, A. J., Mower, W. R., & Hendey, G. W. (2014). Diagnostic yields, charges, and radiation dose of chest imaging in blunt trauma evaluations. *Academic Emergency Medicine: Official Journal of the Society for Academic Emergency Medicine*, 21(6), 644−650. Available from https://doi.org/10.1111/acem.12396.

Rossmann, M., Altomare, M., Pezzoli, I., Abruzzese, A., Spota, A., Vettorello, M., Cioffi, S. P. B., Virdis, F., Bini, R., Chiara, O., & Cimbanassi, S. (2022). Risk Factors for Retained Hemothorax after Trauma: A 10-Years Monocentric Experience from First Level Trauma Center in Italy. *Journal of Personalized Medicine*, 12(10). Available from https://doi.org/10.3390/jpm12101570.

Scott, M. F., Khodaverdian, R. A., Shaheen, J. L., Ney, A. L., & Nygaard, R. M. (2017). Predictors of retained hemothorax after trauma and impact on patient outcomes. *European Journal of Trauma and Emergency Surgery: Official Publication of the European Trauma Society*, 43(2), 179−184. Available from https://doi.org/10.1007/s00068-015-0604-y.

Shorr, R. M., Crittenden, M., Indeck, M., Hartunian, S. L., & Rodriguez, A. (1987). Blunt thoracic trauma. Analysis of 515 patients. *Annals of Surgery*, 206(2), 200−205. Available from https://doi.org/10.1097/00000658-198708000-00013.

Siemens, R., Polk, H. C., Gray, L. A., & Fulton, R. L. (1977). Indications for thoracotomy following penetrating thoracic injury. *The Journal of Trauma, 17*(7), 493−500. Available from https://doi.org/10.1097/00005373-197707000-00002.

Sisley, A. C., Rozycki, G. S., Ballard, R. B., Namias, N., Salomone, J. P., & Feliciano, D. V. (1998). Rapid detection of traumatic effusion using surgeon-performed ultrasonography. *The Journal of Trauma, 44*(2), 291−296. Available from https://doi.org/10.1097/00005373-199802000-00009.

Slessor, D., & Hunter, S. (2015). To be blunt: are we wasting our time? Emergency department thoracotomy following blunt trauma: a systematic review and meta-analysis. *Annals of Emergency Medicine, 65*(3). Available from https://doi.org/10.1016/j.annemergmed.2014.08.020.

Smith, J. A., Secombe, P., & Aromataris, E. (2021). Conservative management of occult pneumothorax in mechanically ventilated patients: A systematic review and meta-analysis. *The Journal of Trauma and Acute Care Surgery, 91*(6), 1025−1040. Available from https://doi.org/10.1097/TA.0000000000003322.

Soldati, G., Testa, A., Pignataro, G., Portale, G., Biasucci, D. G., Leone, A., & Silveri, N. G. (2006). The ultrasonographic deep sulcus sign in traumatic pneumothorax. *Ultrasound in Medicine & Biology, 32*(8), 1157−1163. Available from https://doi.org/10.1016/j.ultrasmedbio.2006.04.006.

Soldati, G., Testa, A., Sher, S., Pignataro, G., La Sala, M., & Silveri, N. G. (2008). Occult traumatic pneumothorax: diagnostic accuracy of lung ultrasonography in the emergency department. *Chest, 133*(1), 204−211. Available from https://doi.org/10.1378/chest.07-1595.

Tanizaki, S., Maeda, S., Sera, M., Nagai, H., Hayashi, M., Azuma, H., Kano, K.-I., Watanabe, H., & Ishida, H. (2017). Small tube thoracostomy (20-22 Fr) in emergent management of chest trauma. *Injury, 48*(9), 1884−1887. Available from https://doi.org/10.1016/j.injury.2017.06.021.

Trupka, A., Waydhas, C., Hallfeldt, K. K., Nast-Kolb, D., Pfeifer, K. J., & Schweiberer, L. (1997). Value of thoracic computed tomography in the first assessment of severely injured patients with blunt chest trauma: results of a prospective study. *The Journal of Trauma, 43*(3), 405−411. Available from https://doi.org/10.1097/00005373-199709000-00003.

Vassiliu, P., Velmahos, G. C., & Toutouzas, K. G. (2001). Timing, safety, and efficacy of thoracoscopic evacuation of undrained post-traumatic hemothorax. *The American Surgeon, 67*(12), 1165−1169. Available from https://doi.org/10.1177/000313480106701210.

Velmahos, G. C., Demetriades, D., Chan, L., Tatevossian, R., Cornwell, E. E., Yassa, N., Murray, J. A., Asensio, J. A., & Berne, T. V. (1999). Predicting the need for thoracoscopic evacuation of residual traumatic hemothorax: chest radiograph is insufficient. *The Journal of Trauma, 46*(1), 65−70. Available from https://doi.org/10.1097/00005373-199901000-00011.

Watkins, J. A., Spain, D. A., Richardson, J. D., & Polk, H. C. (2000). Empyema and restrictive pleural processes after blunt trauma: an under-recognized cause of respiratory failure. *The American Surgeon, 66*(2), 210−214.

Wilson, H., Ellsmere, J., Tallon, J., & Kirkpatrick, A. (2009). Occult pneumothorax in the blunt trauma patient: tube thoracostomy or observation? *Injury, 40*(9), 928−931. Available from https://doi.org/10.1016/j.injury.2009.04.005.

Zhang, M., Teo, L. T., Goh, M. H., Leow, J., & Go, K. T. S. (2016). Occult pneumothorax in blunt trauma: is there a need for tube thoracostomy? *European Journal of Trauma and Emergency Surgery: Official Publication of the European Trauma Society, 42*(6), 785−790. Available from https://doi.org/10.1007/s00068-016-0645-x.

Zhang, M., Liu, Z.-H., Yang, J.-X., Gan, J.-X., Xu, S.-W., You, X.-D., & Jiang, G.-Y. (2006). Rapid detection of pneumothorax by ultrasonography in patients with multiple trauma. *Critical Care (London, England), 10*(4). Available from https://doi.org/10.1186/cc5004.

Chapter 13

Sternal fracture

Korkut Bostanci[1] and Zeynep Bilgi[2]
[1]Department of Thoracic Surgery, Marmara University School of Medicine, Istanbul, Turkey,
[2]Department of Thoracic Surgery, Medeniyet University School of Medicine, Istanbul, Turkey

Epidemiology

Sternal fractures are usually diagnosed after high-energy deceleration injuries and blunt frontal chest trauma. Athletic injuries, falls, and assaults are also commonly seen and their frequency depends on the population makeup of a particular trauma center. Generally, sternal fractures are diagnosed in 3%–6% of motor vehicle accidents (Doyle & Diaz-Gutierrez, 2021).

Sternal fractures are either reported as having slight female predominance or being comparable between genders. Since the position of shoulder restraint is different on a smaller frame, female drivers or passengers may be more prone to sternal fracture when lower bone density is also taken into account (Raghunathan & Porter, 2009). In vehicle-to-vehicle collisions, older cars released to market with previous safety standards without any airbag protection, have been reported to be a cause for this specific injury (Raghunathan & Porter, 2009).

Lateral chest radiograph was considered the gold standard for making the diagnosis of a sternal fracture, but widespread availability and usage of computed tomography have changed the epidemiology (Perez et al., 2015). Since the introduction of pan-CT protocols in the trauma bay, the number of sternal fracture diagnoses has increased, but together with more frequently seen isolated sternal fractures, specific clinical significance has come under debate, as well as the level of care required for this finding alone (Doyle & Diaz-Gutierrez, 2021; Perez et al., 2015).

Mechanism of injury and implications

The thoracic organs can suffer serious harm from a sternum fracture, which can be fatal. These problems frequently need immediate treatment and include lung injuries such as tracheobronchial injuries, pneumothorax, hemothorax, and pulmonary contusions. Resuscitation and treatment of associated injuries are essential in emergencies. Dogrul et al. stated several conditions that are seen concomitantly with a sternal fracture (obstructed airway, tension pneumothorax, open pneumothorax, large hemothorax, flail chest, pericardial tamponade) which need to be worked up urgently, depending on the context (Dogrul et al., 2020).

A cadaver study reported that the sternum and rib cage give the thoracic spine 40% of its stability in flexion and extension, 22% in lateral bending, and 15% in axial rotation. Therefore, thoracic stability is drastically reduced when the sternum and ribs are fractured together, especially when there are multiple rib fractures (Watkins et al., 2005). Previously, in a similar observation, Berg et al. reported that the sternum, together with the ribs, acts as a fourth column to anchor the thoracic spine (Berg, 1993). Many studies show a predilection for the coexistence of sternal fractures and thoracic spinal fractures (which tend to be unstable in the presence of thoracic wall instability). Morgenstern et al reported 62% of the patients who had a fracture of the manubrium sterni, also had concurrent injuries to the thoracic spine (Morgenstern et al., 2016). The authors further note that when sternal and thoracic spinal fractures are at the same segment, spinal fracture instability must be thoroughly accounted for and ruled out as this situation is prone to serious complications (Morgenstern et al., 2016).

Over 80% of patients with sternal fractures (SF) are diagnosed with further chest injuries, the most prevalent of which are rib fractures, mediastinal hemorrhages, and pulmonary contusions. Therefore, the majority of SF patients undergo monitoring in addition to the therapy they would have gotten with or without an SF diagnosis (mainly pain management and respiratory support), and overall mortality and morbidity seem to be unaffected by the presence or absence of a sternal fracture (Doyle & Diaz-Gutierrez, 2021; Perez et al., 2015).

SF may also be a sign of high-energy trauma and is frequently present in deaths at the scene, although it is typically a benign finding on its own when isolated or other associated injuries take primacy. In an autopsy investigation of fatal falls from height, sternal fractures were discovered in 76% of cases involving cardiac injury (among those 16% were multiple) (Türk & Tsokos, 2004). In a different study, of 267 patients out of a >40,000 patient trauma database, 230 (86.1%) with sternal fractures were able to be transported to the hospital, while 30 (11.2%) passed away at the scene. On-scene mortalities had further severe trauma (43% with fatal cranial injury, 32% with fatal thoracic injuries) (Knobloch et al., 2006).

Clinical presentation and diagnosis

Localized sternal discomfort is reported by nearly all patients. Patients with insufficiency fractures may experience more diffuse discomfort, which could necessitate a more thorough work-up to rule out life-threatening injuries, especially in frail patients. Dyspnea may be a result of significant cardiopulmonary contusion. Additional clinical indicators like sternal steps, hematoma, seat belt contusions, and crepitance during cough are intermittently found.

While lateral chest X-ray remains the textbook gold standard for the diagnosis of sternal fracture, the context of trauma usually dictates performing a chest computerized tomography (CT), which is more sensitive (Figs. 13.1 and 13.2). One study performed on over 14,000 patients in a national chest trauma database reported that 94% of sternal fractures were visible on only chest CT (Perez et al., 2015). One should keep in mind that transverse fractures may be missed on CT examinations and a second look imaging may be necessary if the clinical suspicion is high (Huggett & Roszler, 1998).

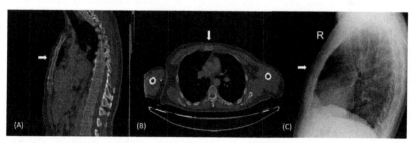

FIGURE 13.1 Fourty five-year-old male, motor vehicle passenger, frontal impact. Discharged after uneventful overnight observation (negative echocardiogram (echo), troponin, and electrocardiogram (ECG)). (A) Saggital CT reconstruction showing sternal fracture line (arrow). (B) Axial CT showing overriding sternal segments (arrow). (C) Lateral chest X-ray of the same patient with visible fracture line (arrow).

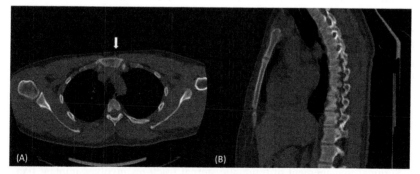

FIGURE 13.2 Twenty six-year-old male, motorcycle accident, high-speed accident with protective equipment. Discharged after uneventful overnight observation (negative echo, troponin, and ECG). (A) Axial CT showing manubrial «tear» (arrow). (B) Same patient, fracture not apparent in sagittal plane (optimal lateral chest was not obtained due to shoulder pain during movement).

While pulse oximetry and ECG monitorization are essential parts of major trauma work-up, repeated electrocardiograms, and cardiac enzyme follow-up are necessary to look for signs of myocardial contusion. Unexplained sinus tachycardia, newly diagnosed arrhythmias, conduction issues, or ST-segment alterations are a few other concerning signs (Khoriati et al., 2013). Echocardiography should be performed to detect wall motion abnormalities in patients with comorbidities or concerning findings.

Management

Patients with isolated sternal fractures do not require extensive testing or routine hospitalization since they are not at high risk for serious cardiac, pulmonary, or mediastinal consequences (Hossain et al., 2010; Perez et al., 2015; Yilmaz et al., 1999). The majority of isolated sternal fractures will heal in 10 weeks on average (Yuan, 2022).

The cornerstone of treatment for solitary sternal fractures is adequate analgesia. Respiratory physiotherapy and early mobilization are essential to prevent further complications in especially elderly and frail patients. Observation admission or hospitalization in the high-frailty patient population should be considered when pain management is difficult, self-ambulation may be unsafe, or adherence to respiratory physiotherapy is doubtful. Non-union or mal-union is rare after sternal fractures but predisposing factors such as malignancy, osteoporosis, diabetes mellitus, chronic steroid therapy, and chest wall instability should be kept in mind. Rarely, osteomyelitis, sternal abscess, and mediastinitis may occur in patients with risk factors (retrosternal hematoma, intravenous drug abuse, etc.) (Doyle & Diaz-Gutierrez, 2021).

Patients with a severe combination of thoracic cage injuries or unstable or considerably displaced sternal fractures should be hospitalized and surgical fixation may be required. The indications for operative management of sternal fractures are reported as unstable anterior chest wall, displaced fracture, nonunion, suboptimal analgesia despite supportive treatment, difficulty weaning from the ventilator (Bauman et al., 2022; Kalberer et al., 2020; Klei et al., 2019; Krinner et al., 2017; Potaris et al., 2002; Queitsch et al., 2011; Richardson et al., 2007; Zhao et al., 2017). Potaris et al.'s report on the outcomes of 239 sternal fractures over 11 years states that while 1.7% of patients ultimately needed surgical fixation of the sternum, 5.4% had at least one trauma complication and the average ward stay was 6.4 days, mainly due to concomitant injuries (Potaris et al., 2002).

Surgical management of traumatic sternal fractures involves implant fixation with or without bone grafts (Fig. 13.3) and studies to date are usually retrospective case series without a control arm with either unreported or short follow-up times. The variety of used devices (absorbable and nonabsorbable plates, internal cemented screws, etc.) and the heterogeneity of the patient population make it difficult to identify patient selection criteria for surgical management. For example, a US study done on the National Trauma Data Bank identified 270 (1.8%) cases of sternal fixation among 14,760 adults encountered with sternal fractures. When patients were propensity score matched, the group who underwent sternal fixation had lower odds of mortality (OR [95%CI]: 0.19 [0.06−0.62], $P = .006$), but pneumonia and respiratory failure risks were comparable. Furthermore, 18 out of 270 operative management cases had concurrent rib fixation as well (Choi et al., 2021). Another US-based study using the Trauma Quality Improvement Program reported a comparable percentage of sternal fracture cases undergoing surgical repair and found overall lower mortality risk after matching but increased median length of stay (16 vs 7 days, $P < .001$), intensive care unit (ICU) stay (9.5 vs 5.5 days, $P = .016$) and ventilator days (8 vs 5, $P = .035$) and similar pulmonary complication rate were evident in patients with surgical management (Christian et al., 2022).

The results of the prominent case series published are summarized in Table 13.1. Overall, sternal fixation surgery is reported as safe and effective, with the majority of the patients achieving chest wall stability, even in cases of sternal non-union as an indication for surgery.

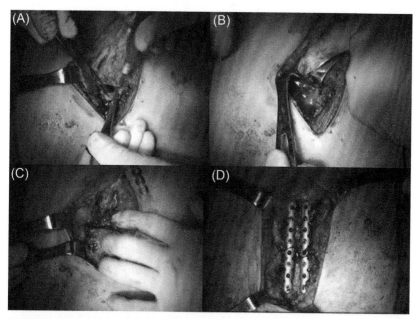

FIGURE 13.3 (A) Sternal fracture is exposed via a midline incision. (B) Displaced sternal fracture is reduced and sharp edges are trimmed if necessary. (C) In situ end-to-end reduction. (D) Fixation with parallel plates.

TABLE 13.1 Outcomes of surgical management of sternal fractures.

Author (year)	Study type	N	Device	Study period	Outcomes	Complications
Zhao et al. (2017)	Case series	64 (45 with combined injuries)	Titanium plate	5 years	Sternal stability in all patients Mean length of stay (LOS) = 45 days Lower postoperative pain score	None observed
Krinner et al. (2017)	Case series	11 (4 isolated, 7 combined with spinal fixation)	Low profile locking titanium plate	4 years	Complete consolidation of SF at 12 weeks	None observed
Queitsch et al. (2011)	Multi-center prospective single arm	12 (10 with non-union, average 8, 2 months from original trauma)	Locked sternum-osteosynthesis plate	4 years	Osteosynthesis in all patients	1 intraoperative (inappropriate device size) 1 wound infection 1 hypertrophic scar and need for device removal
Richardson et al. (2007)	Retrospective case series	35 (31 with severe pain, 4 ventilator dependent)	Titanium mandibular plates	13 years	Chest wall stabilization in all patients	1 removal for possible need for sternotomy 2 removals per patient request 1 mortality (severe chronic obstructive pulmonary disease (COPD) complicating weaning)

Bauman et al. (2022)	Case series (single surgeon)	13 out of 91 diagnosed sternal fractures (average injury severity score at 15.8 ± 10.9, average time to surgery 3.2 ± 2.6 days)	Titanium ladder plate /two titanium parallel straight plates	2 years	ICU days = 2.3 ± 4.1 Length of hospital stay = 10.2 ± 7.1 Full range of motion upper extremity in all Pain score decreased postoperatively (7.08 [2.32] vs 3.54 [2.5], P = .001*)	No postoperative complications
Kalberer et al. (2020)	Retrospective cohort	18 (out of 153 diagnosed) 3 lost to follow up	Double straight locking compression plates	8 years	Osteosynthesis and functional implant in all cases	Pneumonia (20%) respiratory insufficiency (20%) Implant removal in 4 patients (3 due to implant-related irritation, 1 per patient's request)

SF, Sternal fractures, * Statistical significance set at p < 0.05

References

Bauman, Z. M., Yanala, U., Waibel, B. H., Malhotra, G. K., Cemaj, S., Evans, C. H., & Schlitzkus, L. L. (2022). Sternal fixation for isolated traumatic sternal fractures improves pain and upper extremity range of motion. *European Journal of Trauma and Emergency Surgery*, *48*(1), 225−230. Available from https://doi.org/10.1007/s00068-020-01568-x, http://link.springer.com/journal/68.

Berg, E. E. (1993). The sternal-rib complex. A possible fourth column in thoracic spine fractures. *Spine*, *18*(13), 1916−1919.

Choi, J., Khan, S., Syed, M., Tennakoon, L., & Forrester, J. D. (2021). Early national landscape of surgical stabilization of sternal fractures. *World Journal of Surgery*, *45*(6), 1692−1697. Available from https://doi.org/10.1007/s00268-021-06007-5, https://www.springer.com/journal/268.

Christian, A. B., Grigorian, A., Nahmias, J., Duong, W. Q., Lekawa, M., Joe, V., Dolich, M., & Schubl, S. D. (2022). Comparison of surgical fixation and non-operative management in patients with traumatic sternum fracture. *European Journal of Trauma and Emergency Surgery*, *48*(1), 219−224. Available from https://doi.org/10.1007/s00068-020-01527-6, http://link.springer.com/journal/68.

Dogrul, B. N., Kiliccalan, I., Asci, E. S., & Peker, S. C. (2020). Blunt trauma related chest wall and pulmonary injuries: An overview. *Journal of Traumatology − English Edition*, *23*(3), 125−138. Available from https://doi.org/10.1016/j.cjtee.2020.04.003, http://www.elsevier.com/wps/find/journaldescription.cws_home/714951/description#description.

Doyle, J. E., & Diaz-Gutierrez, I. (2021). Traumatic sternal fractures: A narrative review. *Mediastinum*, *5*. Available from https://doi.org/10.21037/MED-21-27, https://med.amegroups.com/article/view/6535/html.

Hossain, M., Ramavath, A., Kulangara, J., & Andrew, J. G. (2010). Current management of isolated sternal fractures in the UK: Time for evidence based practice? A cross-sectional survey and review of literature. *Injury*, *41*(5), 495−498. Available from https://doi.org/10.1016/j.injury.2009.07.072.

Huggett, J. M., & Roszler, M. H. (1998). CT findings of sternal fracture. *Injury*, *29*(8), 623−626. Available from https://doi.org/10.1016/S0020-1383(98)00150-8.

Kalberer, N., Frima, H., Michelitsch, C., Kloka, J., & Sommer, C. (2020). Osteosynthesis of sternal fractures with double locking compression plate fixation: A retrospective cohort study. *European Journal of Orthopaedic Surgery and Traumatology*, *30*(1), 75−81. Available from https://doi.org/10.1007/s00590-019-02526-z, http://link.springer.de/link/service/journals/00590/index.htm.

Khoriati, A. A., Rajakulasingam, R., & Shah, R. (2013). Sternal fractures and their management. *Journal of Emergencies, Trauma and Shock*, *6*(2), 113−116. Available from https://doi.org/10.4103/0974-2700.110763.

Klei, D. S., de Jong, M. B., Öner, F. C., Leenen, L. P. H., & van Wessem, K. J. P. (2019). Current treatment and outcomes of traumatic sternal fractures—A systematic review. *International Orthopaedics*, *43*(6), 1455−1464. Available from https://doi.org/10.1007/s00264-018-3945-4, http://link.springer.de/link/service/journals/00264/index.htm.

Knobloch, K., Wagner, S., Haasper, C., Probst, C., Krettek, C., Otte, D., & Richter, M. (2006). Sternal fractures occur most often in old cars to seat-belted drivers without any airbag often with concomitant spinal injuries: Clinical findings and technical collision variables among 42,055 crash victims. *Annals of Thoracic Surgery*, *82*(2), 444−450. Available from https://doi.org/10.1016/j.athoracsur.2006.03.046.

Krinner, S., Grupp, S., Oppel, P., Langenbach, A., Hennig, F. F., & Schulz-Drost, S. (2017). Do low profile implants provide reliable stability in fixing the sternal fractures as a \fourth vertebral column\ in sternovertebral injuries? *Journal of Thoracic Disease*, *9*(4), 1054–1064. Available from https://doi.org/10.21037/jtd.2017.03.37, http://jtd.amegroups.com/article/download/12625/pdf.

Morgenstern, M., von Rüden, C., Callsen, H., Friederichs, J., Hungerer, S., Bühren, V., Woltmann, A., & Hierholzer, C. (2016). The unstable thoracic cage injury: The concomitant sternal fracture indicates a severe thoracic spine fracture. *Injury*, *47*(11), 2465–2472. Available from https://doi.org/10.1016/j.injury.2016.08.026, http://www.elsevier.com/locate/injury.

Perez, M. R., Rodriguez, R. M., Baumann, B. M., Langdorf, M. I., Anglin, D., Bradley, R. N., Medak, A. J., Mower, W. R., Hendey, G. W., Nishijima, D. K., & Raja, A. S. (2015). Sternal fracture in the age of pan-scan. *Injury*, *46*(7), 1324–1327. Available from https://doi.org/10.1016/j.injury.2015.03.015, http://www.elsevier.com/locate/injury.

Potaris, K., Gakidis, J., Mihos, P., Voutsinas, V., Deligeorgis, A., & Petsinis, V. (2002). Management of sternal fractures: 239 cases. *Asian Cardiovascular and Thoracic Annals*, *10*(2), 145–149. Available from https://doi.org/10.1177/021849230201000212.

Queitsch, C., Kienast, B., Voigt, C., Gille, J., Jürgens, C., & Arndt, S. P. (2011). Treatment of posttraumatic sternal non-union with a locked sternum-osteosynthesis plate (TiFix). *Injury*, *42*(1), 44–46, https://doi.org/10.1016/j.injury.2010.08.013.

Raghunathan, R., & Porter, K. (2009). Sternal fractures. *Trauma*, *11*(2), 77–92. Available from https://doi.org/10.1177/1460408608102007.

Richardson, J. D., Franklin, G. A., Heffley, S., & Seligson, D. (2007). Operative fixation of chest wall fractures: An underused procedure? *American Surgeon*, *73*(6), 591–596.

Türk, E. E., & Tsokos, M. (2004). Blunt cardiac trauma caused by fatal falls from height: An autopsy-based assessment of the injury pattern. *The Journal of Trauma: Injury, Infection, and Critical Care*, *57*(2), 301–304. Available from https://doi.org/10.1097/01.ta.0000074554.86172.0e.

Watkins, R., Watkins, R., Williams, L., Ahlbrand, S., Garcia, R., Karamanian, A., Sharp, L., Vo, C., & Hedman, T. (2005). Stability provided by the sternum and rib cage in the thoracic spine. *Spine*, *30*(11), 1283–1286, https://doi.org/10.1097/01.brs.0000164257.69354.bb.

Yilmaz, E. N., van Heek, N. T., van der Spoel, J. I., Bakker, F. C., Patka, P., & Haarman, H. J. (1999). Myocardial contusion as a result of isolated sternal fractures: A fact or a myth? *European Journal of Emergency Medicine: Official Journal of the European Society for Emergency Medicine*, *6*(4), 293–295. Available from https://doi.org/10.1097/00063110-199912000-00003.

Yuan, S. M. (2022). Sternal fractures due to blunt chest trauma. *Journal of the College of Physicians and Surgeons Pakistan*, *32*(12), 1591–1596. Available from https://doi.org/10.29271/jcpsp.2022.12.1591, https://www.jcpsp.pk/article-detail/psternal-fractures-due-to-blunt-chest-traumaorp.

Zhao, Y., Yang, Y., Gao, Z., Wu, W., He, W., & Zhao, T. (2017). Treatment of traumatic sternal fractures with titanium plate internal fixation: A retrospective study. *Journal of Cardiothoracic Surgery*, *12*(1). Available from https://doi.org/10.1186/s13019-017-0580-x, http://www.cardiothoracicsurgery.org/.

Chapter 14

VATS for delayed hemothorax

Jacie Jiaqi Law[1,2] and Giuseppe Aresu[1,2]
[1]Department of Cardiothoracic Surgery, Royal Papworth Hospital, Cambridge, United Kingdom, [2]Department of Cardiothoracic Surgery, Royal Victoria Hospital, Belfast, United Kingdom

Introduction

Thoracic trauma presenting to the emergency department (ED) secondary to blunt or penetrating chest injuries is a common cause of thoracic surgery consultation. Thoracic trauma accounts for 10%—15% of all trauma cases with blunt chest trauma contributing to 70% of all thoracic injuries (Demirhan et al., 2009). The mortality rate is significant, with chest-related injuries accounting for 20%—25% of all trauma-related deaths (Manlulu et al., 2004). In a retrospective analysis of 515 cases of blunt chest trauma, Shorr et al. demonstrated an overall thoracic morbidity rate of 36% and mortality of 15.5% (Shorr et al., 1987). Approximately 75% of patients are subsequently discharged with minor thoracic injuries (MTI) after comprehensive clinical examination, radiological investigations, and thoracic surgical evaluation (Ziegler & Agarwal, 1994; Liman et al., 2003). However, it is imperative to remain vigilant over delayed complications such as delayed hemothorax (DHTX) which can arise in the following weeks, even in the absence of rib fractures or other thoracic bony fractures (Bundy & Tilton, 2003; Lin et al., 2017).

Definition, epidemiology, and classification of delayed hemothorax

Hemothorax is a frequent consequence of traumatic thoracic injuries and refers to a collection of blood in the pleural space which is a potential space between the parietal and visceral pleura (Boersma et al., n.d.). The core tenets of Advanced Trauma Life Support teachings advocate an aggressive and systematic approach to early recognition and treatment of acute hemothorax (Mahoozi et al., 2016). DHTX on the other hand is an uncommon underrecognized entity that poses life-threatening complications and portends high morbidity. Previous studies report the incidence of delayed hemothorax after blunt chest trauma to range between 2.1% and 33% and up to 40% in elderly patients (Mangram et al., 2016; Muronoi et al., 2020; Sharma et al., 2005). In comparison to other commonly focused delayed complications, such as delayed pneumonia with a reported incidence of less than 2% amongst MTI patients, DHTX occurs more frequently and argues for greater consideration. In a prospective analysis performed by Misthos et al. (2004), 52 out of 709 outpatients (7.4%) who presented to ED with at least 1 rib fracture received a diagnosis of hemothorax within 14 days (Misthos et al., 2004). In an 8-year retrospective analysis by Sharma et al., 7 out of 167 patients (4.2%) had DHTX with an association with older patients with multiple or displaced rib fractures in 92% of cases (Sharma et al., 2005).

Hemothorax can be classified temporally into two conditions. Acute hemothorax which is detected on admission immediately post-trauma on chest x-ray (CXR)manifests as blunting of costophrenic angles or computed tomography (CT) chest demonstrating pleural effusion (Lewis et al., 2021). Existing literature suggests debate over the time definition of DHTX between initial thoracic injury and subsequent diagnosis. This is in part due to substantial institutional practice discrepancy in the way chest trauma patients are follow-up upon hospital discharge. Ritter et al define DHTX as no evidence of hemothorax on first imaging which is subsequently detected on further investigation with a time gap of as little as 2 h (Ritter & Chang, 1995). Masuda et al. and Shorr et al. consider DHTX to arise 24 h or later following initial inciting thoracic injury (Masuda et al., 2013; Shorr et al., 1987). Fig. 14.1 demonstrates real-life radiological images of a 75-year-old elderly gentleman who developed DHTX 48 h after falling down a flight of stairs and sustained multiple right-sided rib fractures.

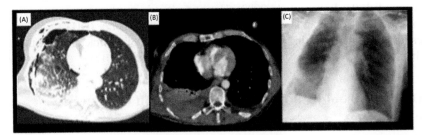

FIGURE 14.1 Chest-x ray and CT chest image of a 75-year-old male patient who developed delayed hemothorax 48 h post-chest injury secondary to a fall. A 75-year-old male patient presented to emergency department after falling down a flight of stairs. (A) demonstrates his CT chest on admission with multiple right-sided rib fractures, traumatic pneumothorax, subcutaneous emphysema, and pulmonary contusion but no evidence of hemothorax. 48 h post-admission, the patient progressively developed respiratory compromise with a noted reduction in systolic blood pressure. (B) demonstrates his repeat CT chest showing the presence of a new hemothorax. The patient underwent fluid resuscitation with 1.5 L of blood drained upon chest drain insertion. (C) demonstrates his post-chest drain insertion chest x-ray.

An overview of existing DHTX literature

Table 14.1 summarizes existing published case reports and observational studies reported in English on DHTX secondary to blunt chest trauma. This table also summarizes the associated mechanism of injury, time of onset of DHTX from initial chest trauma, source of hemothorax (if identified), the extent of rib fractures, other concomitant extra-thoracic injuries, management strategies and patient outcome (Ahn et al., 2016; Brekke et al., 2021; Chang et al., 2018; Chen & Cheng, 2014; Chokshi et al., 2021; Curfman et al., 2015; Igai et al., 2019; Lin et al., 2017; Masuda et al., 2013; Misthos et al., 2004; Muronoi et al., 2020; Ota et al., 2014; Park et al., 2022; Rizk et al., 2020; Sharma et al., 2005; Simon et al., 1998; Sinha & Sarkar, 1998; Verkroost & Hensens, 1998; Yap et al., 2018; Hamamoto et al., 2023).

As observed from the above, the onset of DHTX from the time of injury varies and can range from 12 h to 3 months but is mostly diagnosed during the two weeks after trauma (Plourde et al., 2014). There is also a preponderance of DHTX within the male population as corroborated by larger studies including Misthos et al. (64.3%), Emond et al. (69%), and Gonzalez et al. (85%) (Emond et al., 2015; Gonzalez et al., 2021; Misthos et al., 2004). The most common causes of DHTX are motor vehicle accidents and pedestrian accidents. Sharma et al. further classified DHTX into iatrogenic versus traumatic origins (Sharma et al., 2005). Iatrogenic DHTX, which falls outside the scope of this chapter's discussion, can arise secondary to invasive procedures such as central subclavian access line insertion and post-coronary stent placement (Quinn & Dillard, 1999; Waldman et al., 1984). Traumatic causes of DHTX include blunt or penetrating injuries. Weigelt et al. reported a 9% incident of delayed pneumothorax or hemothorax in 110 asymptomatic stab wounds patients within 48 h of admission (Weigelt et al., 1982). In blunt DHTX, published literature has reported various sources of intra-thoracic bleeding including lung, and intercostal artery injury secondary to multiple displaced rib fractures, internal mammary artery, vena azygous, phrenic artery, and thoracic aorta (Ahn et al., 2016; Baldwin et al., 1984; Ritter & Chang, 1995; Ross & Cordoba, 1986; Simon et al., 1998). Other intrathoracic organ injuries can concomitantly occue whereby isolated pericardial rupture and diaphragmatic injury have been reported. Brekke et al. reports a case of bleeding from herniated omentum (Brekke et al., 2021; Verkroost & Hensens, 1998).

Clinical and radiological predictive factors of delayed hemothorax

In a retrospective analysis of older adults aged ≥ 50 years developing DHTX performed by Choir et al., all 14 patients demonstrated posterolateral fractures between the 6th and 10th ribs with at least 1 offset or displaced

TABLE 14.1 A summary of existing published case reports and observational studies (English) reporting on delayed hemothorax secondary to blunt chest trauma.

Sample size	Year	Author	Age	Sex	Mechanism of injury	Time from onset	Source of hemothorax	Rib fracture	Other extra thoracic injuries	Management	LOS (days)	Outcome
1	1997	Sinha	75	M	Fall	4 days	None	Left 3rd – 8th	None	Thoracotomy	–	Survived
1	1997	Sinha	34	M	Fall	7 days	None	Right 8th – 9th	None	Thoracotomy	–	Survived
12	1998	Simon	–	–	–	18h–6 days	Intercostal artery	–	–	Thoracotomy	–	Survived
1	1998	Verkroost	59	M	Fall	3 months	Isolated pericardial rupture	None	Isolate pericardial rupture	Thoracotomy and mesh repair	7	Survived
13	2004	Misthos	–	64.3% M (9)	–	2–14 days	Intercostal artery	–	–	Thoracotomy	–	Survived
1	2004	Misthos	–	–	–	2–14 days	Intercostal artery	–	–	Thoracotomy	–	Dead
7	2005	Sharma	–	All Male	Fall/MVA	36–60 h	Extra-pleural and diaphragm	–	–	Thoracotomy	–	Survived
1	2013	Masuda	56	M	Fall	30 days	Extra-pleural cavity	Left 10th	None	Thoracotomy	4	Survived
1	2014	Chen	60	M	Fall	6 days	Diaphragm	Right 10th – 11th	None	Thoracotomy	8	Survived

	Year	Author	Age	Sex	Mechanism	Time	Injury	Rib fracture	Associated injuries	Treatment	Hospital stay	Outcome
1	2014	Ota	62	M	Slip	7 days	Diaphragm and right superior phrenic artery	Right 11th – 12th	None	VAT-assisted mini-thoracotomy and diaphragm repair	5	Survived
1	2015	Yamanashi	75	M	MVA	24 h	Diaphragm and inferior phrenic artery	Right 7th – 8th, 10th – 11th	Liver laceration	IVR→VATS	–	Survived
1	2015	Curfman	29	M	Assault	10 days	Intercostal artery	Left 7th	None	Thoracotomy	5	Survived
1	2016	Ahn	24	F	Fall	13 days	Musculophrenic artery	Right 11th–12th	Pelvic fracture	VATS	–	Survived
1	2017	Lin	19	M	Assault	12 h	Diaphragm	None	Left index finger fracture and contusions overhead, chest, abdomen, and extremity	Thoracotomy	7	Survived
1	2017	Yap	52	M	Slip	44 days	–	Right 7th – 8th	–	Chest drain insertion	–	Survived
1	2018	Chang	52	M	Fall	93 h	Diaphragm	Left 4th – 10th	None	Thoracotomy	9	Survived
1	2018	Chang	44	M	Slip	63 h	Diaphragm	Right 8th – 10th	None	VATS→thoracotomy	15	Survived
1	2018	Chang	45	M	MVA	66 h	Diaphragm	Left 10th – 12th	Tibiofibular fracture	Thoracotomy	90	Survived

(Continued)

TABLE 14.1 (Continued)

Sample size	Year	Author	Age	Sex	Mechanism of injury	Time from onset	Source of hemothorax	Rib fracture	Other extra thoracic injuries	Management	LOS (days)	Outcome
1	2018	Chang	59	M	Pedestrian traffic accident	63 h	Diaphragm	Right 1st – 11th	Clavicle fracture, scalp laceration, liver laceration	Thoracotomy, rib fixation, and diaphragm repair	37	Survived
1	2018	Chang	31	M	MVA	33 h	Diaphragm	Right 3rd – 8th	Aortic dissection, pelvic bone fracture, liver laceration, deep knee laceration	VATS→thoracotomy, rib fixation, diaphragm repair, and thoracic endovascular aortic repair	53	Survived
1	2019	Igai	44	F	–	22 days	Diaphragm	Right 9th – 12th	–	VATS, rib fixation and diaphragm repair	3	Survived
1	2019	Igai	55	F	–	30 days	Diaphragm	Left 9th – 11th	–	Thoracotomy, rib fixation, and diaphragm repair	4	Survived
1	2019	Igai	85	F	–	15 days	Diaphragm	Left 9th – 11th	–	Thoracotomy, rib fixation, and diaphragm repair	6	Survived
1	2019	Igai	57	F	–	14 h	Diaphragm	Right 5th – 12th	–	Thoracotomy, rib fixation, and diaphragm repair	28	Survived

N	Year	Author	Age	Sex	Cause	Time	Diaphragm	Ribs	Traumatic SAH	Treatment	Days	Outcome
1	2020	Muronoi	58	M	Fall	17 h	–	Left 11th – 12th	Traumatic SAH	Thoracotomy, rib fixation, and diaphragm repair	30	Survived
44	2020	Hassan	–	65.9% male (29)	MVA	1–5 days	–	–	–	Chest drain insertion/thoracotomy	9	Survived
1	2021	Brekke	63	M	MVA	19 days	Diaphragm and bleeding omentum	Left 2nd – 9th	None	VATS	–	Survived
1	2021	Chokshi	77	M	Fall	11 days	–	Right 3rd – 8th	Shoulder and clavicle fracture	VATS → thoracotomy for decortication and rib fixation	6	Survived
1	2022	Park	59	M	Fall	4th day	–	Right 7th – 11th	Liver contusion	VATS	16	Survived
1	2023	Hamamoto	58	M	MVA	19 days	Intercostal artery	Left 5th – 8th	None	IVR	7	Survived

IVR, Interventional radiology; MVA, motor vehicle accident; –, not applicable/no information available; VATS, video-assisted thoracoscopic surgery.
Modified from Muronoi, T., Kidani, A., Oka, K., Konishi, M., Kuramoto, S., Shimojo, Y., Hira, E., & Watanabe, H. (2020). Delayed massive hemothorax due to diaphragm injury with rib fracture: A case report. International Journal of Surgery Case Reports, 77, 133–137.

fracture. The avarge number of displaced or offset rib fracture per patient was 3.5 with the average number of consecutive rib fractures as defined by the Chest Wall Injury Society estimated to be at 6 consecutive rib fracture per patient. (Choi et al., 2021). The above finding is congruent with the analysis performed by Gonzales et al. which also examined high-risk features on CT scans predictive of DHTX occurrence (Gonzalez et al., 2021). The commonly used "≥ 3 rib fractures" was identified to have low sensitivity (51.5%) and specificity (59.1%) for the prediction of DHTX (Flores-Funes et al., 2020). Rather, it is the magnitude of rib fracture displacement, age, co-morbidity, and trauma type that independently predicts delayed pulmonary complications, days of mechanical ventilation, length of stay, opioid requirement, and poor patient outcomes (Simon et al., 1998; Söderlund et al., 2015; Whitson et al., 2013). Chest wall anatomy would explain the association of displaced posterolateral chest wall injury patterns with DHTX. The 6th– 10th ribs are the largest and the greater bone mass is capable of disrupting longer intercostal arteries naturally accompanying these ribs (Graeber & Nazim, 2007). The posterior intercostal arteries are also derived from the aorta and possess a larger diameter in comparison to the anterior intercostal artery arising from the internal mammary arteries (Kuhlman et al., 2015). In addition, 6th–10th ribs play a greater role in respiratory mechanics in comparison to cranial ribs and floating 11th–12th ribs (Graeber & Nazim, 2007). The greater engagement in respiration movements poses a higher risk of intercostal vessel disruption, especially in the context of displaced and offset rib fractures, resulting in clot disruption or delayed vessel injury. Chang et al. (2018) and Ross and Cordoba (1986) also postulate that DHTX in the setting of diaphragm laceration and lower rib fractures could be temporally related to coughing and maximum ventilatory maneuver associated with nebulization treatments and position change, which causes lower rib fractures with a sharp edge to injure diaphragm. Overall, further studies are required to validate the anatomic hypothesis behind DHTX. It is also prudent to remember that DHTX can still occur in the absence of rib fractures and displaced rib fractures.

The Quebec clinical decision rule

Emond et al. highlighted the importance of triaging patients based on the likelihood of developing DHTX to guide outpatient management and follow-up after discharge from ED as MTI, especially in consideration of resource limitations within healthcare systems (Émond et al., 2017). In a 2017 prospective study, Edmond et al. conceptualized and validated the Quebec clinical decision rule (Table 14.2) to risk stratify patients based on the below 4 variables with a maximum 5-point score to identify those who would benefit most from further follow-up after discharge from ED based on risk of developing DHTX.

TABLE 14.2 The Quebec clinical decision rule.

Characteristic	Points
Age ≥70 years old (highest relative risk)	2
High or mid-rib fracture	2
Age 45–70 years old	1
At least 3 rib fractures	1
Maximum total score	5

The Quebec clinical decision rule has a 90% specificity in identifying patients with high-risk DHTX. A score of "2-3" is considered moderate risk and "4-5" is considered high risk

Adopted from Émond, M., Guimont, C., Chauny, J. M., Daoust, R., Bergeron, É., Vanier, L., Moore, L., Plourde, M., Kuimi, B., Boucher, V., Allain-Boulé, N., & Le Sage, N. (2017). Clinical prediction rule for delayed hemothorax after minor thoracic injury: A multicentre derivation and validation study. *CMAJ Open, 5*(2), E444–E453.

Complications of delayed hemothorax

Emond et al. showed that patients with DHTX, even in the absence of rib fractures, experienced the lowest global physical health score using the 12-Item Short Form Survey (SF-12) with 22.8% of patients experiencing moderate or severe disabilities at 90-days post injury (Emond et al., 2015). Pulmonary infection is also a frequent complication arising from DHTX because the reduction in functional residual capacity results in alveolar consolidation, atelectasis, inflammation, and respiratory impairment. On rare occasions, highly fatal massive DHTX can occur after blunt thoracic injury. In a single institutional study performed by Chang et al., 5 out of 1278 patients (0.4%) experienced massive DHTX (Chang et al., 2018). Massive hemothorax is the evacuation of more than 1500 mL of blood immediately post-tube thoracostomy or 200 mL/h of chest drain output for 4 h (Zeiler et al., 2020). Furthermore, DHTX can progress to become retained clotted hemothorax with a reported incidence of 5%–30% (Mowery et al., 2011). The presence of retained hemothorax can result in lung parenchyma collapse which in combination with lung contusion and post-traumatic pneumonia could potentially lead to an entrapped lung further progressing to become a fibrothorax. Retained pleural collections can also be further contaminated by micro-organisms which leads to the formation of empyema and post-trauma infections. Both entities result in permanent pulmonary dysfunction and can arise in 1%–3.3% of patients with retained hemothorax, necessitating surgical interventions (Milfeld et al., 1978). Given the mortality, morbidity, and spectrum of crucial complications that can arise, DHTX must be regularly considered by the multidisciplinary team in chest trauma.

The diagnostic and therapeutic role of VATS in delayed hemothorax

CT is considered the gold standard diagnostic modality which allows visualization of chest wall, intrathoracic, and intrabdominal injuries. The application of 3D CT is also valuable to preoperative planning of rib fracture fixation while CT angiography is utilized to identify bleeding points and identify extravasation of contrast agent and pseudo-aneurysm (Mahmood et al., 2011). However, it should be noted that in cases of DHTX potentially involving right-sided diaphragmatic rupture, the utility of CT can be limited as the liver can obscure the detection of diaphragmatic lacerations. Only direct visualization with video assisted thoracoscopic surgery (VATS) may confirm or exclude the existence of diaphragmatic injury (Mintz et al., 2007; Ota et al., 2014; Turhan et al., 2008).

Due to the paucity of literature, clear gold standard management guidelines of DHTX have not been established. Examining existing publications reveals a majority of institutions adopting an initial conservative approach such as serial hemodynamic observation and tube thoracostomy before progression to surgical intervention. Overall, the surgical approach adopted to manage DHTX is dependent on a patient's clinical stability and associated underlying injuries, with the ultimate goal aimed at achieving hemostasis, hematoma evacuation, and managing posttraumatic complications.

In cases of massive DHTX, hemodynamic instability, and massive transfusion requirements, an emergency thoracotomy is necessary to achieve good exposure (an anterolateral approach is advocated in instances of unclear source of blood loss), adequate surgeon maneuvering space and expeditious control of underlying bleeding source (Karmy-Jones et al., 2001). Carrillo et al. highlighted the morbidity associated with thoracotomy, especially in comorbid patients who are poor surgical candidates, and also the low detection yield with regard to hemorrhage identification in certain patients (Carrillo et al., 1998). Here, selective angiography and transcatheter embolization (SATE) can be a reliable, less invasive, and accurate alternative (Chemelli et al., 2009). However, caution remains over SATE due to the variable reported efficacy of endovascular treatments. Additionally, tantamount to nonoperative strategies, SATE is also unable to evacuate retained blood within the chest cavity. Empyema arising secondary to hematoma infection has been reported in non-operative management strategies of massive hemothorax in the context of diaphragmatic rupture (Muronoi et al., 2020).

With the rise of minimally invasive surgery, VATS continues to expand its role in the management of patients with chest trauma and has become the preferred method to simultaneously evaluate and treat post-traumatic pleural complications in hemodynamically stable individuals (Divisi et al., 2004; Smith et al., 2011). In the realm of primary diagnosis, Liu et al. demonstrated the safety of directly proceeding to VATS before the bedside placement of chest drains in hemodynamically stable patients to identify injuries that could be missed on CT (Godat et al., 2016; Liu et al., 1997). Apical and basal chest drains can be placed intra-operatively under direct vision with VATS and extra pleural catheters can also be inserted to enhance postoperative pain management. In the context of retained hemothorax and empyema evacuation, Navsaria et al. (2004) and Heniford et al. (1997) advocate for the utilization of VATS, demonstrating its safety and advantages which include reducing pulmonary complications, quicker recovery, and less long-term disability. VATS also provide excellent visualization of the pleural cavity, allowing for much more effective evacuation of pleural collections in comparison to placing secondary tube thoracostomies which inadvertently prolongs the length of hospital stay (Lin et al., 2014). Early VATS (defined as before day 3) also resulted in a statistically significant reduction in operative difficulties with conversion to thoracotomy more likely to occur after 5 days, though it is recognized that successful thoracoscopic evacuation has been achieved as far out as 14 days (Morales Uribe et al., 2008; Mowery et al., 2011). Villaicencio et al. highlights the benefit of thoracoscopy in preventing 62% of trauma patients from having a thoracotomy with a 90% optimal result yield in post-traumatic hemothorax cases, 2% procedure-related complications and 0.8% missed injury rate (Villavicencio et al., 1999). Recent reports have also demonstrated the ability of highly specialized institutions to perform complete VATS rib fracture fixation using an internal memory alloy fixation system (Zhang et al., 2022). Lin et al. further demonstrate the benefit of adding rib fixation during VATS for retained hemothorax, specifically in reducing ventilator dependence, persistent air leaks, hospital length of stay, and opioid consumption (Lin et al., 2019).

Currently, there is no consensus on the use of VATS in hemodynamically unstable individuals although Ota et al. reported the successful application of video-assisted mini-thoracotomy (hybrid VATS) to perform diaphragm repair in a pre-shock patient with DHTX secondary to diaphragmatic laceration involving right superior phrenic artery severance (Ota et al., 2014). Previous studies and our own institutional experience suggest the feasibility of utilizing VATS in cases of massive hemothorax, especially when expert VATS thoracic surgeon and advanced peri-operative care can be provided by highly experienced thoracic anesthesiologist on site (Ahn et al., 2016).

Fig. 14.2 demonstrates the management of a 16-year-old gentleman stab injury victim sustaining hemopneumothorax managed with VATS and no new port sites were created at our institution. (A) illustrates his original injury and chest drain insertion site (performed at his local hospital). (B) shows his CT chest image. The patient was transferred urgently to our institution as he remained hemodynamically unstable with ongoing intra-thoracic bleeding. We first performed a VATS inspection via his original chest drain site and performed a hematoma evacuation and washout (C). We then performed a VATS inspection via his original stab wound site, noting a right-

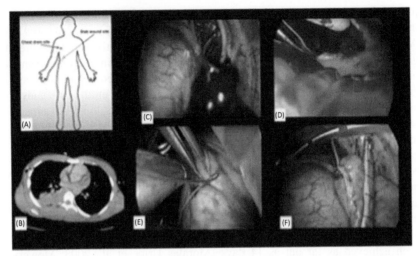

FIGURE 14.2 The Management of Hemopneumothorax in a hemodynamically unstable stab injury patient via video-assisted thoracoscopic surgery from original chest drain insertion and stab injury site. *With permission from Mr. Aman Coonar.*

sided diaphragm injury and bleeding from the intercostal artery (D). (E) demonstrates the repair and closure of the diaphragm injury using heavy ethibond 5 interrupted sutures. (F) shows the insertion of chest drains under VATS via the original chest drain insertion site. Persistent air leak secondary to lung laceration at the inferior part of the right lower lobe was noted and this was stapled and removed. The patient was discharged 3 days post-surgery.

The role of uniportal VATS in blunt chest trauma

Uniportal VATS has undergone substantial growth and uptake in recent years (Salati et al., 2008) since its first introduction by Gonzalez et al. (2011) and Rocco (2016). In experienced hands and high-volume institutions, uniportal VATS can be used as a safe and feasible approach to managing thoracic emergencies as demonstrated by Swierzy et al. (2018).

In our institution, the uniportal VATS peri-operative setup and technical details are summarized below (Box 14.1):

The advantages of uniportal VATS include reduction of postoperative pain and improved postoperative mobilization as one instead of multiple intercostal incisions are performed which reduces intercostal nerve injury (Tamura et al., 2013). Another main strength of this technique is the availability of a better intraoperative view as described by Bertolaccini et al. whereby the operative fulcrum is brought inside the chest which allows surgeons to address the target lesion in a fashion similar to open surgery (Bertolaccini et al., 2016). In cases

> **BOX 14.1 Uniportal video-assisted thoracoscopic surgery setup and technical considerations.**
>
> General anesthesia with double lumen tube insertion to achieve ipsilateral lung collapse. Perform single lung ventilation according to side of injury during surgery.
>
> Position the patient in a lateral decubitus position (left or right) with both arms flexed and stretched towards the head. This permits quick conversion to thoracotomy in the event of failed VATS.
>
> Perform a single 3–4 cm incision in the 5th intercostal space or through chest tube incision.
>
> Placement of an Alexis wound retractor for otptimal visualization.
>
> Utilize a 30° angle thoracoscope with purpose-designed instruments with proximal and distal articulations.
>
> Systematic inspection of the pleural cavity, chest wall, lung parenchyma, diaphragm, and great vessels with additional surgical interventions performed such as blood and clot suction with suction-irrigator system, the release of adhesions, repair of lung laceration with autostapler or wedge resection, resection of fractured rib tip and rib fixation.
>
> Intercostal nerve block was performed for postoperative pain control.
>
> Chest tube insertion at the end of operation.

of emergency, the uniportal incision can be rapidly extended into a lateral thoracotomy to gain rapid and broad exposure. Alternatively, additional port sites can be performed based on surgeon discretion. Further larger prospective or randomized control studies are required to further validate the role of uniportal VATS in thoracic trauma surgery.

Conclusion

DHTX is an under-recognized entity that must not be overlooked by the multidisciplinary team managing blunt chest trauma. Further research and evidence are required to delineate the definition, pathophysiology, diagnosis, and best management of DHTX. With increased clinician awareness and the utilization of clinical risk prediction tools to guide patient follow-up and inform radiological surveillance algorithms, DHTX could be detected in a timely fashion. This could potentially shift the nature of DHTX surgical management from a predominantly emergency resuscitative fashion necessitating urgent maximally invasive thoracotomy to a more controlled elective setting whereby minimally invasive approaches can be adopted based on suitable indication and patient selection criteria. The potential role of utilizing VATS in managing delayed hemothorax lies in its ability to maximize positive postoperative patient outcomes, yield superior long-term outcomes, and gain greater patient satisfaction.

References

Ahn, H. J., Lee, J. W., Kim, K. D., & You, I. S. (2016). Phrenic arterial injury presenting as delayed hemothorax complicating simple rib fracture. *Journal of Korean Medical Science, 31* (4), 641−643. Available from https://doi.org/10.3346/jkms.2016.31.4.641, http://www.jkms.org/Synapse/Data/PDFData/0063JKMS/jkms-31-641.pdf.

Baldwin, J. C., Oyer, P. E., Guthaner, D. F., & Stinson, E. B. (1984). Combined azygous vein and subclavian artery injury in blunt chest trauma. *Journal of Trauma − Injury, Infection and Critical Care, 24*(2), 170−171. Available from https://doi.org/10.1097/00005373-198402000-00018.

Bertolaccini, L., Viti, A., Terzi, A., & Rocco, G. (2016). Geometric and ergonomic characteristics of the uniportal video-assisted thoracoscopic surgery (VATS) approach. *Annals of Cardiothoracic Surgery, 5*(2), 118−122. Available from https://doi.org/10.21037/acs.2015.12.05, http://www.annalscts.com/article/view/9557/pdf.

Boersma, W. G., Stigt, J. A., & Smit, H. J. M. (n.d.). Treatment of haemothorax.

Brekke, I. J., Maidas, P., & Møller, L. (2021). Forsinket pleuraeffusjon etter thoraxtraume. *Tidsskrift for den Norske laegeforening: tidsskrift for praktisk medicin, ny raekke, 141*(3). Available from https://doi.org/10.4045/tidsskr.20.0717.

Bundy, D. W., & Tilton, D. M. (2003). Delayed hemothorax after blunt trauma without rib fractures. *Military Medicine, 168*(6), 501−502. Available from https://doi.org/10.1093/milmed/168.6.501, https://academic.oup.com/milmed/issue.

Carrillo, E. H., Heniford, B. T., Senler, S. O., Dykes, J. R., Maniscalco, S. P., & Richardson, J. D. (1998). Embolization therapy as an alternative to thoracotomy in vascular injuries of the chest wall. *American Surgeon, 64*(12), 1142−1148.

Chang, S. W., Ryu, K. M., & Ryu, J. W. (2018). Delayed massive hemothorax requiring surgery after blunt thoracic trauma over a 5-year period: Complicating rib fracture with sharp edge associated with diaphragm injury. *Clinical and Experimental Emergency Medicine, 5*(1), 60−65. Available from https://doi.org/10.15441/ceem.16.190, https://www.ceemjournal.org/upload/pdf/ceem-16-190.pdf.

Chemelli, A. P., Thauerer, M., Wiedermann, F., Strasak, A., Klocker, J., & Chemelli-Steingruber, I. E. (2009). Transcatheter arterial embolization for the management of iatrogenic and blunt traumatic intercostal artery injuries. *Journal of Vascular Surgery, 49*(6), 1505−1513. Available from https://doi.org/10.1016/j.jvs.2009.02.001.

Chen, C. L., & Cheng, Y. L. (2014). Delayed massive hemothorax complicating simple rib fracture associated with diaphragmatic injury. *American Journal of Emergency Medicine, 32*(7), 818. Available from https://doi.org/10.1016/j.ajem.2013.12.060, http://www.journals.elsevier.com/american-journal-of-emergency-medicine/.

Choi, J., Anand, A., Sborov, K. D., Walton, W., Chow, L., Guillamondegui, O., Dennis, B. M., Spain, D., & Staudenmayer, K. (2021). Complication to consider: Delayed traumatic hemothorax in older adults. *Trauma Surgery & Acute Care Open, 6*(1), e000626. Available from https://doi.org/10.1136/tsaco-2020-000626.

Chokshi, T., Theodosopoulos, A., Wilson, E., Ysit, M., Alhadi, S., & Ranasinghe, L. (2021). A case report of delayed hemothorax complicated by fibrothorax. *Asploro Journal of Biomedical and Clinical Case Reports, 4*(3), 184−190. Available from https://doi.org/10.36502/2021/asjbccr.6252.

Curfman, K. R., Robitsek, R. J., Salzler, G. G., Gray, K. D., Lapunzina, C. S., Kothuru, R. K., & Schubl, S. D. (2015). Massive hemothorax caused by a single intercostal artery bleed ten days after solitary minimally displaced rib fracture. *Case Reports in Surgery, 2015*, 1−4. Available from https://doi.org/10.1155/2015/120140.

Demirhan, R., Onan, B., Oz, K., & Halezeroglu, S. (2009). Comprehensive analysis of 4205 patients with chest trauma: A 10-year experience. *Interactive Cardiovascular and Thoracic Surgery*, 9. 2009/09 https://doi.org/10.1510/icvts.2009.206599 15699293 3 450 453 Turkey http://icvts.ctsnetjournals.org/cgi/reprint/9/3/450?ck = nck 9.

Divisi, D., Battaglia, C., Berardis, D., Vaccarili, M., Di Francescantonio, W., Salvemini, S., & Crisci, R. (2004). Video-assisted thoracoscopy in thoracic injury: Early or delayed indication? *Acta bio-medica: Atenei Parmensis*, 75(3), 158−163.

Émond, M., Guimont, C., Chauny, J. M., Daoust, R., Bergeron, É., Vanier, L., Moore, L., Plourde, M., Kuimi, B., Boucher, V., Allain-Boulé, N., & Le Sage, N. (2017). Clinical prediction rule for delayed hemothorax after minor thoracic injury: A multicentre derivation and validation study. *CMAJ Open*, 5(2), E444. Available from https://doi.org/10.9778/cmajo.20160096.

Emond, M., Sirois, M. J., Guimont, C., Chauny, J. M., Daoust, R., Bergeron, E., Vanier, L., Camden, S., & Le Sage, N. (2015). Functional impact of a minor thoracic injury an investigation of age, delayed hemothorax, and rib fracture effects. *Annals of Surgery*, 262(6), 1115−1122. Available from https://doi.org/10.1097/SLA.0000000000000952, http://journals.lww.com/annalsofsurgery/pages/default.aspx.

Flores-Funes, D., Lluna-Llorens, A. D., Jiménez-Ballester, M. Á., Valero-Navarro, G., Carrillo-Alcaráz, A., Campillo-Soto, Á., & Aguayo-Albasini, J. L. (2020). Is the number of rib fractures a risk factor for delayed complications? A case−control study. *European Journal of Trauma and Emergency Surgery*, 46(2), 435−440. Available from https://doi.org/10.1007/s00068-018-1012-x, http://link.springer.com/journal/68.

Godat, L., Cantrell, E., & Coimbra, R. (2016). Thoracoscopic management of traumatic sequelae. *Current Trauma Reports*, 2(3), 144−150. Available from https://doi.org/10.1007/s40719-016-0047-x, http://www.springer.com/medicine/journal/40719.

Gonzalez, D., de la Torre, M., Paradela, M., Fernandez, R., Delgado, M., Garcia, J., Fieira, E., & Mendez, L. (2011). Video-assisted thoracic surgery lobectomy: 3-year initial experience with 200 cases. *European Journal of Cardio-thoracic Surgery*, 40(1), e21. Available from https://doi.org/10.1016/j.ejcts.2011.02.051.

Gonzalez, G., Robert, C., Petit, L., Biais, M., & Carrié, C. (2021). May the initial CT scan predict the occurrence of delayed hemothorax in blunt chest trauma patients? *European Journal of Trauma and Emergency Surgery*, 47(1), 71−78. Available from https://doi.org/10.1007/s00068-020-01391-4, http://link.springer.com/journal/68.

Graeber, G. M., & Nazim, M. (2007). The anatomy of the ribs and the sternum and their relationship to chest wall structure and function. *Thoracic Surgery Clinics*, 17(4), 473−489. Available from https://doi.org/10.1016/j.thorsurg.2006.12.010.

Heniford, B. T., Carrillo, E. H., Spain, D. A., Sosa, J. L., Fulton, R. L., & Richardson, J. D. (1997). The role of thoracoscopy in the management of retained thoracic collections after trauma. *Annals of Thoracic Surgery*, 63(4), 940−943. Available from https://doi.org/10.1016/S0003-4975(97)00173-2.

Igai, H., Kamiyoshihara, M., Yoshikawa, R., Ohsawa, F., & Yazawa, T. (2019). Delayed massive hemothorax due to a diaphragmatic laceration caused by lower rib fractures. *General Thoracic and Cardiovascular Surgery*, 67(9), 811−813. Available from https://doi.org/10.1007/s11748-018-1033-8, http://www.springer.com/dal/home/generic/search/results?SGWID = 1-40109-70-173671907-0.

Karmy-Jones, R., Jurkovich, G. J., Nathens, A. B., Shatz, D. V., Brundage, S., Wall, M. J., Engelhardt, S., Hoyt, D. B., Holcroft, J., & Knudson, M. M. (2001). Timing of urgent thoracotomy for hemorrhage after trauma: A multicenter study. *Archives of Surgery*, 136(5), 513−518. Available from https://doi.org/10.1001/archsurg.136.5.513, http://archsurg.jamanetwork.com/journal.aspx.

Kuhlman, D. R., Khuder, S. A., & Lane, R. D. (2015). Factors influencing the diameter of human anterior and posterior intercostal arteries. *Clinical Anatomy*, *28*(2), 219−226. Available from https://doi.org/10.1002/ca.22460, http://onlinelibrary.wiley.com/journal/10.1002/(ISSN)1098-2353.

Lewis, B. T., Herr, K. D., Hamlin, S. A., Henry, T., Little, B. P., Naeger, D. M., & Hanna, T. N. (2021). Imaging manifestations of chest trauma. *Radiographics*, *41*(5), 1321−1334. Available from https://doi.org/10.1148/rg.2021210042, https://pubs.rsna.org/doi/pdf/10.1148/rg.2021210042.

Hamamoto, N., Kikuta, S., Takahashi, R., & Ishihara (2023). *Delayed tension hemothorax with nondisplaced rib fractures after blunt thoracic trauma*.

Lin, A. C., Tsai, C. K., Tsai, C. S., & Lin, C. Y. (2017). Blunt chest trauma with diaphragmatic laceration presenting as delayed hemothorax. *Journal of Medical Sciences (Taiwan)*, *37*(4), 175−177. Available from https://doi.org/10.4103/jmedsci.jmedsci_33_17, http://www.jmedscindmc.com/temp/JMedSci374175-2760023_074000.pdf.

Lin, H. L., Huang, W. Y., Yang, C., Chou, S. M., Chiang, H. I., Kuo, L. C., Lin, T. Y., & Chou, Y. P. (2014). How early should VATS be performed for retained haemothorax in blunt chest trauma? *Injury*, *45*(9), 1359−1364. Available from https://doi.org/10.1016/j.injury.2014.05.036, http://www.elsevier.com/locate/injury.

Lin, H. L., Tarng, Y. W., Wu, T. H., Huang, F. D., Huang, W. Y., & Chou, Y. P. (2019). The advantages of adding rib fixations during VATS for retained hemothorax in serious blunt chest trauma − A prospective cohort study. *International Journal of Surgery*, *65*, 13−18. Available from https://doi.org/10.1016/j.ijsu.2019.02.022, http://www.elsevier.com/wps/find/journaldescription.cws_home/705107/description#description.

Liu, D. W., Liu, H. P., Lin, P. J., & Chang, C. H. (1997). Video-assisted thoracic surgery in treatment of chest trauma. *Journal of Trauma - Injury, Infection and Critical Care*, *42*. 1 1997/01 https://doi.org/10.1097/00005373-199704000-00015 00225282 4 670 674 Lippincott Williams and Wilkins Taiwan http://www.jtrauma.com.

Mahmood, I., Abdelrahman, H., Al-Hassani, A., Nabir, S., Sebastian, M., & Maull, K. (2011). Clinical management of occult hemothorax: A prospective study of 81 patients. *The American Journal of Surgery*, *201*(6), 766−769. Available from https://doi.org/10.1016/j.amjsurg.2010.04.017.

Mahoozi, H. R., Volmerig, J., & Hecker, E. (2016). Modern management of traumatic hemothorax. *Journal of Trauma & Treatment*, *5*(3). Available from https://doi.org/10.4172/2167-1222.1000326.

Mangram, A. J., Zhou, N., Sohn, J., Moeser, P., Sucher, J. F., Hollingworth, A., Ali Osman, F. R., Moyer, M., Johnson, V. A., Jr, & Dzandu, J. K. (2016). Pleural effusion following rib fractures in the elderly: Are we being aggressive enough? *Journal of Gerontology & Geriatric Research*, *5*(5). Available from https://doi.org/10.4172/2167-7182.1000341.

Manlulu, A. V., Lee, T. W., Thung, K. H., Wong, R., & Yim, A. P. C. (2004). Current indications and results of VATS in the evaluation and management of hemodynamically stable thoracic injuries. *European Journal of Cardio-thoracic Surgery*, *25*(6), 1048−1053. Available from https://doi.org/10.1016/j.ejcts.2004.02.017.

Masuda, R., Ikoma, Y., Oiwa, K., Nakazato, K., Takeichi, H., & Iwazaki, M. (2013). Delayed hemothorax superimposed on extrapleural hematoma after blunt chest injury: A case report. *Tokai Journal of Experimental and Clinical Medicine*, *38*(3), 97−102. Available from http://mj.med.u-tokai.ac.jp/pdf/380301.pdf.

Milfeld, D. J., Mattox, K. L., & Beall, A. C. (1978). Early evacuation of clotted hemothorax. *The American Journal of Surgery*, *136*(6), 686−692. Available from https://doi.org/10.1016/0002-9610(78)90336-7.

Mintz, Y., Easter, D. W., Izhar, U., Edden, Y., Talamini, M. A., & Rivkind, A. I. (2007). Minimally invasive procedures for diagnosis of traumatic right diaphragmatic tears: A method for correct diagnosis in selected patients. *The American Surgeon*, 73(4), 388−392. Available from https://doi.org/10.1177/000313480707300416.

Misthos, P., Kakaris, S., Sepsas, E., Athanassiadi, K., & Skottis, I. (2004). A prospective analysis of occult pneumothorax, delayed pneumothorax and delayed hemothorax after minor blunt thoracic trauma. *European Journal of Cardio-thoracic Surgery*, 25. 5 2004/05 https://doi.org/10.1016/j.ejcts.2004.01.044 5 859 864 Greece.

Morales Uribe, C. H., Villegas Lanau, M. I., & Petro Sánchez, R. D. (2008). Best timing for thoracoscopic evacuation of retained post-traumatic hemothorax. *Surgical Endoscopy and Other Interventional Techniques*, 22(1), 91−95. Available from https://doi.org/10.1007/s00464-007-9378-6.

Mowery, N. T., Gunter, O. L., Collier, B. R., Diaz, J. J., Haut, E., Hildreth, A., Holevar, M., Mayberry, J., & Streib, E. (2011). Practice management guidelines for management of hemothorax and occult pneumothorax. *Journal of Trauma − Injury, Infection and Critical Care*, 70(2), 510−518. Available from https://doi.org/10.1097/TA.0b013e31820b5c31.

Muronoi, T., Kidani, A., Oka, K., Konishi, M., Kuramoto, S., Shimojo, Y., Hira, E., & Watanabe, H. (2020). Delayed massive hemothorax due to diaphragm injury with rib fracture: A case report. *International Journal of Surgery Case Reports*, 77, 133−137. Available from https://doi.org/10.1016/j.ijscr.2020.10.125, http://www.elsevier.com/wps/find/journaldescription.cws_home/723449/description#descriptionElsevier.

Navsaria, P. H., Vogel, R. J., & Nicol, A. J. (2004). Thoracoscopic evacuation of retained post-traumatic hemothorax. *Annals of Thoracic Surgery*, 78(1), 282−285. Available from https://doi.org/10.1016/j.athoracsur.2003.11.029, http://www.elsevier.com/locate/athoracsur.

Ota, H., Kawai, H., & Matsuo, T. (2014). Video-assisted minithoracotomy for blunt diaphragmatic rupture presenting as a delayed hemothorax. *Annals of Thoracic and Cardiovascular Surgery*, 20(Supplement), 911−914. Available from https://doi.org/10.5761/atcs.cr.13-00201.

Park, C. H., Kim, K. E., Chae, M. C., & Lee, J. W. (2022). Delayed massive hemothorax after blunt thoracic trauma requiring thoracotomy by VATS: A case report. *Journal of Surgical Case Reports*, 2022(1). Available from https://doi.org/10.1093/jscr/rjab537, https://academic.oup.com/jscr.

Plourde, M., mond, M. E., Lavoie, A., Guimont, C., Le Sage, N., Chauny, J. M., Bergeron, E., Vanier, L., Moore, L., Allain-Boulé, N., Fratu, R. F., & Dufresne, M. (2014). Cohort study on the prevalence and risk factors for delayed pulmonary complications in adults following minor blunt thoracic trauma. *Canadian Journal of Emergency Medicine*, 16(2), 136−143. Available from https://doi.org/10.2310/8000.2013.131043, http://journals.deckerpublishing.com/pubs/CJEM/volume%2016,%202014/issue%2002,%20March/CJEM_2014_131043_English/CJEM_2014_131043_English.pdf.

Quinn, M. W., & Dillard, T. A. (1999). Delayed traumatic hemothorax on ticlopidine and aspirin for coronary stent. *American College of Chest Physicians, United States Chest*, 116(1), 257−260. Available from https://doi.org/10.1378/chest.116.1.257, http://www.chestjournal.org/.

Ritter, D. C., & Chang, F. C. (1995). Delayed hemothorax resulting from stab wounds to the internal mammary artery. *Journal of Trauma − Injury, Infection and Critical Care*, 39(3), 586−589. Available from https://doi.org/10.1097/00005373-199509000-00032, http://www.jtrauma.com.

Rizk, W., Kadry, A., Awad, G., Abdelmohty, H., & Elsaid, A. (2020). Delayed Hemothorax: Don't miss it. *Zagazig University Medical Journal*, 0−0. Available from https://doi.org/10.21608/zumj.2020.47844.1984.

Rocco, G. (2016). Fact checking in the history of uniportal video-assisted thoracoscopic surgery. *Journal of Thoracic Disease, 8*(8), 1849–1850. Available from https://doi.org/10.21037/jtd.2016.07.77, http://jtd.amegroups.com/article/download/8571/pdf.

Ross, R. M., & Cordoba, A. (1986). Delayed life-threatening hemothorax associated with rib fractures. *Journal of Trauma – Injury, Infection and Critical Care, 26*(6), 576–578. Available from https://doi.org/10.1097/00005373-198606000-00018.

Salati, M., Brunelli, A., & Rocco, G. (2008). Uniportal video-assisted thoracic surgery for diagnosis and treatment of intrathoracic conditions. *Thoracic Surgery Clinics, 18*(3), 305–310. Available from https://doi.org/10.1016/j.thorsurg.2008.04.005.

Sharma, O. P., Hagler, S., & Oswanski, M. F. (2005). Prevalence of delayed hemothorax in blunt thoracic trauma. *American Surgeon, 71*, 12 2005/12 00031348 6 481 486 United States.

Shorr, R. M., Crittenden, M., Indeck, M., Hartunian, S. L., & Rodriguez, A. (1987). Blunt thoracic trauma. Analysis of 515 patients. *Annals of Surgery, 206*(2), 200–205. Available from https://doi.org/10.1097/00000658-198708000-00013.

Simon, B. J., Chu, Q., Emhoff, T. A., Fiallo, V. M., & Lee, K. F. (1998). Delayed hemothorax after blunt thoracic trauma: An uncommon entity with significant morbidity. *Journal of Trauma – Injury, Infection and Critical Care, 45*(4), 673–676. Available from https://doi.org/10.1097/00005373-199810000-00005, http://www.jtrauma.com.

Sinha, P., & Sarkar, P. (1998). Late clotted haemothorax after blunt chest trauma. *Emergency Medicine Journal, 15*(3), 189–191. Available from https://doi.org/10.1136/emj.15.3.189, http://emj.bmj.com/.

Smith, J. W., Franklin, G. A., Harbrecht, B. G., & Richardson, J. D. (2011). Early VATS for blunt chest trauma: A management technique underutilized by acute care surgeons. *Journal of Trauma – Injury, Infection and Critical Care, 71*(1), 102–107. Available from https://doi.org/10.1097/TA.0b013e3182223080.

Söderlund, T., Ikonen, A., Pyhältö, T., & Handolin, L. (2015). Factors associated with in-hospital outcomes in 594 consecutive patients suffering from severe blunt chest trauma. *Scandinavian Journal of Surgery, 104*(2), 115–120. Available from https://doi.org/10.1177/1457496914543976, http://sjs.sagepub.com/content/104/2/115.full.pdf+html.

Swierzy, M., Faber, S., Nachira, D., Günsberg, A., Rückert, J. C., & Ismail, M. (2018). Uniportal video-assisted thoracoscopic surgery for the treatment of thoracic emergencies. *Journal of Thoracic Disease, 10*. 11 1 2018/11/01 https://doi.org/10.21037/jtd.2018.08.126 20776624 S3720-S3725S3720 AME Publishing Company Germany http://www.jthoracdis.com/.

Tamura, M., Shimizu, Y., & Hashizume, Y. (2013). Pain following thoracoscopic surgery: Retrospective analysis between single-incision and three-port video-assisted thoracoscopic surgery. *Journal of Cardiothoracic Surgery, 8*(1). Available from https://doi.org/10.1186/1749-8090-8-153Japan, http://www.cardiothoracicsurgery.org/content/8/1/153.

Turhan, K., Makay, O., Cakan, A., Samancilar, O., Firat, O., Icoz, G., & Cagirici, U. (2008). Traumatic diaphragmatic rupture: Look to see. *European Journal of Cardio-thoracic Surgery, 33*(6), 1082–1085. Available from https://doi.org/10.1016/j.ejcts.2008.01.029.

Verkroost, M. W., & Hensens, A. G. (1998). Isolated pericardial rupture with left-sided haematothorax after blunt chest trauma. *European Journal of Cardio-thoracic Surgery, 14*(5), 517–519. Available from https://doi.org/10.1016/S1010-7940(98)00235-8.

Liman, Kuzucu, A., Tastepe, Ulasan, G.N., & Topcu (2003). Chest injury due to blunt trauma.

Waldman, R. P., Donner, M., Bilsky, A. C., Stom, M. C., & Narins, R. G. (1984). Delayed onset of hemothorax: An unusual complication of subclavian access for hemodialysis. *Nephron, 37*(4), 270–272. Available from https://doi.org/10.1159/000183262.

Weigelt, J. A., Aurbakken, C. M., Meier, D. E., & Thal, E. R. (1982). Management of asymptomatic patients following stab wounds to the chest. *Journal of Trauma — Injury, Infection and Critical Care*, 22(4), 291–294. Available from https://doi.org/10.1097/00005373-198204000-00005.

Whitson, B. A., Mcgonigal, M. D., Anderson, C. P., & Dries, D. J. (2013). Increasing numbers of rib fractures do not worsen outcome: An analysis of the national trauma data bank. *American Surgeon*, 79(2), 140–150. Available from http://www.ingentaconnect.com/search/download?pub = infobike%3a%2f%2fsesc%2ftas%2f2013%2f00000079%2f00000002%2fart00024&mimetype = application%2fpdf&exitTargetId = 1366712312890.

Yap, D., Ng, M., Chaudhury, M., & Mbakada, N. (2018). Longest delayed hemothorax reported after blunt chest injury. *American Journal of Emergency Medicine*, 36(1), 171. Available from https://doi.org/10.1016/j.ajem.2017.10.025, http://www.journals.elsevier.com/american-journal-of-emergency-medicine/.

Villavicencio, R. T., Aucar, J. A., & Wall, M. J. (1999). *Analysis of thoracoscopy in trauma.*

Zhang, J., Hong, Q., Mo, X., & Ma, C. (2022). Complete video-assisted thoracoscopic surgery for rib fractures: Series of 35 cases. *Annals of Thoracic Surgery*, 113(2), 452–458. Available from https://doi.org/10.1016/j.athoracsur.2021.01.065, http://www.elsevier.com/locate/athoracsur.

Ziegler, D. W., & Agarwal, N. N. (1994). The morbidity and mortality of RIB fractures. *Journal of Trauma — Injury, Infection and Critical Care*, 37(6), 975–979. Available from https://doi.org/10.1097/00005373-199412000-00018.

Zeiler, J., Idell, S., Norwood, S., & Cook, A. (2020). *Hemothorax: A review of the literature.*

Chapter 15

Chest drains: which, where, when

Tanzil Rujeedawa[1], Ahmed Mohamed Osman[2] and Adam Peryt[3]
[1]School of Clinical Medicine, University of Cambridge, Cambridge, United Kingdom,
[2]Department of Thoracic Surgery, Nottingham University Hospital, Nottingham, United Kingdom, [3]Department of Thoracic Surgery, Royal Papworth Hospital, Cambridge, United Kingdom

Introduction

Insertion of chest drains is one of the most common procedures performed to manage different diseases of the pleura, to evacuate air or fluid from the cavity, to restore lung function (Porcel, 2018).

Chest tube insertion and drainage have a long history, dating back to the 5th century when Hippocrates first described using a drain to treat empyema. The next recorded account of pus evacuation from the chest to enable healing is documented by Mitchell in Medicine in the Crusades—where the treatment of draining pus from the wound was based on the experimental treatment of a bear with a similar injury (Mitchell, 2004).

In more recent history, the first documented descriptions of using waterseal drainage systems can be attributed to Playfair in 1873, for treating a child with pleural empyema (Mitchell, 2004), and Hewitt in 1876 where they described the use of closed drainage systems attached to chest tubes. Furthermore, for patients undergoing modern thoracic surgery, chest tubes were documented in the postoperative care of patients in 1922, and were also used throughout World War II after traumatic thoracotomies, and the Korean War. From these experiences, chest drains were the standard of care for pleural space drainage following trauma during the Vietnam War (Monaghan & Swan, 2008). Since then, chest tube design and drainage systems have evolved to accommodate modern needs, but the principles have remained the same—to evacuate intrapleural air or fluid to restore intrathoracic lung function.

Chest tubes can be classified by their size, either small-bore chest tubes (SBCTs) or large-bore chest tubes (LBCTs) (Anderson et al., 2022). They can also be classified according to the technique of insertion, including blunt dissection, Seldinger technique (with or without radiological guidance), and the trocar technique (Filosso et al., 2016; Mahmood & Wahidi, 2013). Some authors advocate against using the trocar technique due to the higher risk of misplacement and organ perforation (John et al., 2014).

Which

Chest drain components

Chest drainage systems are composed of a chest tube and a collection system with or without an integrated suction unit.

Chest tubes are available in differing sizes Fig. 15.1 and materials to best suit the clinical needs of the patient. Generally speaking, the size is measured by the internal diameter of the tube in units of French (F, or Fr), where 1 increment of the French scale corresponds to a third of a millimeter diameter (Anderson et al., 2022)—in other words, F refers to the outer diameter of a cylindrical tube and it is equivalent to 0.333 mm (Filosso et al., 2016). For example, a size 24 F chest tube has an 8-mm caliber, and a 32 F chest tube has a 10.7 mm. The typical size of chest tubes ranges from 8Fr to 36Fr, with

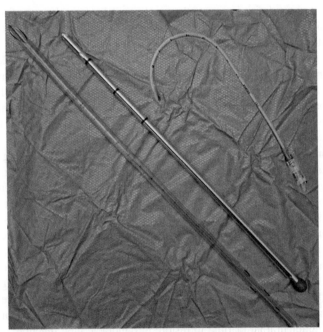

FIGURE 15.1 **Different size chest tubes.** Two straight large-bore chest tubes and a curved small-bore chest tube. *From Porcel, J. M. (2018). Chest tube drainage of the pleural space: A concise review for pulmonologists. Tuberculosis and Respiratory Diseases, 81(2), 106–115. https://doi.org/10.4046/trd.2017.0107.*

those less than 20Fr classified as SBCTs, and those equal to or larger than 20Fr classified as LBCTs (Anderson et al., 2022). However, some authors classify LBCTs as those being larger than 14 F (Mahmood & Wahidi, 2013). When choosing which size chest tube to insert, it is important to account for the rate at which air or liquid flows through the chest tube, which is determined by its internal diameter and length (Venuta et al., 2017), and the viscosity of the fluid to be drained—where highly viscous fluids (such as blood, complex exudative effusions and empyemas) need larger bore tubes for drainage when compared to air or simple effusions (Baumann, 2003).

With regards to chest drainage collection systems, one of the following four types can be used in Fig. 15.2:

1. **One-way Heimlich valve**—which consists of a one-way rubber flutter valve that occludes during inspiration and opens during expiration. These valves are usually used for the ambulatory management of pneumothorax (Gogakos et al., 2015).
2. **Analog three-container systems**—being the most common collection systems, they consist of (1) a collection chamber, into which fluid or air drains, (2) a water-seal chamber that prevents atmospheric air from entering into the pleural space with inspiration, and (3) a suction control chamber which uses either wet or dry suction—this component of the drainage system can be attached to a continuous, low-pressure, wall suction or placed on "water seal" with no active suction mechanism (i.e., placed on "gravity") (Porcel, 2018). Another commonly used drainage collection system is the Rocket® Underwater Seal Single Bottle System, which consists of a single bottle containing a prescribed amount of water, with the chest tube from the patient submerged approximately 2cm below the surface of the water, and one short tube leading to the outside atmosphere which functions as a vent.
3. **Digital collection units**—these provide a continuous digital recording of air leaks, fluid drainage, and intrapleural pressure (Arai et al., 2018; Kiefer, 2017) and systems such as the Thopaz + have an in-built, regulated suction pump with digital display.
4. **Simple vacuum bottles**—these typically drain indwelling pleural catheters when connected to their external one-way valve (Porcel, 2018).

When

Indications of chest tube insertion and drainage

As stated earlier in this chapter, chest tubes are inserted to drain the pleural cavity of air or fluid (or both). Specifically, the indications include (Asciak et al., 2023):

- Pneumothorax failing other treatments.

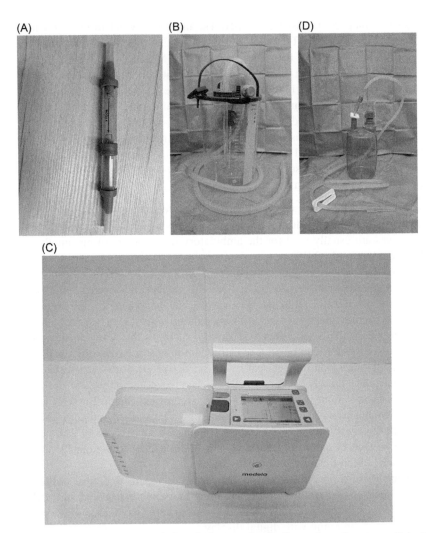

FIGURE 15.2 Different types of chest drainage and collection systems. Four types of chest drainage and collection systems. *From Porcel, J. M. (2018). Chest tube drainage of the pleural space: A concise review for pulmonologists. Tuberculosis and Respiratory Diseases, 81(2), 106−115. https://doi.org/10.4046/trd.2017.0107.*

- Simple drainage of large benign or malignant pleural effusions.
- Symptomatic pleural effusions in patients on mechanical ventilation.
- Talc pleurodesis.
- Pleural infection.
- Traumatic hemothorax and/or pneumothorax.

- Post-thoracic cavity procedures (i.e., medical thoracoscopy, thoracic, esophageal, or cardiac surgery).
- Bronchopleural fistula (Venuta et al., 2017).

SBCTs are typically sufficient to drain pneumothoraxes and pleural infections (including empyema). However, in secondary spontaneous pneumothorax with substantial air leaks, some clinicians favor LBCTs (Iepsen & Ringbæk, 2013; Mahmood & Wahidi, 2013; Rahman et al., 2010). In traumatic pneumothorax, studies have demonstrated that 14 F drains are equally as effective as 28 F drains, with no increase in complications (Kulvatunyou et al., 2011, 2014). Similarly, small-bore drains have been used successfully in the drainage of traumatic hemothorax in stable patients without increased rates of failure or complications (Kulvatunyou et al., 2011; Rivera et al., 2009).

The use of LBCTs is indicated in certain clinical situations such as post-thoracic surgery, or cases of hemothorax in unstable patients (Mahmood & Wahidi, 2013). An ex vivo study simulating massive hemothorax drainage demonstrated that 28 F drains provided an optimal balance between flow rate efficiency, reduced risk of occlusion, and manageable tube size. Therefore, sizes exceeding 32 F are rarely warranted (Chestovich et al., 2020), and a consensus from four international thoracic surgery societies supports the use of 28–32 F chest drains post-thoracotomy (Brunelli et al., 2011). Furthermore, in cases of barotrauma-induced pneumothorax from mechanical ventilation, LBCTs are preferred due to the lower success rate observed with SBCTs (Lin et al., 2010)

Once the indications of chest tube insertion are met, and prior to the insertion of a chest tube, every effort should be made to minimize the potential risks and complications. Written consent should be obtained where possible, and a thorough review of radiological images should be undertaken prior to chest tube insertion, as well as ensuring there isn't an underlying coagulopathy which may increase the risk of bleeding (Laws et al., 2003).

Precautions to chest tube insertion

While there are no absolute contraindications for chest drain insertion especially in emergencies, it is important to note that in some situations, surgical intervention is preferred to chest drainage. These include blunt or penetrating traumas with extensively destroyed chest walls, impaled objects, and suspicion of cardiac or great vessel injury (Molnar, 2017). Other contraindications for chest drains may include uncorrected coagulopathy and insertion over an infected skin area (Porcel, 2018). Furthermore, giant bullae may mimic a pneumothorax on imaging (Waseem et al., 2005) Fig. 15.3, necessitating caution when interpreting imaging.

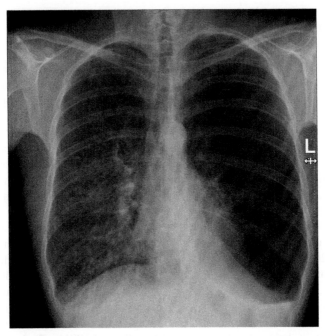

FIGURE 15.3 Plain film chest x-ray showing a right-sided giant bulla. A plain chest x-ray demonstrating a right-sided giant bulla. *Courtesy: Hacking, C. (2024). Giant pulmonary bulla. Case study. https://doi.org/10.53347/rID-73253.*

Where

Chest tube insertion

Standard insertion technique

Patient positioning and landmarks for insertion

Position the patient supine and raise the head of the bed until it is about 45 degrees (or between 30–60 degrees). Abduct the ipsilateral arm of the patient and secure it above the patient's head, making a note of the safety triangle, which is bounded by the lateral borders of the latissimus dorsi and pectoralis major, and extends from the base of the axilla to the 5th intercostal space. The standard chest tube insertion site is the 4th or 5th intercostal space, between the anterior and midaxillary lines (roughly corresponding to the level of the nipple for men and above the inframammary fold for women). This area of the chest allows for chest tube insertion with reduced risk of injury to vessels, nerves, and muscles. Alternatively, the patient could be positioned in the lateral decubitus position, with the side of the lesion uppermost, however, this should not be used for the drainage of the fluid in the postpneumonectomy space when a bronchopleural fistula (BPF) is suspected.

Steps for insertion of a large-bore chest tubes using blunt dissection technique

Following the appropriate positioning of the patient, prep the area with an antiseptic solution and administer local anesthesia along the anticipated tract, from the skin down to the intercostal muscle space. Once the anesthetic settles, a 3–5 cm transverse incision is made through the skin and subcutaneous tissues, and blunt dissection is carried out down to the pleura, along the superior border of the rib using a Kelly clamp. Penetrate the pleura by applying firm pressure and opening the tip of the clamp to widen the pleural opening.

Remove the Kelly clamp and perform a finger sweep in the pleural space to ensure there aren't vital structures around the created hole. Clamp the distal end of the tube using the Kelly clamp and insert it into the pleural cavity in a controlled manner along the created tract. Once all the side holes of the tube are within the pleural space, make a note of the immediate "fogging" of the chest tube in the case of a pneumothorax, or the gush of fluid in effusions or hemothorax, indicating intrapleural insertion. Secure the chest tube using a 0 or 1 silk "drain stitch" and connect the tube to the prepared chest drainage system. A post-procedural chest x-ray (CXR) is warranted to confirm satisfactory positioning of the tube.

Steps for insertion of a small-bore chest tubes using the Seldinger technique

SBCTs are usually inserted utilizing the Seldinger technique with the aid of a guidewire. After positing the patient, utilizing a sterile technique, and ensuring adequate analgesia, the introducer needle is advanced over the superior border of the rib in the safety triangle. Once the pleural space is breached, air or fluid is aspirated to confirm the correct position. The guidewire is then advanced through the needle without resistance. Once the guidewire is in the pleural space, the needle is removed, taking care to leave the guidewire in place. Dilators are then passed over the guidewire into the pleural space, and once adequate tract dilatation is achieved, the chest tube assembly is passed over the guidewire into the pleural space, ensuring that all the holes are in the pleural space. The guidewire and chest tube inserter are removed once the SBCT is in the pleural cavity. Once inserted, it should be secured in the usual fashion and connected to the chest drainage system of choice.

Apical/anterior chest tube insertion

As for the standard lateral insertion, the patient for an apical drain insertion is laid supine with the head of the bed raised between 30 and 60 degrees. The ipsilateral arm is adducted to the side of the patient. The landmark for insertion is at the intersection of the second intercostal space and midclavicular line (identify the sternal angle of Louis and move laterally on the corresponding intercostal space, which is the second intercostal space, stopping at the midclavicular line). The indications for this approach include decompressing a tension pneumothorax via needle thoracocentesis, and insertion of either a SBCT or LBCT for an anteriorly located (Fig. 15.4) or apical pneumothorax. The steps for insertion in this anatomical landmark follow the same steps as the standard insertion technique above. Care should be taken not to insert the drain medial to the midclavicular line to avoid injuring the internal mammary vessels, leading to a hemothorax.

Chest tube removal

Several factors come into play when determining the correct time to remove a chest tube. In cases of pneumothorax, ensure that there is no air leak or bubbling into the water-seal drainage system in an otherwise patent chest tube. With regard to fluid drainage, make a note of both the quality and quantity of the fluid. It should be non-purulent, and free of chyle, or blood. There is no consensus with regard to the quantity of fluid that is acceptable before the removal of a chest tube. The recommended volume thresholds for chest tube removal range from 200 mL to 500 mL in 24 hours (Cerfolio & Bryant, 2008; Cerfolio et al., 2013; Grodzki, 2008). A CXR should confirm proper lung re-expansion with resolution of pneumothorax, effusion, or hemothorax, before removal of the chest tube. Removal of the chest tube at the end of inspiration or expiration is another widely debated

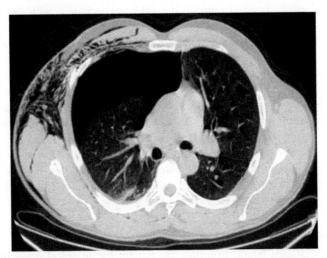

FIGURE 15.4 Computed tomography (CT) in the lung window at the level of the pulmonary trunk showing a right-sided large pneumothorax. CT scan of the chest showing a large right pneumothorax with associated surgical emphysema. *Courtesy: Palas, J., Matos, A., Mascarenhas, V., Herédia, V., & Ramalho, M. (2014). Multidetector computer tomography: Evaluation of blunt chest trauma in adults. Radiology Research and Practice. 864369. https://doi.org/10.1155/2014/864369.*

question (French, 2023; Novoa & Fuentes, 2022), but regardless of the respiratory phase in which it is removed, the chest tube should be removed swiftly and the defect in the chest wall closed with either a suture that was placed at the time of chest tube placement or with a properly occlusive dressing (in selective patients where SBCT's are used).

Potential pitfalls and complications

The average rate of complications associated with chest tube insertion is less than 10%. It mainly depends on the operator's experience, the size of the tube inserted, and whether or not imaging was used to guide insertion. Fewer complications appear when experienced operators insert SBCT under image guidance. The most frequent immediate complications documented following SBCT insertion (with or without image guidance) were pain (4.1%), failure to place the drain (2.4%), and vasovagal reactions (2.1%), while delayed complications included pain (18%), drain blockage (7.4%), accidental dislodgment (7.3%), and subcutaneous emphysema (3.4%). Commonly reported post-insertion complications associated with LBCT (≥ 20 F) were malposition (6.5%), drain blockage (5.2%), organ injuries (1.4%), and empyema (1.4%).

In an attempt to standardize reporting of complications surrounding chest tubes, Aho et al. proposed five complication categories to allow for easier recording and collection of data. These were insertional, positional, removal, infectious, and malfunction (Aho et al., 2015):

- Insertional complications: these include injury to intrathoracic or extrathoracic organs within 24 hours of chest tube insertion. This complication is most common with chest tubes inserted using the trocar technique.
- Positional complications: defined as occurring 24 hours after insertion, include erosions into adjacent organs or any tube kinking, obstruction, or being entrapped in the fissure after lung expansion.
- Removal complications: encompass failure to seal the chest defect after the chest tube is removed, resulting in entraining atmospheric air, or the retention of any foreign objects after removal.
- Infectious complications: any external (i.e., wound-related from improper sterilizing techniques) or internal (empyema) infections related to chest tube insertion.
- Malfunction complications: these include problems that may arise from the health care clinician managing the chest tube or equipment issues.

Summary

Several types and sizes of chest tubes are available. It is imperative to have a thorough understanding of the principles, indications, and rationale of choosing different tube sizes when dealing with trauma and disease processes of the chest and pleura, to greatly improve patient outcomes and minimize morbidity. The benefits of SBCTs include patient comfort and ease of placement, and most tubes can be placed using ultrasound guidance and the Seldinger technique. LBCTs may be necessary for more complex intrathoracic collections or barotrauma-associated pneumothoraces in mechanically ventilated patients and the postoperative setting.

References

Aho, J. M., Ruparel, R. K., Rowse, P. G., Brahmbhatt, R. D., Jenkins, D., & Rivera, M. (2015). Tube Thoracostomy: A Structured Review of Case Reports and a Standardized Format for Reporting Complications. *World J Surg, 39*(11), 2691−2706. Available from https://doi.org/10.1007/s00268-015-3158-6.

Anderson, D., Chen, S. A., Godoy, L. A., Brown, L. M., & Cooke, D. T. (2022). Comprehensive review of chest tube management: A review. *JAMA Surgery, 157*(3), 269−274. Available from https://doi.org/10.1001/jamasurg.2021.7050, http://archsurg.jamanetwork.com/issues.aspx.

Arai, H., Tajiri, M., Kameda, Y., Shiino, K., Ando, K., Okudela, K., & Masuda, M. (2018). Evaluation of a digital drainage system (Thopaz) in over 250 cases at a single site: A retrospective case-control study. *The Clinical Respiratory Journal, 12*(4), 1454−1459. Available from https://doi.org/10.1111/crj.12683.

Asciak, R., Bedawi, E. O., Bhatnagar, R., Clive, A. O., Hassan, M., Lloyd, H., Reddy, R., Roberts, H., & Rahman, N. M. (2023). British Thoracic Society Clinical Statement on pleural procedures. *Thorax, 78*(Suppl 3), s43. Available from https://doi.org/10.1136/thorax-2022-219371.

Baumann, M. H. (2003). What size chest tube? What drainage system is ideal? And other chest tube management questions. *Current Opinion in Pulmonary Medicine, 9*(4), 276−281. Available from https://doi.org/10.1097/00063198-200307000-00006.

Brunelli, A., Beretta, E., Cassivi, S. D., Cerfolio, R. J., Detterbeck, F., Kiefer, T., Miserocchi, G., Shrager, J., Singhal, S., Van Raemdonck, D., & Varela, G. (2011). Consensus definitions to promote an evidence-based approach to management of the pleural space. A collaborative proposal by ESTS, AATS, STS, and GTSC. *European Journal of Cardio-thoracic Surgery, 40*(2), 291−297. Available from https://doi.org/10.1016/j.ejcts.2011.05.020.

Cerfolio, R. J., & Bryant, A. S. (2008). Results of a prospective algorithm to remove chest tubes after pulmonary resection with high output. *Journal of Thoracic and Cardiovascular Surgery, 135*(2), 269−273. Available from https://doi.org/10.1016/j.jtcvs.2007.08.066.

Cerfolio, R. J., Bryant, A. S., Skylizard, L., & Minnich, D. J. (2013). Optimal technique for the removal of chest tubes after pulmonary resection. *Journal of Thoracic and Cardiovascular Surgery, 145*(6), 1535−1539. Available from https://doi.org/10.1016/j.jtcvs.2013.02.007.

Chestovich, P. J., Jennings, C. S., Fraser, D. R., Ingalls, N. K., Morrissey, S. L., Kuhls, D. A., & Fildes, J. J. (2020). Too big, too small or just right? Why the 28 French chest tube is the best size. *Journal of Surgical Research, 256*, 338−344. Available from https://doi.org/10.1016/j.jss.2020.06.048, http://www.elsevier.com/inca/publications/store/6/2/2/9/0/1/index.htt.

Filosso, P. L., Sandri, A., Guerrera, F., Ferraris, A., Marchisio, F., Bora, G., Costardi, L., Solidoro, P., Ruffini, E., & Oliaro, A. (2016). When size matters: Changing opinion in the management of pleural space - the rise of small-bore pleural catheters. *Journal of Thoracic Disease, 8*(7), E503. Available from https://doi.org/10.21037/jtd.2016.06.25, http://www.jthoracdis.com/.

French, D. (2023). Contributing to drainology: Removing chest drains after pulmonary resection based on air leak alone. *Journal of Thoracic Disease*, *15*(11), 5885–5888. Available from https://doi.org/10.21037/jtd-23-1433.

Gogakos, A., Barbetakis, N., Lazaridis, G., Papaiwannou, A., Karavergou, A., Lampaki, S., Baka, S., Mpoukovinas, I., Karavasilis, V., Kioumis, I., Pitsiou, G., Katsikogiannis, N., Tsakiridis, K., Rapti, A., Trakada, G., Zissimopoulos, A., Tsirgogianni, K., Zarogoulidis, K., & Zarogoulidis, P. (2015). Heimlich valve and pneumothorax. *Annals of Translational Medicine*, *3*(4). Available from https://doi.org/10.3978/j.issn.2305-5839.2015.03.25, http://atm.amegroups.com/.

Grodzki, T. (2008). Prospective algorithm to remove chest tubes after pulmonary resection with high output – is it valid everywhere? *The Journal of Thoracic and Cardiovascular Surgery*, *136*(2), 536. Available from https://doi.org/10.1016/j.jtcvs.2008.04.017.

Iepsen, U. W., & Ringbæk, T. (2013). Small-bore chest tubes seem to perform better than larger tubes in treatment of spontaneous pneumothorax. *Danish Medical Journal*, *60*(6). Available from http://www.danmedj.dk/portal/pls/portal/!PORTAL.wwpob_page.show?_docname = 10109043.PDF.

John, M., Razi, S., Sainathan, S., & Stavropoulos, C. (2014). Is the trocar technique for tube thoracostomy safe in the current era? *Interactive Cardiovascular and Thoracic Surgery*, *19*(1), 125–128. Available from https://doi.org/10.1093/icvts/ivu071, http://icvts.oxfordjournals.org/.

Kiefer, T. (2017). *Chest drains in daily clinical practice*. Springer Nature.

Kulvatunyou, N., Erickson, L., Vijayasekaran, A., Gries, L., Joseph, B., Friese, R. F., O'Keeffe, T., Tang, A. L., Wynne, J. L., & Rhee, P. (2014). Randomized clinical trial of pigtail catheter versus chest tube in injured patients with uncomplicated traumatic pneumothorax. *British Journal of Surgery*, *101*(2), 17–22. Available from https://doi.org/10.1002/bjs.9377.

Kulvatunyou, N., Vijayasekaran, A., Hansen, A., Wynne, J. L., O'Keeffe, T., Friese, R. S., Joseph, B., Tang, A., & Rhee, P. (2011). Two-year experience of using pigtail catheters to treat traumatic pneumothorax: A changing trend. *Journal of Trauma - Injury, Infection and Critical Care*, *71*(5), 1104–1107. Available from https://doi.org/10.1097/TA.0b013e31822dd130.

Laws, D., Neville, E., & Duffy, J. (2003). BTS guidelines for the insertion of a chest drain. *Thorax*, *58*(2), ii53. Available from https://doi.org/10.1136/thx.58.suppl_2.ii53, http://thorax.bmj.com/.

Lin, Y. C., Tu, C. Y., Liang, S. J., Chen, H. J., Chen, W., Hsia, T. C., Shih, C. M., & Hsu, W. H. (2010). Pigtail catheter for the management of pneumothorax in mechanically ventilated patients. *American Journal of Emergency Medicine*, *28*(4), 466–471. Available from https://doi.org/10.1016/j.ajem.2009.01.033.

Mahmood, K., & Wahidi, M. M. (2013). Straightening out chest tubes: What size, what type, and when. *Clinics in Chest Medicine*, *34*(1), 63–71. Available from https://doi.org/10.1016/j.ccm.2012.11.007.

Mitchell, P. (2004). *Medicine in the crusades: Warfare, wounds and the medieval surgeon*. Cambridge University Press.

Molnar, T. F. (2017). Thoracic trauma: Which chest tube when and where? *Thoracic Surgery Clinics*, *27*(1), 13–23. Available from https://doi.org/10.1016/j.thorsurg.2016.08.003, http://www.elsevier.com/wps/find/journaldescription.cws_home/702699/description#description.

Monaghan, S. F., & Swan, K. G. (2008). Tube thoracostomy: The struggle to the "Standard of Care". *Annals of Thoracic Surgery*, *86*(6), 2019–2022. Available from https://doi.org/10.1016/j.athoracsur.2008.08.006.

Novoa, N. M., & Fuentes, M. G. (2022). Digital drainage systems: Reminding what is important. *Journal of Thoracic Disease*, *14*(9), 3103–3104. Available from https://doi.org/10.21037/jtd-22-940, https://jtd.amegroups.com/article/view/67090/html.

Porcel, J. M. (2018). Chest tube drainage of the pleural space: A concise review for pulmonologists. *Tuberculosis and Respiratory Diseases*, *81*(2), 106−115. Available from https://doi.org/10.4046/trd.2017.0107, https://www.e-trd.org/Synapse/Data/PDFData/0003TRD/trd-81-106.pdf.

Rahman, N. M., Maskell, N. A., Davies, C. W. H., Hedley, E. L., Nunn, A. J., Gleeson, F. V., & Davies, R. J. O. (2010). The relationship between chest tube size and clinical outcome in pleural infection. *Chest*, *137*(3), 536−543. Available from https://doi.org/10.1378/chest.09-1044. Available from: http://chestjournal.chestpubs.org/content/137/3/536.full.pdf + html.

Rivera, L., O'Reilly, E. B., Sise, M. J., Norton, V. C., Sise, C. B., Sack, D. I., Swanson, S. M., Iman, R. B., Paci, G. M., & Antevil, J. L. (2009). Small catheter tube thoracostomy: Effective in managing chest trauma in stable patients. *Journal of Trauma - Injury, Infection and Critical Care*, *66*(2), 393−399. Available from https://doi.org/10.1097/TA.0b013e318173f81e.

Venuta, F., Diso, D., Anile, M., Rendina, E. A., & Onorati, I. (2017). Chest tubes: Generalities. *Thoracic Surgery Clinics*, *27*(1), 1−5. Available from https://doi.org/10.1016/j.thorsurg.2016.08.001, http://www.elsevier.com/wps/find/journaldescription.cws_home/702699/description#description.

Waseem, M., Jones, J., Brutus, S., Munyak, J., Kapoor, R., & Gernsheimer, J. (2005). Giant bulla mimicking pneumothorax. *The Journal of Emergency Medicine*, *29*(2), 155−158. Available from https://doi.org/10.1016/j.jemermed.2005.04.004.

Chapter 16

Chest tube placement: options and guidelines

Kunal Bhakhri and Anuj Wali
Department of Thoracic Surgery, University College Hospital, London, United Kingdom

Introduction

All patients with blunt chest trauma should be assessed with a systematic approach to identify and treat life-threatening complications in a timely manner. This assessment could and should happen in a pre-hospital setting by relevantly trained first responders or on arrival at the emergency department (ED). Insertion of a chest tube is an integral part of this initial assessment and can be a lifesaving intervention.

Advanced Trauma Life Support (ATLS) guidelines have become the gold standard for acute trauma management (Advanced Trauma Life Support Course for Physicians, 1987). In addition to this, many countries and societies have released their own guidance and algorithms for decision-making in trauma management.

Most guidance surrounding the insertion and management of chest drains in chest trauma comes from first principles, expert opinion, and common-sense practice, and there is a paucity of high-level evidence in the literature. There are few prospective randomized trials and most of the guideline's recommendations are based on retrospective studies.

In this chapter, we will explore the guidelines currently available for the management of chest tubes in the context of blunt chest trauma. Then we will consider the available literature that explores the rationale behind the current common practices in more detail.

Indications

A chest tube is inserted into the pleural space to drain air or fluid from the chest cavity to assist in total re-expansion of the lung and to reduce the risk of immediate and delayed complications.

Indications for Chest Drain insertion can be simplified into two main categories:

- Pneumothorax — air in the pleural space
- Hemothorax — blood in the pleural space

The decision-making and clinical reasoning surrounding chest drain management depend on the indication for chest drain insertion.

The American College of Surgeons ATLS 10th Edition has become the most ubiquitous international guideline available for the management of trauma patients. Other institutions at a regional and national level have also produced guidelines and algorithms to guide clinicians in the management of trauma. Western Trauma Association (USA), Eastern Trauma Association (USA), NICE guidelines (UK), and BOAST guidelines (UK) are all readily available and generally well regarded.

Most trauma centers will now also have local guidelines that supplement these international guidelines. A patient with a chest tube needs input from multiple members of the multidisciplinary team and access to these essential treatments depends on local protocols and availability of resources.

Local guidelines offer practical advice regarding local resources and procedural protocol, and we recommend that all clinicians who are expected to manage chest trauma patients are familiar with their local practice.

Chest tubes for pneumothorax

Pneumothorax is very common after blunt chest trauma. Most commonly, lung laceration secondary to rib fracture is the mechanism for air to enter the potential space between the visceral and parietal pleura.

Pneumothorax can be classified as follows:

- Simple
- Tension
- Open
- Complex (Haemopneumothorax, Hydropneumothorax, etc.)

and the indications for chest tubes vary accordingly.

Do you need a chest drain for pneumothorax?

A **tension pneumothorax** is defined as a pneumothorax associated with hemodynamic instability.

It is an emergency presentation and needs immediate treatment based on clinical findings. If there is a clinical suspicion of tension pneumothorax, treatment should be initiated immediately and in any setting.

A tension pneumothorax develops when more air enters the pleural space that can leave. Usually, this is due to a one-way flow of air from the lung parenchyma to the pleural space but can also occur through the chest wall from an open wound. The build-up of pressure in the pleural space causes the ipsilateral lung to collapse and the mediastinum to shift towards the contralateral side. As the thoracic cavity pressure builds, this can exceed the pressure of venous return to the heart (from both vena cava & pulmonary veins) causing catastrophic loss of pre-load to the heart and circulatory arrest.

If there is clinical evidence of tension pneumothorax, (reduced breath sounds, tympanic percussion, tracheal deviation & hemodynamic instability), then radiological confirmation of the diagnosis should not delay immediate decompression and chest tube insertion. Decompression should be performed according to ATLS guidelines. For children, the Second intercostal space midclavicular line is recommended; for adults, the 4th or 5th intercostal space anterior to the midaxillary line is the first line but either location is acceptable. Following decompression, a chest tube should be inserted for ongoing management.

If there is an **open pneumothorax** chest tube insertion at a separate site to the injury is mandatory. After chest tube insertion the open chest wound should be covered with a bio-occlusive dressing prior to formal closure.

For the management of **simple pneumothorax**, there is a paucity of evidence in trauma literature. This is reflected in the current guidelines where there are few robust recommendations and instead, the clinical judgment of experienced clinicians is promoted. Chest tube insertion can be associated with serious complications; therefore we must weigh up the potential risks and benefits of intervention (Bailey, 2000). Indeed, ATLS states: "Any pneumothorax is best treated with a chest tube placed in the fifth intercostal space, just anterior to the midaxillary line. Observation and aspiration of a small, asymptomatic pneumothorax may be appropriate, but a qualified doctor should make this treatment decision."

The Western trauma guidelines algorithm suggests a 35 mm cut-off for intervention with a chest tube (Moya, 2022, 2022). Measurement should be calculated on a chest radiograph or computed tomography (CT) scan perpendicular to the chest wall at the deepest point of the pneumothorax. Evidence for this recommendation comes from single-center retrospective reviews of pneumothorax management where those with PTX >35 mm are more likely to fail observant management (Bou Zein Eddine et al., 2019; Figueroa et al., 2022).

Increased prevalence of whole-body trauma scanning in the assessment of the polytrauma patient has resulted in the identification of previously missed pneumothorax and has provided a treatment dilemma. **Occult pneumothorax** is defined as a sub-clinical pneumothorax not seen on chest radiographs but visible on subsequent CT imaging. Occult pneumothorax is found in approximately 5% of hospitalized trauma patients (Kirkpatrick et al., 2013). Most pneumothoraces are missed on initial chest radiograph assessment due to the supine position required for the x-ray to facilitate spinal precautions. In a supine position, the pneumothorax often redistributes to lie anteriorly, and therefore on a supine chest radiograph the air is not visible as it overlies the lung markings. As CT scanning becomes the first-line imaging modality, the distinction of occult pneumothorax becomes less relevant as small pneumothoraces will be readily identified on first-line imaging.

It is important to attempt to distinguish the features of those with simple pneumothorax that will progress and those that can be safely observed. Attempts at producing scoring systems for the prediction of progression have been attempted but are not yet adequately validated with prospective randomized trials (De Moya et al., 2007). Large retrospective studies and case studies describe that conservative management of simple pneumothorax is possible in a certain selection of patients (Berliner, 2018; Walker et al., 2018). A randomized control trial investigating the conservative management of small traumatic pneumothoraces would offer more robust evidence for this practice.

There is increasing evidence from clinical trials that **spontaneous** simple pneumothorax can be managed conservatively irrespective of size with similar re-expansion rates but with fewer serious chest tube-related complications (Brown et al., 2020).

The theory behind this is that if there is a defect in the lung parenchymal surface allowing air to escape into the pleural cavity, then the quickest recovery is made if the lung defect heals and closes. As the lung deflates, it is proposed that the free edges of the parenchymal defect are better apposed to one another which promotes healing of the defect. If the lung is prematurely re-expanded by insertion of a chest drain, then this could interfere with the healing process prolong the need for a chest tube, and prolong hospital stays. These findings have been described in the context of spontaneous pneumothorax; however, the principles are likely to be transferrable to a trauma context.

Ultimately, the decision to insert a chest tube depends on the clinician's judgment on whether the simple asymptomatic pneumothorax will progress to respiratory distress or tension. This judgment should be based on the size of the pneumothorax, severity of the injuries, clinical signs of respiratory distress, and comorbid status.

If the patient has symptoms of respiratory distress or an injury pattern with a high risk for deterioration, then the pragmatic approach would be to insert a chest tube to improve the re-expansion of the lung and rectify ventilation-perfusion mismatch. If the patient is deemed low risk for progression, then the risks of inserting a chest tube may outweigh the benefits and a conservative approach should be attempted with monitoring for respiratory deterioration.

Positive pressure ventilation

Patients with simple traumatic pneumothorax are at theoretical risk of developing tension pneumothorax with **positive pressure ventilation**

(**PPV**). ATLS guidelines state that "Ideally, a patient with a known pneumothorax should not undergo general anesthesia or receive positive pressure ventilation without having a chest tube inserted." The evidence behind this recommendation is mixed, as two early small-size randomized trials concluded opposing findings (Brasel et al., 1999; Enderson et al., 1993). A more recent clinical trial was stopped early for protocol violations which highlights the lack of equipoise in clinicians who would rather follow the "safety first" approach to management of this patient group (Clements et al., 2021).

Ultimately, with this lack of firm evidence, there should be a discussion between the surgeon and anesthetist/intensivist before PPV being initiated in a patient with pneumothorax.

Chest tubes for haemothorax

Hemothorax describes a collection of blood in the pleural space and is a common consequence of blunt chest trauma. Injury to vessels in the chest wall (intercostals, internal thoracic, etc.), diaphragm, or bleeding from injury to the lung parenchyma are the most common cause of a hemothorax requiring chest tube management. Injuries to the great vessels and the myocardium are more commonly associated with penetrating chest trauma and cause catastrophic bleeding requiring emergency surgical intervention.

In the context of chest tube insertion for traumatic hemothorax, the purpose of the intervention is threefold:

- To drain the chest cavity to aid re-expansion of the lung for ventilation.
- To prevent secondary infection of the pleural space, empyema, and trapped lung.
- To monitor for active bleeding and the volume of blood loss.

Classically, the severity of hemothorax is classified according to the volume of blood present in the pleural cavity (Zeiler et al., 2020):

- <400 mL — minimal
- $400-1000$ mL — moderate
- >1000 mL — massive

CT scan is considered the gold standard for radiological assessment of the volume of pleural collections.

Does any volume of blood loss mandate surgery?

Approximately 15% of cases of blunt chest trauma require emergency thoracotomy. Criteria for consideration of urgent surgical thoracotomy from international guidelines (Advanced Trauma Life Support Course for Physicians, 1987; Mowery et al., 2011; Moya, 2022a, 2022b):

- More than 1500 mL of blood was immediately evacuated by a chest tube.
- Persistent bleeding — 150 to 200 mL/h for 2–4 h measured from chest tube.
- Persistent Blood transfusion is required to maintain hemodynamic instability.

These criteria are based on consensus agreement, but the evidence is poor and mainly comes from retrospective studies focusing on penetrating chest trauma rather than blunt trauma. Mortality rates have been shown to increase linearly in relation to the volume of blood lost. Patients who lose at least 1500 mL have three times higher mortality rates than those who lost 500 mL (Karmy-Jones et al., 2001).

The decision to operate should be focused on sound clinical judgment based upon the clinical picture, chest tube output with the mechanism, and timing of the injury. For example, it must be appreciated that the high-energy car accident patient who bleeds 1500 mL in a few hours post-injury is different from the patient who had a low-energy rib fracture that may have accumulated 1500 mL over 24 h and is drained all in one go on arrival to ED.

Do you need a chest drain for "small" haemothorax?

All available guidelines recommend chest tube drainage for moderate to large-size hemothorax. For smaller hemothoraces, the evidence is less clear and the recommendations vary between guidelines (Advanced Trauma Life Support Course for Physicians, 1987; Moya, 2022a, 2022b; Patel et al., 2021).

The rationale for draining moderate to large hemothorax is to prevent infection of the pleural space, empyema, trapped lung, and Fibrothorax. Blood is a nutrient-rich substrate that easily becomes infected if in contact with infective material. If the collection is large, then this is more likely to result in atelectasis and collapsed airways which are at high risk of bronchoalveolar infection which can easily migrate to the pleural space. Early drainage is key, before the activation of the clotting cascade and clot formation which can hinder chest tube drainage and lead to retained collections.

There is no consensus definition for what the exact cut-off is for small hemothorax, but all the relevant literature generally suggests that a collection below 300–500 mL in volume is considered small. ATLS guidelines suggest chest tube drainage for any hemothorax visible on chest x-rays. This is a simple and practical way of estimating hemothorax volume in the emergency room and in resource-poor settings. It must be remembered that large volumes of fluid can collect in the pleural space and chest x-ray does not always visualize this.

Small hemothorax can be managed expectantly as small collections can be spontaneously reabsorbed by the pleura. In a normal person, pleural fluid turnover is estimated to be approximately 0.15 mL/kg/h and this rate is estimated to increase in pathological states (Miserocchi, 1997). We must therefore carefully weigh up the risks of injury and introducing infection to the pleural space when there is a good chance of spontaneous resolution. Retrospective reviews have suggested that small (<300 mL) hemothorax can be managed conservatively and that there is the little additional clinical risk to those patients who fail conservative management (Demetri et al., 2018).

If an observation approach is used, then repeated imaging to assess for progression and a high index of suspicion for infection is necessary.

Measuring the size of hemothorax on chest radiography is inaccurate and CT imaging is the gold standard. However, with the increasing use of Focused Assessment with Sonography for Trauma scanning in the trauma setting, there is an increased body of evidence suggesting that ultrasound is a fast and accurate method of deciding whether a chest tube is required (Chung et al., 2018). Ultrasound assessment can be used as a safe and radiation-free adjunct if there is availability.

Multiple formulae are validated for use in estimating the volume of pleural collections using CT imaging and ultrasound imaging. One simple formula is as follows (Mergo et al., 1999):

$$volume = d^2 \times l$$

where d = largest depth of the collection on a single CT image, l = greatest length of the effusion.

Management of retained haemothorax

Retained hemothorax is defined as any hemothorax that remains in the pleural space after initial drainage with a chest tube. Retained hemothorax Is associated with high rates of development of empyema (around 20%), pneumonia (around 20%), and fibrothorax (Zeiler et al., 2020).

Retained hemothorax can be managed conservatively if the retained collection is small. For larger collections, there are multiple options:

1. A second chest tube could be inserted, targeting the residual collection using ultrasound-guided methods (pigtail catheter or surgical chest tube).
2. Instillation of thrombolytic agents (alteplase, streptokinase, etc.) to the pleural space to encourage the breakdown of clots and free drainage of the residual collection
3. Video assisted thoracic surgery (VATS) procedure for washout.

In all but the smallest retained hemothoraces (which should spontaneously resolve), the guidelines recommend that early intervention with VATS washout is superior to a second drain or thrombolytic agents. Early VATS washout is associated with shorter hospital stays and fewer further interventions (Kumar et al., 2015; Meyer et al., 1997; Oğuzkaya et al., 2005). Currently, intrapleural thrombotic agents should be considered a second-line agents behind surgical intervention for patients for who VATS surgery +/− thoracotomy is too risky.

Early VATS is defined as surgical drainage and washout within 4 days of injury. It is associated with reduced need for conversion to thoracotomy, reduced rates of empyema, and shortened length of stay in hospital. All available guidelines recommend early intervention, if necessary, for the best outcomes.

Chest tube management

Suction versus no suction

Once a chest tube is successfully placed, we can either manage it with an underwater seal or with low-pressure suction. There is no consensus view on the use of suction in the chest tube management of traumatic pneumothorax.

In general, there are two schools of thought regarding the use of negative-pressure chest tube therapy. One theory suggests that suction expedites the re-expansion of the lung through the accelerated removal of air and fluid from the pleural space which therefore shortens the duration of treatment. On the other hand, it could be argued that the mechanical negative pressure in the pleural space encourages more air leak from the lung parenchyma and therefore prolongs the air leak and time of chest tube management. There is evidence for both phenomena in the literature, however, this evidence comes from observation of spontaneous pneumothorax or post-surgical air leak. The evidence for this suction in the trauma setting is scarce.

In the context of hemothorax, there is less controversy, as there are few negative consequences of suction therapy. A review and meta-analysis of three randomized trials suggest some evidence that suction decreases the duration of chest tube management, length of hospital stay, and persistent air leak (Feenstra et al., 2018). However, the authors of the meta-analysis concede that the trials are of poor quality, with low numbers, and susceptible to bias.

Do you need prophylactic antibiotics?

The routine use of prophylactic antibiotics when inserting a chest tube for chest trauma is controversial. Proponents of antibiotic use cite the increased risk of introducing infection when inserting a chest drain. Patients with chest tubes are also more likely to be at risk of pneumonia because of reduced ventilation caused by pain resulting in pooling of secretions and causing airway collapse.

Despite the perceived increased risk of infection, the increased focus on antimicrobial stewardship means there needs to be a convincing clinical benefit to justify the use of prophylactic antibiotics. Unfortunately, the current literature suffers from underpowered clinical trials and observational studies susceptible to bias. The most recent guidelines from the Eastern Association for the Surgery of Trauma (EAST) reflect the lack of convincing evidence: "There is insufficient published evidence to support any recommendation either for or against the use of presumptive antibiotics to reduce the incidence of empyema or pneumonia in chest tubes for traumatic hemopneumothorax."

Interestingly, a meta-analysis showed a protective effect against empyema in penetrating chest trauma but no effect in blunt trauma (Elnahla et al., 2021). This may reflect the 'sterile' nature of the blunt injury compared to a penetrating injury which may introduce external pathogens.

In blunt trauma, there is insufficient evidence to endorse prophylactic antibiotics after chest drain insertion. However, there should be a high clinical suspicion for infection and early antimicrobial treatment of infective complications associated with chest tube insertion.

Summary

Whenever we perform any intervention, we must weigh up the benefits of the procedure with the potential harms. The decision to insert a chest drain should depend on the patient's clinical picture and the prognosis of the injuries sustained.

There is a requirement for more prospective research into the potential conservative management of small simple hemothorax or pneumothorax which would help to guide future recommendations.

References

Advanced Trauma Life Support Course for Physicians. (1987).

Bailey, R. C. (2000). Complications of tube thoracostomy in trauma. *Emergency Medicine Journal*, *17*(2), 111–114. Available from https://doi.org/10.1136/emj.17.2.111.

Berliner, T. (2018). Conservative management in traumatic pneumothoraces: An observational study. *The Journal of Emergency Medicine*, *55*(2), 301–302. Available from https://doi.org/10.1016/j.jemermed.2018.06.031.

Bou Zein Eddine, S., Boyle, K. A., Dodgion, C. M., Davis, C. S., Webb, T. P., Juern, J. S., Milia, D. J., Carver, T. W., Beckman, M. A., Codner, P. A., Trevino, C., & de Moya, M. A. (2019). Observing pneumothoraces: The 35-millimeter rule is safe for both blunt and penetrating chest trauma. *Journal of Trauma and Acute Care Surgery*, *86*(4), 557–564. Available from https://doi.org/10.1097/ta.0000000000002192.

Brasel, K. J., Stafford, R. E., Weigelt, J. A., Tenquist, J. E., & Borgstrom, D. C. (1999). Treatment of occult pneumothoraces from blunt trauma. *Journal of Trauma – Injury, Infection and Critical Care*, *46*(6), 987–991. Available from https://doi.org/10.1097/00005373-199906000-00001, http://www.jtrauma.com.

Brown, S. G. A., Ball, E. L., Perrin, K., Asha, S. E., Braithwaite, I., Egerton-Warburton, D., Jones, P. G., Keijzers, G., Kinnear, F. B., Kwan, B. C. H., Lam, K. V., Lee, Y. C. G., Nowitz, M., Read, C. A., Simpson, G., Smith, J. A., Summers, Q. A., Weatherall, M., & Beasley, R. (2020). Conservative versus interventional treatment for spontaneous pneumothorax. *New England Journal of Medicine*, *382*(5), 405−415. Available from https://doi.org/10.1056/NEJMoa1910775, http://www.nejm.org/medical-index.

Chung, M. H., Hsiao, C. Y., Nian, N. S., Chen, Y. C., Wang, C. Y., Wen, Y. S., Shih, H. C., & Yen, D. H. T. (2018). The benefit of ultrasound in deciding between tube thoracostomy and observative management in hemothorax resulting from blunt chest trauma. *World Journal of Surgery*, *42*(7), 2054−2060. Available from https://doi.org/10.1007/s00268-017-4417-5, link.springer.de/link/service/journals/00268/index.htm.

Clements, T. W., Sirois, M., Parry, N., Roberts, D. J., Trottier, V., Rizoli, S., Ball, C. G., Xiao, Z. J., & Kirkpatrick, A. W. (2021). OPTICC: A multicentre trial of Occult Pneumothoraces subjected to mechanical ventilation: The final report. *American Journal of Surgery*, *221*(6), 1252−1258. Available from https://doi.org/10.1016/j.amjsurg.2021.02.012, www.elsevier.com/locate/amjsurg.

De Moya, M. A., Seaver, C., Spaniolas, K., Inaba, K., Nguyen, M., Veltman, Y., Shatz, D., Alam, H. B., & Pizano, L. (2007). Occult pneumothorax in trauma patients: Development of an objective scoring system. *Journal of Trauma − Injury, Infection and Critical Care*, *63*(1), 13−17. Available from https://doi.org/10.1097/TA.0b013e31806864fc.

Demetri, L., Martinez Aguilar, M. M., Bohnen, J. D., Whitesell, R., Yeh, D. D., King, D., & de Moya, M. (2018). Is observation for traumatic hemothorax safe? *Journal of Trauma and Acute Care Surgery*, *84*(3), 454−458. Available from https://doi.org/10.1097/ta.0000000000001793.

Elnahla, A., Iuliucci, K. R., Toraih, E., Duchesne, J. C., Nichols, R. L., & Kandil, E. (2021). The efficacy of the use of presumptive antibiotics in tube thoracostomy in thoracic trauma-results of a meta-analysis. *The American Journal of Surgery*, *222*(5), 1017−1022. Available from https://doi.org/10.1016/j.amjsurg.2021.05.003.

Enderson, B. L., Abdalla, R., Frame, S. B., Casey, M. T., Gould, H., & Maull, K. (1993). Tube thoracostomy for occult pneumothorax: A prospective randomized study of its use. *Journal of Trauma − Injury, Infection and Critical Care*, *35*(5), 726−730. Available from https://doi.org/10.1097/00005373-199311000-00013.

Feenstra, T. M., Dickhoff, C., & Deunk, J. (2018). Systematic review and meta-analysis of tube thoracostomy following traumatic chest injury; suction versus water seal. *European Journal of Trauma and Emergency Surgery*, *44*(6), 819−827. Available from https://doi.org/10.1007/s00068-018-0942-7, http://link.springer.com/journal/68.

Figueroa, J. F., Karam, B. S., Gomez, J., Milia, D., Morris, R. S., Dodgion, C., Carver, T., Murphy, P., Elegbede, A., Schroeder, M., & De Moya, M. A. (2022). The 35-mm rule to guide pneumothorax management: Increases appropriate observation and decreases unnecessary chest tubes. *Journal of Trauma and Acute Care Surgery*, 92, 6 1 2022/06/01 10.1097/TA.0000000000003573 21630763 6 951 957 Lippincott Williams and Wilkins United States http://journals.lww.com/jtrauma.

Karmy-Jones, R., Jurkovich, G. J., Nathens, A. B., Shatz, D. V., Brundage, S., Wall, M. J., Engelhardt, S., Hoyt, D. B., Holcroft, J., & Knudson, M. M. (2001). Timing of urgent thoracotomy for hemorrhage after trauma: A multicenter study. *Archives of Surgery*, *136*(5), 513−518. Available from https://doi.org/10.1001/archsurg.136.5.513, http://archsurg.jama-network.com/journal.aspx.

Kirkpatrick, A. W., Rizoli, S., Ouellet, J. F., Roberts, D. J., Sirois, M., Ball, C. G., Xiao, Z. J., Tiruta, C., Meade, M., Trottier, V., Zhu, G., Chagnon, F., & Tien, H. (2013). Occult pneumothoraces in critical care: A prospective multicenter randomized controlled trial of pleural drainage for mechanically ventilated trauma patients with occult pneumothoraces. *Journal of Trauma and Acute Care Surgery*, *74*(3), 747–755. Available from https://doi.org/10.1097/TA.0b013e3182827158.

Kumar, S., Rathi, V., Rattan, A., Chaudhary, S., & Agarwal, N. (2015). VATS versus intrapleural streptokinase: A prospective, randomized, controlled clinical trial for optimum treatment of post-traumatic Residual Hemothorax. *Injury*, *46*(9), 1749–1752. Available from https://doi.org/10.1016/j.injury.2015.02.028.

Mergo, P. J., Helmberger, T., Didovic, J., Cernigliaro, J., Ros, P. R., & Staab, E. V. (1999). New formula for quantification of pleural effusions from computed tomography. *Journal of Thoracic Imaging*, *14*(2), 122–125. Available from https://doi.org/10.1097/00005382-199904000-00011, http://journals.lww.com/thoracicimaging.

Meyer, D. M., Jessen, M. E., Wait, M. A., & Estrera, A. S. (1997). Early evacuation of traumatic retained hemothoraces using thoracoscopy: A prospective, randomized trial. *Annals of Thoracic Surgery*, *64*(5), 1396–1401. Available from https://doi.org/10.1016/S0003-4975(97)00899-0.

Miserocchi, G. (1997). Physiology and pathophysiology of pleural fluid turnover. *European Respiratory Journal*, *10*(1), 219–225. Available from https://doi.org/10.1183/09031936.97.10010219.

Mowery, N. T., Gunter, O. L., Collier, B. R., Diaz, J. J., Haut, E., Hildreth, A., Holevar, M., Mayberry, J., & Streib, E. (2011). Practice management guidelines for management of hemothorax and occult pneumothorax. *Journal of Trauma – Injury, Infection and Critical Care*, *70*(2), 510–518. Available from https://doi.org/10.1097/TA.0b013e31820b5c31.

Moya (2022) *2023 Western trauma Hemothorax Algorithm*. https://www.westerntrauma.org/wp-content/uploads/2021/03/b21435d708c8497f9a77cf92125c319b1.pdf.

Moya. (2022). Evaluation and management of traumatic pneumothorax: A Western Trauma Association critical decisions algorithm. *Journal of Trauma and Acute Care Surgery*, *92*, 2022.

Oğuzkaya, F., Akçali, Y., & Bilgin, M. (2005). Videothoracoscopy versus intrapleural streptokinase for management of post traumatic reta haemothorax: A retrospective study of 65 cases. *Injury*, *36*(4), 526–529. Available from https://doi.org/10.1016/j.injury.2004.10.008, www.elsevier.com/locate/injury.

Patel, N. J., Dultz, L., Ladhani, H. A., Cullinane, D. C., Klein, E., McNickle, A. G., Bugaev, N., Fraser, D. R., Kartiko, S., Dodgion, C., Pappas, P. A., Kim, D., Cantrell, S., Como, J. J., & Kasotakis, G. (2021). Management of simple and retained hemothorax: A practice management guideline from the Eastern Association for the Surgery of Trauma. *The American Journal of Surgery*, *221*(5), 873–884. Available from https://doi.org/10.1016/j.amjsurg.2020.11.032.

Walker, S. P., Barratt, S. L., Thompson, J., & Maskell, N. A. (2018). Conservative management in traumatic pneumothoraces. *Chest*, *153*(4), 946–953. Available from https://doi.org/10.1016/j.chest.2017.10.015.

Zeiler, J., Idell, S., Norwood, S., & Cook, A. (2020). Hemothorax: A review of the literature. *Clinical Pulmonary Medicine*, *27*(1), 1–12. Available from https://doi.org/10.1097/cpm.0000000000000343.

Chapter 17

Diaphragmatic rupture and herniation following blunt trauma: minimally invasive versus open approach

Duaa Ali Faruqi[1,2], Waad Attafi[2], M. Yousuf Salmasi[1,3] and Nizar Asadi[3]
[1]*Faculty of Medicine, Imperial College London, London, United Kingdom,* [2]*Faculty of Medicine, Kings College London, London, United Kingdom,* [3]*Department of Thoracic Surgery, Harefield Hospital, London, United Kingdom*

Introduction

The diaphragm plays an integral role in respiratory mechanics as well as the maintenance of intra-abdominal pressure (IAP). As such, diaphragmatic rupture poses a significant clinical challenge. Traumatic diaphragmatic rupture occurs in 1%–7% of patients presenting with blunt trauma (Lim & Park, 2018).

The incidence may be subtle in the initial presentation, however, undiagnosed, and untreated diaphragmatic ruptures can culminate in life-threatening complications, such as the herniation of abdominal viscera into the thoracic cavity, respiratory distress, and strangulation of herniated organs. Recognizing and promptly addressing diaphragmatic injuries is of paramount importance.

While smaller defects may remain asymptomatic, larger ruptures necessitate surgical intervention to prevent potential complications and ensure anatomical and functional restoration. There are three primary approaches for the repair of diaphragmatic ruptures:

- Laparotomy (Rajaretnam et al., 2024)

    A midline incision through the abdomen to access the abdominal cavity has traditionally been the primary approach for diaphragmatic repair. It provides wide exposure and direct access to all abdominal and thoracic organs, ensuring comprehensive management of associated thoracic injuries. As contemporary surgical paradigms evolve, the role of laparotomy continues to be critically evaluated against emerging minimally invasive techniques.

- Laparoscopy
    The application of laparoscopy, involving the insertion of instruments through the abdominal cavity, has revolutionized the field of surgical intervention, owing to its minimally invasive nature and improved postoperative outcomes. Laparoscopy offers a clear view of both hemidiaphragms and it is particularly preferred when there is a high suspicion of diaphragmatic tear, even with inconclusive imaging results.
- Thoracotomy
    A chest wall incision through the intercostal spaces grants direct access to the diaphragm and is particularly useful for patients with delayed presentations. Furthermore, video-assisted thoracoscopic surgery (VATS) can also be used as a minimally invasive approach.
    The aim of this chapter is to explore the mechanisms of diaphragmatic injury and evaluate the different surgical approaches to repair diaphragmatic rupture and herniation.

Anatomy of the diaphragm

The diaphragm, anatomically a broad, dome-shaped musculotendinous structure, separates the thoracic and abdominal cavities, establishing a critical boundary within the human body. The diaphragm originates from various structures, including the xiphoid process, the lower six costal cartilages, and the first three lumbar vertebrae via the right and left crus. The muscular periphery converges on a resilient, central tendinous region, making it ideally suited to withstand the constant tension imposed during the respiratory cycle (Nason et al., 2012).

Anatomically, notable are three major apertures — the caval opening at the T8 level, the esophageal hiatus at T10, and the aortic hiatus at T12, which facilitate the passage of the inferior vena cava (IVC), esophagus, vagus nerves, and descending aorta. These openings allow the diaphragm to maintain its role as a dynamic partition, enabling the passage of critical anatomical structures while preserving the pressure gradient between the thoracic and abdominal cavities.

The caval opening, on the right side of the diaphragm, allows the passage of the IVC and branches of the right phrenic nerve. On the right side, the diaphragm is associated with the liver and gallbladder, which rest in the falciform recess. The bare area of the liver and diaphragm are in apposition, without any intervening peritoneum. Superiorly, the diaphragm borders the lower right lung and pleura. depicts the relationship of the diaphragm to surrounding structures.

The esophageal hiatus, an opening in the left diaphragm, permits the passage of the esophagus and the vagal trunks. The left diaphragm forms a surface for the heart, creating the diaphragmatic surface of the pericardium. It is also closely associated with the stomach, with the left dome forming the

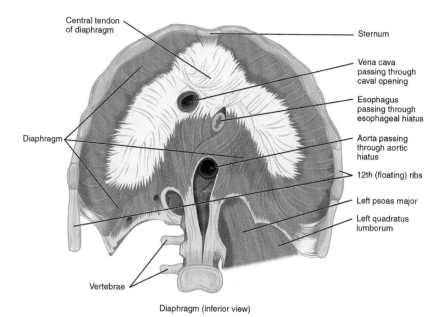

FIGURE 17.1 A detailed depiction of the diaphragm and its relationship to surrounding structures (Betts, 2022).

anterior wall of the lesser sac of the peritoneal cavity. The spleen is positioned posterolateral to the stomach and rests against the diaphragm.

The diaphragm is innervated by the phrenic nerves (C3-C5). These nerves provide sensory and motor supply, enabling the contraction and relaxation of the diaphragm (Jones, 2024) (Fig. 17.1).

Physiology of the diaphragm

The diaphragm's core function is to act as the principal muscle of respiration, contributing to about 75% of the total work of breathing at rest. During inspiration, contraction of the diaphragm muscle leads to a downward displacement of the central tendon, expanding the vertical diameter of the thoracic cavity. This decreases the intra-thoracic pressure, allowing for an influx of air into the lungs. Conversely, during expiration, the diaphragm relaxes, and the elastic recoil of the chest wall and lungs compresses the thoracic cavity, allowing for the expulsion of air (Jones, 2024). briefly illustrates the role of the diaphragm in respiration.

Beyond respiration, the diaphragm also plays a significant role in maintaining IAP, aiding in defecation, urination, vomiting, and childbirth, as well as serving as a secondary pump to facilitate venous return to the heart (Fig. 17.2).

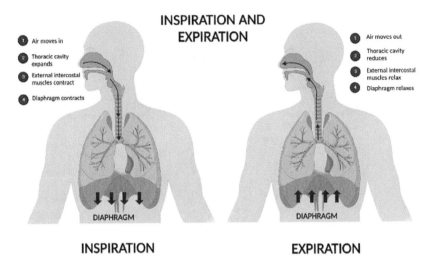

FIGURE 17.2 Diagram illustrating the process of inspiration and expiration in the respiratory system and the role of the diaphragm in this process.

Mechanism of diaphragmatic injury

When subjected to significant trauma, the diaphragm can sustain injury, leading to functional impairment and potentially life-threatening complications. In this section, we will explore the mechanism of underlying diaphragmatic injuries and their clinical implications. TDI can be broadly classified into blunt and penetrating.

Blunt diaphragmatic trauma

Blunt diaphragmatic trauma represents the most common source of TDI, primarily associated with high-impact injuries from falls or motor vehicle accidents (Morgan et al., 2010). Blunt injury to the thorax accounts for approximately 25% of trauma-related mortality and is often the result of the pressure gradient between the pleural and peritoneal cavities. When subjected to such forceful impact, the raised pressure within the abdominal cavity causes the tear of the diaphragm and the negative intrathoracic pressure may draw abdominal viscera into the thoracic cavity. This displacement of viscera disrupts normal respiratory mechanics and potentially introduces fatal complications, such as the strangulation of herniated organs or tension pneumothorax (Chughtai et al., 2009). Importantly, the diaphragm's vulnerability to rupture lies in its left posterolateral segments, where the embryological fusion line is located (Al-Refaie et al., 2009). Consequently, the majority of blunt diaphragmatic ruptures occur on the left side (Rashid et al., 2009). Right-sided diaphragmatic ruptures tend to be less common due to the protective effect of the liver and the increased strength of the right hemi-diaphragm.

Penetrating trauma

In cases of penetrating trauma, sharp objects such as knives, bullets, or shrapnel, can directly lacerate the diaphragmatic tissue and breath the integrity of the diaphragm. This form of injury can happen in the context of stab wounds, gunshot wounds, or any sharp object trauma (Jones, 2024). The precise mechanism of these injuries is determined by several factors, including the trajectory and velocity of the penetrating object, its caliber, and the angle of impact. Due to the close anatomical relation between the diaphragm and other organs, penetrating injuries to the diaphragm often coexist with injuries to the other organs which can further complicate diagnosis and management.

Clinical presentation and diaphragmatic rupture

A salient challenge of diaphragmatic rupture lies in their non-specific presentation and the presence of other concomitant injuries which can obfuscate diagnosis. The clinical presentation of diaphragmatic rupture has often been characterized by spanning "both ends of the spectrum," underscoring the complexity and challenge clinicians face in establishing a diagnosis. This section aims to provide an overview of the clinical manifestations associated with diaphragmatic rupture.

Acute presentation

In an acute setting, patients with diaphragmatic rupture can present with signs of severe trauma, which may mask underlying diaphragmatic injury. Patients may exhibit hypotension, tachycardia, and altered mental state, alongside other signs listed in Table 17.1 (Simon et al., 2024).

Trauma injuries with a significant amount of force to the anterior abdomen should prompt immediate concerns for potential diaphragmatic injury, especially if coupled with the auscultation of bowel sounds or nasogastric suction over the left thorax.

Delayed presentation

As many as two-thirds of patients may experience a delay in diagnosis due to their latent nature and the presence of concurrent injuries. Moreover, rupture of devitalized diaphragmatic muscle can occur days after the injury and in the presence of a diaphragmatic tear without visceral herniation, patients may be asymptomatic for years (Rashid et al., 2009). As such, a high clinical index of suspicion is required.

TABLE 17.1 Physical examination for diaphragmatic rupture.

Physical Examination

Kirschner's Sign: Characterized by increased breath sounds over the abdomen
Kehr's Sign: Shoulder pain due to irritation of the phrenic nerve
Decrease in breath sounds on the side of the injury
Bowel sounds in the chest due to herniation of abdominal organs into the thoracic cavity
Paradoxical movement of the abdomen and chest
Hypotension: Due to decreased venous return secondary to compression from herniated abdominal contents
Abdominal pain
Chest pain
Tachycardia
Altered mental status

Delayed presentation generally coincides with visceral or bowel herniation into the chest cavity, potentially leading to strangulation, incarceration, or even cardiac tamponade. Patients with abdominal contents herniated into the thoracic cavity may exhibit recurrent chest infections, chronic dyspnea, and gastrointestinal symptoms such as vomiting or constipation (Rashid et al., 2009). When abdominal contents herniate, bowel sounds can often be auscultated in the chest. Given that these symptoms are non-specific, diagnosis is often delayed or missed, emphasizing the importance of a comprehensive patient history and thorough physical examination.

Diagnostic imaging

Chest X-ray

Chest radiography is the preferred initial diagnostic method for diaphragmatic rupture, though their sensitivity in detecting suspected ruptures is only about 25%–50%. While obvious findings such as bowel contents or a coiled nasogastric tube in the chest can be seen, up to 40% of chest radiographs may be non-diagnostic. This is especially true in intubated patients, where positive pressure ventilation can prevent herniation of abdominal contents into the chest. Additionally, the use of chest X-ray in diagnosing diaphragmatic ruptures has shown variable accuracy, being as high as 62% for left-sided injuries but only 17% for right-sided injuries (Morgan et al., 2010). Table 17.2 lists key clinical diagnostic signs, however, despite these specific diagnostic findings chest radiographs may not be visible in up to 50% of diaphragmatic rupture cases.

Ultrasound

The use of abdominal or pleural ultrasound for detecting diaphragmatic ruptures appears to have relatively low accuracy, with various studies reporting rates between 0% and 26% (Hofmann et al., 2012). Additionally, the Focused Assessment with Sonography in Trauma (FAST) examination is incorporated to detect free intraperitoneal blood following trauma, particularly in hemodynamically unstable patients. The use of ultrasound to detect diaphragmatic rupture is not yet standardized, and consequently, a negative study does not definitively exclude the diagnosis. Key clinical diagnostic signs are listed in Table 17.3.

Computed tomography scan

The introduction of image reformatting has significantly increased the sensitivity of computed tomography (CT) scans in diagnosing diaphragmatic

TABLE 17.2 Chest X-ray clinical diagnostic signs.

Chest X-ray clinical diagnostic signs
Elevated diaphragm
Basilar atelectasis
Pleural effusion
Unclear hemidiaphragm
Abnormal nasogastric tube positioning
Hemothorax
Inability to trace the normal hemidiaphragm contour
Intrathoracic herniation of a hollow viscus (such as the stomach, colon, or small bowel) with or without focal constriction of the viscus at the tear site (known as the collar sign)
Contralateral mediastinal shift due to a large positive mass effect
Visualization of a nasogastric tube above the left hemidiaphragm
Left hemidiaphragm is positioned significantly higher than the right

TABLE 17.3 Ultrasound clinical diagnostic signs.

Ultrasound clinical diagnostic signs
Discontinuity of the diaphragm
Herniation of the liver
Bowel loops through a diaphragmatic defect
Floating diaphragm
Non-visualization of the diaphragm
Indirect findings: Pleural effusion or subphrenic fluid collection

ruptures and key clinical signs are listed in Table 17.4. CT scans are known to have higher diagnostic accuracy, ranging from 14% to 82%, and are the imaging modality of choice in hemodynamically stable patients (Ganie et al., 2013). Multi-slice spiral CT scans have demonstrated sensitivities as high as 80% and specificities of 100% in various studies, making them the diagnostic tool of choice in cases of blunt or penetrating trauma, or polytraumatized patients, due to their ability to swiftly, and reliably, scan all concomitant injuries. The utility of these scans is increased further with newer generation multidetector machines, which may detect even subtle injuries with a sensitivity of around 66.7%. Despite the high diagnostic accuracy, certain sections of the diaphragm might still be obscured due to similarities in attenuation levels with adjacent structures like the liver or spleen. CT scan detects diaphragm injury and is more useful for assessing the posterior lumbar elements of the diaphragm (crura and arcuate ligaments) compared with the anterior leaflets Table 17.5.

TABLE 17.4 Clinical diagnostic signs in computed tomography scans.

Direct signs	Direct discontinuity of the diaphragm
	Dangling diaphragm sign: A comma-shaped fragment of the diaphragm
	Hump Sign: Presence of a rounded portion of the liver herniating through the diaphragm, forming a hump-shaped mass
	Organ herniation sign: Presence of an intra-abdominal organ within the thoracic cavity through a defect in the diaphragm
	Collar Sign: The presence of a focal, waist-like constriction of the herniating abdominal viscus at the level of the torn diaphragm
	Dependent Viscera Sign: When a patient with a ruptured diaphragm lies supine at CT examination, the herniated viscera are no longer supported posteriorly by the injured diaphragm, and descend to a position against the posterior ribs
	Contiguous Sign: The ability to track an injury on both sides of the diaphragm even in the absence of seeing discontinuation of the diaphragm
	Band Sign: The presence of a band-like density across the thoracic cavity, along the torn edges of the hemidiaphragm, representing the herniated viscera, extending through the diaphragmatic defect
Signs of uncertain origin	Focal Thickening of the Diaphragm: Thickening of the diaphragm at the site of injury, with or without the retraction of edges
	Hypo-attenuated Diaphragm: Decrease in attenuation of the diaphragm compared to surrounding structures

CT, Computed tomography.

Commonly associated injuries

Commonly associated injuries are visceral lacerations of the thoracal or abdominal organs and bone fractures. Further details are listed in Table 17.6. Most likely associated fractures are pelvic and rib fractures, followed by organ injuries involving the spleen, lungs, and liver. The most frequent visceral lacerations are splenic ruptures, pulmonary contusions, pneumothorax, liver lacerations, and cardiac contusions.

TABLE 17.5 Imaging mimics of diaphragmatic injury on computed tomography imaging.

Imaging mimics of diaphragmatic injury
Diaphragmatic hernia
Bochdalek hernia
Morgagni hernia
Hiatus hernia
Eventration of the diaphragm
Discontinuity of the lateral arcuate ligament

TABLE 17.6 Key clinical signs in the acute and delayed presentations of patients with diaphragmatic rupture.

Acute presentation	Delayed presentation	Commonly associated injuries
Hypotension due to decreased venous return secondary to compression from herniated abdominal contents	Recurrent chest infections	Rib fractures
	Chronic dyspnea	Spleen rupture
	Vomiting	Liver laceration
	Constipation	Abdominal wall laceration
Tachycardia	Recurrent abdominal pain	Gastric perforation
Altered mental status	Features of bowel obstruction (nausea, vomiting, abdominal distension, constipation)	Pneumothorax
Difficulty breathing		Hemothorax
Chest pain		Pulmonary contusion
Abdominal pain		Abdominal organ herniation
Pneumothorax		Pelvic fractures
Decreased breath sounds on the affected side		Spinal injuries
Bowel sounds in the chest due to herniation of abdominal organs into the thoracic cavity		
Shoulder pain (Kehr's sign) due to irritation of the phrenic nerve		

Rib fractures

Rib fractures are frequently associated with diaphragmatic rupture, occurring in approximately 60.7% of patients. The mechanism of injury that leads to rib fractures often involves significant blunt force trauma, which can also cause rupture of the diaphragm. Management of rib fractures is crucial in the overall care of these patients as it can lead to further complications like pneumothorax and hemothorax (Weber et al., 2022).

Spleen rupture

Spleen rupture is a commonly associated injury, particularly for patients who present with left-sided diaphragmatic rupture due to its anatomic proximity to the diaphragm. The splenic injury can range from a hematoma to a spleen rupture, which may necessitate splenectomy. The management of spleen rupture should be timely and appropriate to prevent life-threatening hemorrhage (Hofmann et al., 2012).

Liver laceration

In right-sided diaphragmatic ruptures, liver laceration can be a concurrent injury due to its close anatomical relationship with the diaphragm. The severity of liver injuries can vary widely, and these patients may require operative intervention for hemostasis and repair.

Gastric perforation

In cases of high-impact trauma or penetrating injuries, gastric perforation may occur, particularly if the stomach herniates through a diaphragmatic rupture and becomes strangulated. Prompt recognition and surgical management are vital to prevent sepsis and further complications.

Pneumothorax and hemothorax

Air and blood in the pleural space can accumulate, especially when accompanied by rib fractures or other chest injuries. It is noted that when diaphragmatic hernia and tension pneumothorax co-exist, the contents of the visceral sac may be fully retracted, which effectively hides the presence of the hernia. If a significant volume of serous fluid is drained along with air when dealing with tension pneumothorax, it may indicate a connection with the peritoneal cavity. These conditions can lead to respiratory compromise and require prompt management, often involving chest thoracostomy (Ramdass et al., 2006) (Fig. 17.3).

Abdominal organ herniation

Herniation of abdominal organs into the thoracic cavity can lead to the strangulation and ischemia of the herniated organs. The surgical repair of the diaphragmatic defect and careful evaluation of herniated organs are vital to prevent evisceration.

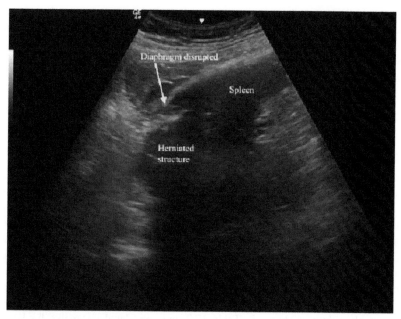

FIGURE 17.3 An ultrasound image depicting a disrupted diaphragm with loop structures (Tjhia & Noor, 2018).

Sepsis

Sepsis can be a serious complication associated with diaphragmatic rupture. The rupture can serve as a gateway for bacteria and other pathogens from the abdominal cavity to migrate into the thoracic cavity, thereby increasing the risk of infection. If left untreated, this infection can proliferate and culminate in sepsis. Moreover, any suspected infections in patients with a history of diaphragmatic rupture should be promptly addressed to reduce such risks.

Mechanisms of repair

The management of traumatic diaphragmatic injuries (TDI) includes a range of surgical approaches, including laparotomy, laparoscopy, and thoracoscopy. In addition, innovative techniques such as robotic transthoracic repair and VATS have proven successful in diaphragm repair. The choice of approach is contingent upon various factors including the patient's hemodynamic status, presence of comorbidities, surgeon's expertize and skills, and availability of hospital resources and facilities.

Laparoscopic approach

The laparoscopic approach to diaphragmatic repair has been demonstrated to serve as an exceptional diagnostic and therapeutic modality for mending suspected diaphragmatic ruptures irrespective of the presence or absence of abdominal organ herniation. This approach is particularly favored in the event of strong suspicions of a diaphragmatic rupture, despite inconclusive imaging findings by providing the distinct advantage of clear visualization of both hemidiaphragms. This enables thorough assessment of diaphragmatic integrity, alongside accurate identification of associated injuries within the abdominal cavity, facilitating simultaneous exploration and repair of the rupture site.

Who should be offered a laparoscopic approach?

Patient selectivity helps ensure procedural safety, minimizing complications and avoiding the need for converting to an open surgical approach. Ideally, patients who are in a stable hemodynamic condition and have undergone minimal previous abdominal surgeries are considered ideal candidates for a laparoscopic approach. During the procedure, carbon dioxide (CO_2) pneumoperitoneum is artificially created which increases IAP and aids in obtaining a clear view of the surgical site. This can lead to significant hemodynamic and ventilatory challenges (Srivastava & Niranjan, 2010). Patients who are classified as having Class II, III, or IV hypovolemic shock according to the Advanced Trauma Life Support algorithm; as shown in Table 17.7, are considered unsuitable for laparoscopy (Spahn et al., 2013) (Table 17.7).

TABLE 17.7 Classes of hypovolemic shock according to the Advanced Trauma Life Support algorithm.

	Class I	Class II	Class III	Class IV
Blood loss (mL)	Up to 750	750–1500	1500–2000	>2000
Blood loss (% blood volume)	Up to 15%	15%–30%	30%–40%	>40%
Pulse rate (bpm)	<100	100–120	120–140	>140
Systolic blood pressure	Normal	Normal	Decreased	Decreased
Pulse pressure (mmHg)	Normal or increased	Decreased	Decreased	Decreased
Respiratory rate	14–20	20–30	30–40	>35
Urine output (mL/h)	>30	20–30	5–15	Negligible
Mental status	Slightly anxious	Mildly anxious	Anxious, confused	Confused, lethargic
Initial fluid replacement	Crystalloid	Crystalloid	Crystalloid and blood	Crystalloid and blood

Patients with suspected central venous injuries, particularly traumatic brain injuries are also excluded from laparoscopic surgery, due to the elevated risk of increased intracranial pressure as a result of CO_2 insufflation. CO_2 has higher solubility in the blood which can exacerbate the adverse effects on the central nervous system in these patients.

Additionally, patients with severe respiratory failure, such as those with chronic obstructive pulmonary disease, and severe hypercapnia (excess carbon dioxide in the blood), are considered relative contraindications to pneumoperitoneum. The risk of developing malignant hypercapnia and toxic shock syndrome is increased in these individuals (Atkinson et al., 2017).

Laparoscopic repair of diaphragmatic ruptures raises a theoretical concern regarding the possibility of cardiac tamponade resulting from CO_2 insufflation. As a precautionary measure, in the majority of cases involving a laparoscopic approach, the level of pneumoperitoneum is carefully regulated to maintain a range of 10–12 cmH_2O. This controlled pneumoperitoneum pressure aims to prevent any potential adverse impacts on the heart that could lead to cardiac tamponade (Table 17.8).

TABLE 17.8 Data for 23 patients identified as having an exclusively laparoscopic approach for diaphragmatic rupture repair (Aborajooh & Al-Hamid, 2020; Bairwa & Anand, 2023; Dushnov et al., 2022; Gaillard et al., 2021; Gielis et al., 2022; Jain et al., 2022; Kori et al., 2022; Kuy et al., 2014; Marappan et al., 2017; Nguyen et al., 2017; Nho et al., 2021; Nishikawa et al., 2021; Safdar et al., 2013; Sala et al., 2017; Shah et al., 2013; Singh et al., 2020; Siow et al., 2016; Toh et al., 2020; Xenaki et al., 2014; Zarzavadjian Le Bian et al., 2014; Zubaidah et al., 2015).

Number of patients identified	23
Acute intervention	39.1%
Delayed intervention	60.9%
Gender	
Male	65.2%
Female	34.8%
Age	
20–30	21.7%
31–40	13.1%
41–50	17.4%
51–60	26.1%

(Continued)

TABLE 17.8 (Continued)

Number of patients identified	23
61 +	21.7%
Imaging used	
Chest X-ray (CXR)	65.2%
CT scan	95.7%
Other	8.7%
Unspecified	4.3%
Symptoms	
Trauma (including fractures)	26.1%
Chest pain	13.1%
Abdominal pain	47.8%
Dyspnea	13.1%
Constipation	4.3%
Cough	8.7%
Site swelling	8.7%
Vomiting	13.1%
Bruising	17.4%
Herniated organs	
Stomach	43.5%
Small intestine	39.1%
Large intestine	65.2%
Spleen	13.1%
Pancreas	13.1%
Omentum	34.8%
Liver	21.7%
Gallbladder	4.3%
Site of defect	
Left hemidiaphragm	73.9%
Right hemidiaphragm	26.1%

CT, Computed tomography.

Procedure

Typically, the patient is placed in a supine position on the operating table with a reverse Trendelenburg (head-up) tilt to 15 degrees—45 degrees. This positioning allows the patient's abdominal organs to move away from the diaphragm, facilitating better access and visualization of the defect. This creates a spacious area for maneuvering the laparoscopic trocars during surgical repair. Moreover, this position improves exposure by creating an angle that minimizes interference with the surgical field, thereby reducing the risk of injury to the abdominal organs during insertion and manipulation of laparoscopic instruments. The operating surgeon will evaluate the individual circumstances of the patient and determine the most appropriate degree of angulation for optimal access and safety during the procedure.

The number and positioning of the trocars can vary depending on surgeon preference and the specific site of diaphragmatic rupture with accessibility to the right hemidiaphragm laparoscopically, posing increased difficulty compared to the left. generally, 3—5 trocars are used with the first access best achieved at the umbilicus, with an open technique using a 10—12 mm trocar. This is the primary trocar and serves as the main access point for the laparoscopy, providing visual guidance during surgery. Two to four additional trocars are inserted (5 or 12 mm) under direct vision, generally achieving a configuration of an inverted V-shape pattern, according to the perioperative injury suspected location. For instance, in a left hemi-diaphragmatic rupture repair, the most common site for diaphragmatic ruptures, left lateral trocars may be used, usually in the midclavicular line or slightly anterior to it, to allow for manipulation of the left side of the diaphragm. This can be accompanied by a right lateral trocar to provide access and manipulation of the repair site. Additional trocars may be inserted if needed, depending on procedure complexity and the surgeon's judgment (Fig. 17.4).

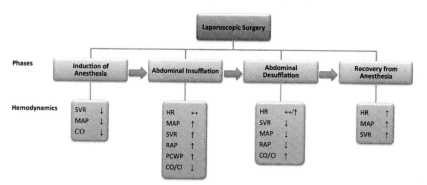

FIGURE 17.4 Phases of laparoscopic surgery and physiological effects on the body (Atkinson et al., 2017). The 4 phases; induction of anesthesia, abdominal insufflation, abdominal desufflation, and recovery from anesthesia each encompass unique hemodynamic and ventilatory changes. *CO*, Cardiac output; *CO/CI*, cardiac output/cardiac index; *HR*, heart rate; *MAP*, mean arterial pressure; *PCWP*, pulmonary capillary wedge pressure; *RAP*, right atrial pressure; *SVR*, systematic vascular resistance.

Right-sided diaphragmatic laparoscopic repair

The laparoscopic repair of right-sided diaphragmatic ruptures poses significant surgical challenges, mainly due to the presence of the liver, which frequently obstructs access to the site of injury. As mentioned previously, the liver may act as a protective barrier, resulting in fewer cases of right-sided diaphragmatic hernias in general. However, when these hernias do occur, they often carry high mortality rates. Right-sided repairs have a greater likelihood of converting to an open approach or resorting to a combined thoracoscopic approach to effectively repair the defect. This is necessary to ensure optimal surgical outcomes and reduce the risks associated with the repair process.

In the context of repairing right-sided diaphragmatic ruptures, it is more common to utilize additional trocars compared to left-sided repairs. Shichirireported a specific technique for the repair of acute right-sided diaphragmatic ruptures without liver herniation (hepatothorax) (Shichiri et al., 2020). This included a 5 mm port in the seventh intercostal space along the anterior axillary line for the right working hand, an 11 mm port in the eighth intercostal space along the anterior axillary line for the left working hand, and a 5 mm port in the seventh intercostal space along the mid-axillary line for a 5 mm endoscope (Fig. 17.5).

These carefully chosen trocar positions allowed for a minimally invasive approach to repairing the right-sided defect. Significantly, this technique deviated from the commonly employed reverse Trendelenburg position, by positioning the patient in lithotomy. Through this alternative positioning, coupled with precise placement of trocars, successful minimally invasive repair of the right-sided diaphragmatic defect was repaired in the acute setting.

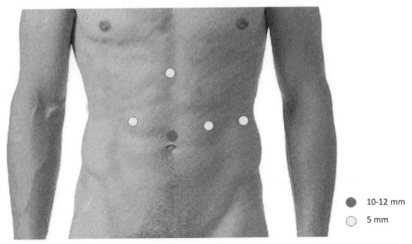

FIGURE 17.5 Example port placements of trocars for diaphragmatic rupture repair for left hemidiaphragm rupture (Gielis et al., 2022).

In comparison, Gaillard reported the successful use of the laparoscopic approach in repairing delayed right-sided diaphragmatic ruptures, accompanied by hepatothorax through a 3×3 cm defect (Gaillard et al., 2021). This achievement was made possible through the complete mobilization of the right liver lobe, which helped visualize the rupture. This allows full access to the anterior, lateral, and posterior aspects of the right hemidiaphragm. Similarly, Jain highlighted the advantages of a laparoscopic approach in repairing delayed right diaphragmatic ruptures (Jain et al., 2022). This approach was particularly beneficial in cases where standby procedures like cholecystectomy were required due to inflammation and wall-thickening of the gallbladder observed on imaging.

Left-sided diaphragmatic laparoscopic repair

Herniation of intra-abdominal structures can occur both acutely and in delayed left-sided diaphragmatic ruptures, with severe cases leading to complications such as bowel strangulation, incarceration, and obstruction. This further complicates the situation at hand. In cases of trauma, the abdominal approach is commonly chosen as it allows for the simultaneous management of concurrent intra-abdominal injuries, which are often prevalent, especially in cases with immediate presentation.

There are documented cases of abdominal viscera herniating into the pericardial sac, necessitating surgical intervention. Without intervention, the risk of life-threatening cardiac tamponade increases. Although these cases often present with non-specific symptoms, and exhibit delayed onset, successful repairs have been achieved using a laparoscopic approach. Repair methods for intrapericardial involvement have been reported to include thoracotomy, laparotomy, and abdominal laparoscopy with either primary or patch repair (Aborajooh & Al-Hamid, 2020; Dushnov et al., 2022; El Bakouri et al., 2021; Kuy et al., 2014; Shabhay et al., 2020).

Other minimally invasive approaches

Video-assisted thoracoscopic surgery

VATS may be considered for diaphragmatic rupture repair in specific cases, depending on factors such as rupture size, location size, patient stability, and surgeon expertize. VATS is particularly suitable for small ruptures in stable patients, allowing repair through small incisions while providing enhanced visualization of the surgical field. Lee et al. described a single-port VATS technique using a laparoscopic surgery port placed at a 4 cm incision in the anterior axillary line on the right-hand side in the seventh intercostal space (Lee et al., 2019). CO_2 insufflation at a pressure ranging from 8 to 12 mmHg was used to reduce the liver herniation. This approach is suitable in patients who have an absence of intra-abdominal injury with weak adhesions surrounding herniated organs. The technique, adapted from esophagectomy and diaphragmatic plication, relies on the surgical center's expertize and experience. It demonstrates that CO_2-assisted reduction of herniated organs eliminates the need for additional instruments, usually required for repositioning during VATS, minimizing organ damage, compared to manual reduction in open surgery. In cases of delayed diagnosis in stable patients, VATS can be a less invasive choice of repair.

In addition to the previously mentioned VATS approach, other cases in the literature have reported the use of a 3-port CO_2 VATS approach for diaphragmatic rupture repair (Nardini et al., 2017). This approach involves a 5 mm camera port and two instrument ports, either both 5 mm or one 5 mm and one 10 mm port positioned posterior and anterior to the first port. These ports serve as a flexible thoracoscope and grasping forceps. Patients are placed in a semi-lateral position under general anesthesia with endotracheal intubation. In some instances, VATS can be employed even in the presence of herniated abdominal contents. Kunz et al. reported a case where the VATS approach was unable to reduce a hernia or repair the defect directly, however, the approach provided direct visualization of the defect, facilitating subsequent smaller thoracotomy incisions (Kunz et al., 2017). This approach resulted in reduced postoperative hospital stay for the patient, potentially even shorter than if a laparoscopic approach had been attempted.

These variations in the VATS technique demonstrate the adaptability of the procedure to different scenarios and highlight its potential advantages in terms of reduced invasiveness and direct visualization (Kamiyoshihara et al., 2016). VATS can serve as a valuable tool in cases of diagnostic uncertainty, allowing surgeons to visually assess the presence and extent of the rupture (Brekke et al., 2021). It is crucial to acknowledge that VATS is not suitable for all diaphragmatic ruptures. Larger, complex, or unstable ruptures may require an open surgical approach or laparoscopy to provide better exposure and perform more extensive repairs. Additionally, the availability of VATS may vary amongst healthcare facilities and surgeons, as not all hospitals may offer this approach.

Robotic transthoracic approach

The robotic transthoracic approach for diaphragmatic defect repair has been adapted from the correction of congenital diaphragmatic abnormalities (Counts et al., 2018). It has been successful in cases where patients did not have intra-abdominal herniation or significant injuries that would deter them from this procedure but instead experienced herniation of the liver. The repair of diaphragmatic defects using this approach resulted in tension-free re-approximation and plication of redundant diaphragm. It has proven to be a reproducible and teachable operation.

The patient is usually positioned in the lateral decubitus position. Between 3 and 5 port sites are usually used, including a 12 mm robotic camera port incision and an additional 8 mm robotic trocars. The Da Vinci Si robot is usually positioned posterior to the patient, with the docking angle approximately 30 degrees from the feet. Through the robotic console, the surgeon can control the robotic arms, which offer improved dexterity and precision during the procedure.

This novel procedure translates the surgeon's hand movements into precise movements of the surgical instruments inside the patient's body. To repair the diaphragmatic rupture, the surgeon first identifies the location and extent of the rupture by visualization through the robotic camera. The edges of the rupture are then carefully dissected and cleaned to prepare for repair in which depending on the size and characteristics of the rupture, different techniques may be used for repair.

The surgeon performs meticulous repair by manipulating the tissues with precision, resulting in a fully intact and functional diaphragm. Successful utilization of this approach has been reported by Kim, in the repair of right-sided diaphragmatic defects, including a defect measuring 9×11 cm (Kim et al., 2020). This technique provides optimized three-dimensional visualization and improved suturing dexterity due to the technical superiority of instruments with multiple degrees of freedom.

Open approach

Laparotomy

Laparotomy has traditionally been the primary approach for repairing diaphragmatic ruptures, both in the acute and delayed setting, classically constituting a midline incision in the abdomen. Since the emergence of minimally invasive techniques, there has been a shift in the surgical landscape with laparotomy now used selectively to minimize the potential complications associated with its invasiveness. The approach is increasingly reserved for cases where its specific advantages are required while opting for less invasive procedures where possible.

Who would be offered a laparotomy approach?

In the acute phase of diaphragmatic rupture, a laparotomy approach is used, when the patient exhibits hemodynamic instability and there is either evidence of additional intra-abdominal injuries, abdominal contents herniating into the thorax, as evidenced on imaging, or associated injuries indicated by the patient's signs and symptoms (Alsuwayj et al., 2021; Beshay et al., 2021; Conti et al., 2021). Laparotomy provides a comprehensive exploration of the abdominal organs, facilitating the reduction of herniated tissue and repair of the diaphragm simultaneously. Laparotomy is also particularly advantageous in cases involving large or complex ruptures. In situations where there is bilateral diaphragmatic rupture, the accessibility to both hemidiaphragms becomes even more crucial. However, in cases of chronic presentations with adhesions between abdominal contents and pleura, combined approaches such as thoracotomy with laparotomy may be necessary (Chatzoulis et al., 2013; Haranal et al., 2017; Mubashir et al., 2022; Pace et al., 2021; Quadrozzi et al., 2016).

In specific situations, a combined approach involving VATS followed by open surgery may be utilized when managing patients who are hemodynamically unstable but exhibit strong suspicions of diaphragmatic rupture. The combined approach ensures careful and thorough assessment of the diaphragm before making a definitive decision, thus minimizing risks of unnecessary laparotomies and optimizing patient management.

Intraoperative considerations

During the acute phase, a meticulous exploration of the abdominal viscera is performed to identify associated injuries and determine the extent of the diaphragmatic rupture. Any herniated contents are carefully reduced. The repair of the diaphragmatic rupture involves utilizing sutures or mesh reinforcement, depending on the size and complexity of the defect. If additional injuries or complications are present, they are addressed within the surgical procedure or may require subsequent interventions scheduled accordingly.

Laparotomy in delayed cases becomes more likely when there is evidence of adhesions between the abdominal contents and the pleura (Alsuwayj et al., 2021; Sauer Durand et al., 2021; Soomro et al., 2022). This scenario would then require careful dissection and exposure of the diaphragmatic defect to ensure repair. Over time, negative pressure in the thoracic cavity creates suction forces that gradually lead to the adherence of abdominal contents with the pleura and/or thoracic contents. To achieve optimal repair and gain access to sites of adhesions at the interface between the abdominal and thoracic cavities, it is important to note that the specific approach and interventions may vary based on the individual patient factors, expertize of the surgical teams, and institutional protocols.

Thoracotomy

A thoracotomy approach is a type of open approach whereby an incision is created through the chest wall to directly access the diaphragm to repair injuries. It is particularly advantageous for injuries located in the upper region of the diaphragm as the incision allows for better exposure in this area, aiding accurate repair. In cases of significant bleeding accompanying the diaphragmatic rupture, the thoracotomy approach facilitates effective control of bleeding. This approach, like a laparotomy, is also favorable in patients with hemodynamic instability (Chen & Cheng, 2014).

Thoracotomy can be indicated in patients with delayed presentations, whilst laparotomy is usually favored in patients presenting immediately after trauma (Conti et al., 2021; Maldonado-Chaar et al., 2022). Successful cases utilizing a thoracotomy approach include patients with a right-sided diaphragmatic rupture, involving the liver herniating into the thoracic space. Due to the liver obstructing the optimal field of view in an open approach, thoracotomy can allow for better visualization of right-sided ruptures, without the need for a laparotomy (Shabhay et al., 2020).

In some cases, additional thoracic injuries can occur in conjunction with a diaphragmatic rupture, such as lung contusions or chest wall fractures. This approach allows for simultaneous, focused repair of multiple injuries, minimizing overall surgical trauma, and expediting recovery.

Primary repair versus mesh

Primary repair

Primary repair involves direct suturing of lacerated or breached tissue, devoid of patches or grafts. Meticulous suturing of the torn margins of the diaphragmatic musculature can be reinforced with additional sutures in layers (Chatzoulis et al., 2013). Non-absorbable sutures such as polypropylene, or Nylon are commonly used due to their inherent strengths.

Smaller defects tend to be repaired with primary repair, whilst larger, more anatomically complex defects tend to be repaired with mesh. Reinforcement with pledgets is sometimes utilized in the event of further diaphragmatic muscular injury or tearing during repair (Counts et al., 2018). This aims to reinforce the sutured defect and minimize tearing when tightening during repair.

Mesh

Comparatively, mesh repair allows for tension-free closure of larger defects. Synthetic mesh can be used to bridge the gap created by the torn diaphragmatic defect to strengthen and stabilize the repair and reduce the risk of adhesions and erosion of nearby viscera. The mesh acts as a scaffold, supporting tissue regeneration and provides long-term reinforcement to prevent recurrence (Aborajooh & Al-Hamid, 2020; Ahmed et al., 2018; De Nadai et al., 2015; Iadicola et al., 2019). The choice of mesh is dependent on various factors including the size and location of the diaphragmatic defect. Four main types of mesh can be used for repair. Polypropylene (Prolene), Polyester (Dacron), polytetrafluoroethylene (Gore-Tex), and composite meshes.

Mesh reinforcement, although favorable in larger ruptures can introduce risks of infection, mesh migration, or adhesion formation with neighboring tissue. Therefore, careful consideration is required to be tailored to individual patient's needs, the clinical picture taking into account surgical expertize and postoperative complications.

References

Aborajooh, E. A., & Al-Hamid, Z. (2020). Case report of traumatic intrapericardial diaphragmatic hernia: Laparoscopic composite mesh repair and literature review. *International Journal of Surgery Case Reports*, *70*, 159−163. Available from https://doi.org/10.1016/j.ijscr.2020.04.077, http://www.elsevier.com/wps/find/journaldescription.cws_home/723449/description#description.

Ahmed, H. H., Afzal, S., & Jamaluddin, M. (2018). Traumatic diaphragmatic hernia in a 40-year-old lady after 34 years of trauma. *Annals of Abbasi Shaheed Hospital and Karachi Medical & Dental College*, *23*(3), 161−166.

Al-Refaie, R. E., Awad, E., & Mokbel, E. M. (2009). Blunt traumatic diaphragmatic rupture: A retrospective observational study of 46 patients. *Interactive Cardiovascular and Thoracic Surgery*, *9*(1), 45−49. Available from https://doi.org/10.1510/icvts.2008.198333.

Alsuwayj, A. H., Al Nasser, A. H., Al Dehailan, A. M., Alburayman, A. Z., Alhuwaiji, K. A., Binsifran, K. F., Almulhim, I. M., Almulhim, A. F., Al Amer, M. A., Almulhim, M. A., Almulhim, A. Y., Almulhim, A. A., Alhazoom, I. A., Albakheet, A. A., & Al-Hawaj, F. (2021). Giant traumatic diaphragmatic hernia: A report of delayed presentation. *Cureus*. Available from https://doi.org/10.7759/cureus.20315.

Atkinson, T. M., Giraud, G. D., Togioka, B. M., Jones, D. B., & Cigarroa, J. E. (2017). Cardiovascular and ventilatory consequences of laparoscopic surgery. *Circulation*, *135*(7), 700−710. Available from https://doi.org/10.1161/CIRCULATIONAHA.116.023262, http://circ.ahajournals.org.

Bairwa, B. L., & Anand, A. (2023). Laparoscopic management of traumatic diaphragmatic rupture with herniation of abdominal contents into left hemithorax. *Clinical Case Reports*, *11*(5). Available from https://doi.org/10.1002/ccr3.7385.

Beshay, M., Krüger, M., Singh, K., Borgstedt, R., Benhidjeb, T., Bölke, E., Vordemvenne, T., & Schulte am Esch, J. (2021). Grave thoraco-intestinal complication secondary to an undetected traumatic rupture of the diaphragm: A case report. *European Journal of Medical Research*, *26*(1). Available from https://doi.org/10.1186/s40001-021-00488-9.

Betts, G. (2022). *Anatomy and physiology 2e* (p. 2022) OpenStax.

Brekke, I. J., Maidas, P., & Møller, L. (2021). Forsinket pleuraeffusjon etter thoraxtraume. *Tidsskrift for Den norske legeforening*, *141*(3). Available from https://doi.org/10.4045/tidsskr.20.0717.

Chatzoulis, G., Papachristos, I. C., Daliakopoulos, S. I., Chatzoulis, K., Lampridis, S., Svarnas, G., & Katsiadramis, I. (2013). Septic shock with tension fecothorax as a delayed presentation of a gunshot diaphragmatic rupture. *Journal of Thoracic Disease*, *5*(5), E195. Available from https://doi.org/10.3978/j.issn.2072-1439.2013.08.63Greece, http://www.jthoracdis.com/article/download/1519/pdf.

Chen, C.-L., & Cheng, Y.-L. (2014). Delayed massive hemothorax complicating simple rib fracture associated with diaphragmatic injury. *The American Journal of Emergency Medicine*, *32*(7), 818.e3. Available from https://doi.org/10.1016/j.ajem.2013.12.060.

Chughtai, T., Ali, S., Sharkey, P., Lins, M., & Rizoli, S. (2009). Update on managing diaphragmatic rupture in blunt trauma: A review of 208 consecutive cases. *Canadian Medical Association, Canada Canadian Journal of Surgery*, *52*(3), 177−181. Available from http://www.cma.ca/multimedia/staticContent/HTML/N0/l2/cjs/vol-52/issue-3/pdf/pg182.pdf.

Conti, L., Grassi, C., Delfanti, R., Cattaneo, G. M., Banchini, F., & Capelli, P. (2021). Left diaphragmatic rupture in vehicle trauma: Report of surgical treatment and complications of two consecutive cases. *Acta Biomedica*, *92*(1). Available from https://doi.org/10.23750/abm.v92iS1.10931, https://www.mattioli1885journals.com/index.php/actabiomedica/article/download/10931/9696.

Counts, S. J., Saffarzadeh, A. G., Blasberg, J. D., & Kim, A. W. (2018). Robotic transthoracic primary repair of a diaphragmatic hernia and reduction of an intrathoracic liver. *Innovations: Technology and Techniques in Cardiothoracic and Vascular Surgery*, *13*(1), 54–55. Available from https://doi.org/10.1097/imi.0000000000000455, https://journals.sagepub.com/home/INV.

De Nadai, T. R., Lopes, J. C. P., Inaco Cirino, C. C., Godinho, M., Rodrigues, A. J., & Scarpelini, S. (2015). Diaphragmatic hernia repair more than four years after severe trauma: Four case reports. *International Journal of Surgery Case Reports*, *14*, 72–76. Available from https://doi.org/10.1016/j.ijscr.2015.07.014, http://www.elsevier.com/wps/find/journaldescription.cws_home/723449/description#description.

Dushnov, V., McNicholas, M., Su, M., Keita, P., & Moore, H. (2022). Small bowel does not belong in the pericardium: Case of a traumatic intrapericardial diaphragmatic hernia. *Cureus*. Available from https://doi.org/10.7759/cureus.32966.

El Bakouri, A., El Karouachi, A., Bouali, M., El Hattabi, K., Bensardi, F. Z., & Fadil, A. (2021). Post-traumatic diaphragmatic rupture with pericardial denudation: A case report. *International Journal of Surgery Case Reports*, *83*, 105970. Available from https://doi.org/10.1016/j.ijscr.2021.105970.

Gaillard, M., Tranchart, H., & Dagher, I. (2021). Laparoscopic repair for delayed right-sided traumatic diaphragmatic rupture (with video). *Surgery Open Digestive Advance*, *1*, 100004. Available from https://doi.org/10.1016/j.soda.2021.100004.

Ganie, F. A., Lone, H., Lone, G. N., Wani, M. L., Ganie, S. A., Wani, N. U. D., & Gani, M. (2013). Delayed presentation of traumatic diaphragmatic hernia: A diagnosis of suspicion with increased morbidity and mortality. *Trauma Monthly*, *18*(1), 12–16. Available from https://doi.org/10.5812/traumamon.7125, http://traumamon.com/?page=download&file_id=17696.

Gielis, M., Bruera, N., Pinsak, A., Olmedo, I., Fabián, P. W., & Viscido, G. (2022). Laparoscopic repair of acute traumatic diaphragmatic hernia with mesh reinforcement: A case report. *International Journal of Surgery Case Reports*, *93*, 106910. Available from https://doi.org/10.1016/j.ijscr.2022.106910.

Haranal, M. Y., Buggi, S., & Sanjeevaiah, S. (2017). Tension fecopneumothorax secondary to unrecognized delayed traumatic diaphragmatic hernia. *Indian Journal of Thoracic and Cardiovascular Surgery*, *33*(1), 58–60. Available from https://doi.org/10.1007/s12055-016-0465-y, http://www.springer.com/medicine/surgery/journal/12055.

Hofmann, S., Kornmann, M., Henne-Bruns, D., & Formentini, A. (2012). Traumatic diaphragmatic ruptures: Clinical presentation, diagnosis and surgical approach in adults. *GMS Interdisciplinary Plastic and Reconstructive Surgery DGPW*, *1*. Available from https://doi.org/10.3205/iprs0000022012.

Iadicola, D., Branca, M., Lupo, M., Grutta, E. M., Mandalà, S., Cocorullo, G., & Mirabella, A. (2019). Double traumatic diaphragmatic injury: A case report. *International Journal of Surgery Case Reports*, *61*, 82–85. Available from https://doi.org/10.1016/j.ijscr.2019.07.030, http://www.elsevier.com/wps/find/journaldescription.cws_home/723449/description# description.

Jain, N., Raju, B. P., Dhanda, S., Johri, V., Reddy, P. K., & Jameel, J. K. A. (2022). Delayed presentation of a post-traumatic large right diaphragmatic hernia displacing liver and gallbladder – A case report. *Asian Journal of Endoscopic Surgery*, *15*(2), 388–392. Available from https://doi.org/10.1111/ases.13015, https://onlinelibrary.wiley.com/loi/17585910.

Jones, O. (2024). *The diaphragm*. Teach Me Anatomy, 2024.

Kamiyoshihara, M., Igai, H., Kawatani, N., & Ibe, T. (2016). A minimally invasive technique for stabilizing the diaphragm on the thoracic wall after blunt chest trauma: The "lifting-up method". *Surgery Today*, *46*(7), 872–875. Available from https://doi.org/10.1007/s00595-015-1249-5.

Kim, J. K., Desai, A., Kunac, A., Merchant, A. M., & Lovoulos, C. (2020). Robotic transthoracic repair of a right-sided traumatic diaphragmatic rupture. *The Surgery Journal*, *06*(03), e164. Available from https://doi.org/10.1055/s-0040-1716330.

Kori, M., Endo, H., Yamamoto, K., Awano, N., & Takehana, T. (2022). Laparoscopic repair and total gastrectomy for delayed traumatic diaphragmatic hernia complicated by intrathoracic gastric perforation with tension empyema: A case report. *Surgical Case Reports*, *8*(1). Available from https://doi.org/10.1186/s40792-022-01477-8.

Kunz, S., Goh, S. K., Stelmach, W., & Seevanayagam, S. (2017). Traumatic rupture of the diaphragm resulting in the sub-acute presentation of an incarcerated intra-thoracic transverse colon. *Journal of Surgical Case Reports*, *2017*(3). Available from https://doi.org/10.1093/jscr/rjx057, https://academic.oup.com/jscr.

Kuy, S., Juern, J., & Weigelt, J. A. (2014). Laparoscopic repair of a traumatic intrapericardial diaphragmatic hernia. *Journal of the Society of Laparoendoscopic Surgeons*, *18*(2), 333–337. Available from https://doi.org/10.4293/108680813X13753907290955, http://docserver.ingentaconnect.com/deliver/connect/sls/10868089/v18n2/s27.pdf.

Lee, J. H., Han, K. N., Hong, J. I., & Kim, H. K. (2019). A single-port video-assisted thoracoscopic surgery with CO2 insufflation for traumatic diaphragmatic hernia. *Interactive Cardiovascular and Thoracic Surgery*, *29*(5), 808–810. Available from https://doi.org/10.1093/icvts/ivz173, http://icvts.oxfordjournals.org/.

Lim, K. H., & Park, J. (2018). Blunt traumatic diaphragmatic rupture Single-center experience with 38 patients. *Medicine (United States)*, *97*(41). Available from https://doi.org/10.1097/MD.0000000000012849, https://journals.lww.com/md-journal/pages/default.aspx.

Maldonado-Chaar, S. M., Miró-González, Á. A., Ramírez, N., & Ramirez-Ferrer, L. O. (2022). Delayed hepatothorax: An unusual presentation case report. *International Journal of Surgery Case Reports*, *94*. Available from https://doi.org/10.1016/j.ijscr.2022.107017, http://www.elsevier.com/wps/find/journaldescription.cws_home/723449/description#description.

Marappan, A., Burud, I., & Tata, M. D. (2017). Laparoscopic repair of acquired abdominal intercostal hernia. *Rawal Medical Journal*, *42*(3), 444–445. Available from https://www.ejmanager.com/mnstemps/27/27-1469716686.pdf?t = 1506153407.

Morgan, B., Watcyn-Jones, T., & Garner, J. (2010). Traumatic diaphragmatic injury. *Journal of the Royal Army Medical Corps*, *156*(3), 139–144. Available from https://doi.org/10.1136/jramc-156-03-02.

Mubashir, M., Barron, J. O., Mubashir, H., DeMare, A., Raja, S., Murthy, S., & Schraufnagel, D. P. (2022). Thoracoabdominal approach for traumatic diaphragmatic hernia in a hemodynamically unstable patient. *European Surgery – Acta Chirurgica Austriaca*, *54*(6), 331–334. Available from https://doi.org/10.1007/s10353-022-00782-8, https://www.springer.com/journal/10353.

Nardini, M., Jayakumar, S., Elsaegh, M., & Dunning, J. (2017). Left video-assisted thoracoscopic surgery for hemidiaphragm traumatic rupture repair. *Interactive Cardiovascular and Thoracic Surgery*, *24*(5), 815–816. Available from https://doi.org/10.1093/icvts/ivw448.

Nason, L. K., Walker, C. M., Mcneeley, M. F., Burivong, W., Fligner, C. L., & Godwin, J. D. (2012). Imaging of the diaphragm: Anatomy and function. *Radiographics*, *32*(2), E51. Available from https://doi.org/10.1148/rg.322115127, http://radiographics.rsna.org/content/32/2/E51.full.pdf#page = 1&view = FitH.

Nguyen, P., Davis, B., & Tran, D. D. (2017). Laparoscopic repair of diaphragmatic rupture: A Case report with radiological and surgical correlation. *Case Reports in Surgery*, *2017*, 1–4. Available from https://doi.org/10.1155/2017/4159108.

Nho, W. Y., Kim, J. O., Nam, S. Y., & Kee, S. K. (2021). Laparoscopic repair of transdiaphragmatic intercostal hernia following blunt trauma. *Trauma (United Kingdom)*, *23*(4), 351−355. Available from https://doi.org/10.1177/14604086211005211, https://journals.sagepub.com/home/TRA.

Nishikawa, S., Miguchi, M., Nakahara, H., Urushihara, T., Egi, H., Shorin, D., Fukuda, H., & Itamoto, T. (2021). Laparoscopic repair of traumatic diaphragmatic hernia with colon incarceration: A case report. *Asian Journal of Endoscopic Surgery*, *14*(2), 258−261. Available from https://doi.org/10.1111/ases.12843.

Pace, M., Vallati, D., Belloni, E., Cavallini, M., Ibrahim, M., Rendina, E. A., & Nigri, G. (2021). Blunt trauma associated with bilateral diaphragmatic rupture: A case report. *Frontiers in Surgery*, *8*. Available from https://doi.org/10.3389/fsurg.2021.772913, http://journal.frontiersin.org/journal/surgery.

Quadrozzi, F., Favoriti, P., Favoriti, M., & Cofini, G. (2016). Unusual repair in a rare case of hepatothorax due to right-sided diaphragmatic rupture: Case report. CIC Edizioni Internazionali s.r.l., Italy. *Giornale di Chirurgia*, *37*(2), 84−85. Available from https://doi.org/10.11138/gchir/2016.37.2.084, http://www.giornalechirurgia.it/common/php/portiere.php?ID = a92d509ae64af87bb94909e2da28cd24.

Rajaretnam, N., Okoye, E., & Burns, B. (2024). *Laparotomy*, 2024.

Ramdass, M. J., Kamal, S., Paice, A., & Andrews, B. (2006). Traumatic diaphragmatic herniation presenting as a delayed tension faecopneumothorax. *Emergency Medicine Journal: EMJ*, *23*(10), e54. Available from https://doi.org/10.1136/emj.2006.039438.

Rashid, F., Chakrabarty, M. M., Singh, R., & Iftikhar, S. Y. (2009). A review on delayed presentation of diaphragmatic rupture. *World Journal of Emergency Surgery*, *4*(1). Available from https://doi.org/10.1186/1749-7922-4-32.

Safdar, G., Slater, R., & Garner, J. P. (2013). Laparoscopically assisted repair of an acute traumatic diaphragmatic hernia. *BMJ Case Reports*, *2013*(June 24), 1. Available from https://doi.org/10.1136/bcr-2013-009415, bcr2013009415.

Sala, C., Bonaldi, M., Mariani, P., Tagliabue, F., & Novellino, L. (2017). Right post-traumatic diaphragmatic hernia with liver and intestinal dislocation. *Journal of Surgical Case Reports*, *2017*(3). Available from https://doi.org/10.1093/jscr/rjw220.

Sauer Durand, A. M., Nebiker, C. A., Hartel, M., & Kremer, M. (2021). Bilateral delayed traumatic diaphragmatic injury. *Journal of Surgical Case Reports*, *2021*(4). Available from https://doi.org/10.1093/jscr/rjab052, https://academic.oup.com/jscr.

Shabhay, A., Horumpende, P., Shabhay, Z., Van Baal, S. G., Lazaro, E., Chilonga, K., & Kirshtein, B. (2020). Surgical approach in management of posttraumatic diaphragmatic hernia: Thoracotomy versus laparotomy. *Case Reports in Surgery*, *2020*, 1−4. Available from https://doi.org/10.1155/2020/6694990.

Shah, N., Fernandes, R., Thakrar, A., & Rozati, H. (2013). Diaphragmatic hernia: An unusual presentation. *Case Reports*, *2013*(April 23), 1. Available from https://doi.org/10.1136/bcr-2013-008699, bcr2013008699.

Shichiri, K., Imamura, K., Takada, M., & Anbo, Y. (2020). Minimally invasive repair of right-sided blunt traumatic diaphragmatic injury. *BMJ Case Reports*, *13*(11), e235870. Available from https://doi.org/10.1136/bcr-2020-235870.

Simon, L. V., Lopez, R. A., & Burns, B. (2024). *Diaphragm rupture*, 2024.

Singh, D., Aggarwal, S., & Vyas, S. (2020). Laparoscopic repair of recurrent traumatic diaphragmatic hernia. *Journal of Minimal Access Surgery*, *16*(2), 166−168. Available from https://doi.org/10.4103/jmas.JMAS_298_18, http://www.journalofmas.com.

Siow, S. L., Wong, C. M., Hardin, M., & Sohail, M. (2016). Successful laparoscopic management of combined traumatic diaphragmatic rupture and abdominal wall hernia: A case report. *Journal of Medical Case Reports*, *10*(1). Available from https://doi.org/10.1186/s13256-015-0780-8, http://www.jmedicalcasereports.com/articles/browse.asp.

Soomro, F. H., Hassan, A., Nazir, I., Azam, S., & Yasmin, A. (2022). Intra-thoracic symptomatic gallstones in a right-sided post-traumatic diaphragmatic hernia: A case report. *Cureus*. Available from https://doi.org/10.7759/cureus.32824.

Spahn, D. R., Bouillon, B., Cerny, V., Coats, T. J., Duranteau, J., Fernández-Mondéjar, E., Filipescu, D., Hunt, B. J., Komadina, R., Nardi, G., Neugebauer, E., Ozier, Y., Riddez, L., Schultz, A., Vincent, J. L., & Rossaint, R. (2013). Management of bleeding and coagulopathy following major trauma: An updated European guideline. *Critical Care*, *17*(2). Available from https://doi.org/10.1186/cc12685, http://ccforum.com/content/17/2/R76.

Srivastava, A., & Niranjan, A. (2010). Secrets of safe laparoscopic surgery: Anaesthetic and surgical considerations. *Journal of Minimal Access Surgery*, *6*(4), 91−94. Available from https://doi.org/10.4103/0972-9941.72593.

Tjhia, J., & Noor, J. M. (2018). Beyond E-FAST scan in trauma: Diagnosing of traumatic diaphragmatic rupture with bedside ultrasound. *Journal of Emergency Medicine*, *25*(3), 163−165. Available from https://doi.org/10.1177/102490791774523, https://journals.sagepub.com/home/HKJ.

Toh, P. Y., Parys, S., & Watanabe, Y. (2020). Traumatic diaphragmatic rupture: Delayed presentation following a SCUBA dive. *BMJ Case Reports*, *13*(9). Available from https://doi.org/10.1136/bcr-2019-234040, http://casereports.bmj.com/.

Weber, C., Willms, A., Bieler, D., Schreyer, C., Lefering, R., Schaaf, S., Schwab, R., Kollig, E., & Güsgen, C. (2022). Traumatic diaphragmatic rupture: Epidemiology, associated injuries, and outcome—An analysis based on the Trauma Register DGU®. *Langenbeck's Archives of Surgery*, *407*(8), 3681−3690. Available from https://doi.org/10.1007/s00423-022-02629-y, https://www.springer.com/journal/423.

Xenaki, S., Lasithiotakis, K., Andreou, A., Chrysos, E., & Chalkiadakis, G. (2014). Laparoscopic repair of posttraumatic diaphragmatic rupture. Report of three cases. *International Journal of Surgery Case Reports*, *5*(9), 601−604. Available from https://doi.org/10.1016/j.ijscr.2014.07.007.

Zarzavadjian Le Bian, A., Costi, R., & Smadja, C. (2014). Delayed right-sided diaphragmatic rupture and laparoscopic repair with mesh fixation. *Annals of Thoracic and Cardiovascular Surgery: Official Journal of the Association of Thoracic and Cardiovascular Surgeons of Asia*, *20*(Supplement), 550−553. Available from https://doi.org/10.5761/atcs.cr.12.02065.

Zubaidah, N. H., Azuawarie, A., Ong, K. W., & Gee, T. (2015). Combined laparoscopic and thoracoscopic repair of a large traumatic diaphragmatic hernia: A case report. *The Medical Journal of Malaysia*, *70*(1), 57−58.

Chapter 18

Traumatic pericardial effusion and cardiac tamponade

Filippo Antonacci[1], Ahmed Mohamed Osman[2] and Piergiorgio Solli[3]
[1]*Unit of Thoracic Surgery, IRCCS Az Univ Ospedaliera Policlinico S. Orsola, Bologna, Italy,*
[2]*Department of Thoracic Surgery, Nottingham University Hospital, Nottingham, United Kingdom,* [3]*Unit of Thoracic Surgery, Royal Papworth Hospital NHS Foundation Trust, Cambridge, United Kingdom*

Introduction

Trauma represents a significant global health concern. It is the third leading cause of death across all age groups, following cardiovascular diseases and cancer. Furthermore, it is an important cause of morbidity and long-term health problems. It is also important to note that the impact of trauma on mortality and morbidity varies across different age groups (Dogrul et al., 2020; Marro et al., 2019), in among the younger population (those under 40 years of age), trauma stands as the primary cause of death as this demographic is particularly vulnerable to accidents, violence, and other traumatic incidents. Moreover, of all the trauma-related injuries, approximately 25% of trauma patients succumb to thoracic injuries or complications arising from them (Dogrul et al., 2020; Mattox et al., n.d.).

Multiple mechanisms of chest injuries have been described and generally speaking, 5 mechanisms of chest injuries have been observed, including (1) penetrating thoracic trauma, (2) blunt thoracic trauma, (3) crush mechanism, (4) deceleration thoracic trauma, (5) blast mechanism (Lasek & Jadczuk, 2020).

Furthermore, with regards to blunt chest trauma, four mechanisms have been identified: (1) direct impact injuries to the chest, (2) chest compression or crush injuries, (3) acceleration/deceleration injuries, and (4) blast injuries (Dogrul et al., 2020). The damage attributed to thoracic trauma can be severe, causing damage to various thoracic structures, leading to cardiac/pulmonary contusions, rib fractures, and thoracic spine fractures, among others.

In cases of polytrauma, where multiple injuries are sustained simultaneously, around 60% of patients experience chest trauma, and when assessing patients with thoracic trauma, a primary survey to identify and address life-threatening conditions promptly must be carried out. These life-threatening injuries include airway obstruction, tracheobronchial tree injury, tension pneumothorax, open pneumothorax, massive hemothorax, cardiac tamponade, and traumatic circulatory arrest (ATLS, 2018; Dogrul et al., 2020).

Blunt traumatic pericardial effusion, while rare in trauma cases (accounting for around 0.5% of trauma admissions) carries a high mortality if it results from cardiac rupture, with an overall mortality rate of approximately 90%. Cardiac rupture is not common, accounting for 0.16%−2% of trauma admissions. However, around 80% succumb to the injury before reaching the hospital. Among patients who survive in the hospital after experiencing blunt cardiac rupture, the reported mortality rate is 89%. These patients present with shock and signs of tamponade, highlighting the need for swift diagnosis and treatment (Huang et al., 2018; Tanizaki et al., 2018).

Cardiac tamponade occurs when the pericardium fills with fluid leading to compression of the heart and impairment of cardiac function (Dogrul et al., 2020; Mattox et al., n.d.). With regards to mortality rates among patients with thoracic trauma following motor vehicle accidents (10%−70%), cardiac tamponade plays a significant contributory role in these deaths.

The effective management of traumatic pericardial effusions and cardiac tamponade relies on the understanding and recognition of their clinical signs and symptoms, as well as proper and timely utilization of appropriate diagnostic tools, and early involvement of experienced thoracic and cardiac surgeons.

Clinical findings

Pericardial effusion is the accumulation of fluid in the pericardium. The amount of fluid required to classify as an effusion typically ranges from 15 to 50 milliliters (Alerhand et al., 2022).

Cardiac tamponade, on the other hand, is a clinical condition characterized by hemodynamic instability, often presenting with symptoms such as hypotension or dyspnea. Its development depends on the rate at which fluid accumulates. In cases of chronic effusion, where fluid builds up over time, the pericardium can gradually stretch and accommodate up to two liters of fluid. However, in acute settings such as trauma, the pericardium's capacity for stretch is limited, and even a smaller amount of fluid, such as 50 mL, can lead to the development of tamponade (Alerhand et al., 2022).

In tamponade, the right atrium and right ventricle are the first to collapse. This is because the right side of the heart typically operates at lower pressure compared to the left side. The collapse of the right-sided chambers leads to decreased preload on the right side, which subsequently affects the left side of the heart, resulting in reduced cardiac output and shock (Alerhand et al., 2022).

The clinical appearance can vary from mild elevation of intrapericardial pressure with minimal hemodynamic consequences in blunt traumatic pericardial effusion to severely increased intrapericardial pressure with hemodynamic collapse and cardiogenic shock, potentially leading to cardiac arrest, as with cardiac tamponade (Alerhand et al., 2022).

The classic clinical presentation of tamponade is known as Beck's triad, which includes hypotension, jugular vein distention, and muffled heart sounds. Other examination findings may include tachycardia (heart rate above 100 beats per minute), tachypnea (respiratory rate exceeding 20 breaths per minute), and low urine output (Alerhand et al., 2022).

Based on autopsy data, blunt cardiac trauma with chamber rupture most often affects the left ventricle. However, patients who arrive at the hospital alive, most often suffer from right atrial disruption. In a recent series published by Huang, patients with traumatic blunt pericardial effusion who survived until reaching the hospital were more likely to have injuries in the low-pressure cardiovascular system. In a review of blunt cardiac rupture reported by Braithwaite, among 32 patients with cardiac rupture, 13 had a right atrial rupture, 8 left atrial rupture, 10 had a right ventricular, 4 left ventricular and 3 had a rupture of two heart chambers. Twelve patients arrived at the hospital with vital signs and 50% of them survived (5 had a right atrial rupture, 1 had a left atrial rupture) (Brathwaite et al., 1990; Huang et al., 2018).

Investigations

To confirm or rule out pericardial effusion or tamponade, urgent investigations may be necessary. These typically include an electrocardiogram (ECG), chest X-ray, and bedside echocardiography, which can be performed as a focused assessment by sonography in trauma (FAST), often done by the emergency department (ED) physician, or a standard bedside transthoracic echocardiogram (Alerhand et al., 2022), and computed tomography (CT). Additionally, the elevation of cardiac Troponin T has shown a good correlation with the severity of blunt cardiac injury (Shoar et al., 2021).

Typical ECG findings in cases of pericardial effusion or tamponade may include tachycardia, low voltage QRS complexes, ST-segment elevation, electrical alternans (alternating amplitude of QRS complexes), PR segment depression, and atrial arrhythmias (Alerhand et al., 2022).

To classify a pericardial effusion, its hemodynamic impact, size, distribution, and composition should be considered. The size of the effusion can be estimated by measuring the end-diastolic diameter between the visceral and parietal pericardium during echocardiography. It is important to differentiate pericardial fat pads from intrapericardial fluid and be aware that left pleural effusion can mimic pericardial effusion (Alerhand et al., 2022).

Echocardiographic findings that suggest tamponade include diastolic collapse of the right ventricle, systolic collapse of the right atrium (which may be the first sign of tamponade), plethoric and non-collapsible inferior vena cava, and sonographic pulsus paradoxus (an abnormal drop in blood pressure during inspiration) (Alerhand et al., 2022). In cases where an occult pericardial effusion is not initially detected due to other associated surgical emergencies, intraoperative transesophageal echocardiography can be a useful adjunct to exclude the diagnosis (Huang et al., 2010).

CT imaging has become the preferred modality for the early evaluation of patients with blunt chest trauma who are hemodynamically stable. CT scans can identify very small pericardial effusions, pericardial tears, and less common entities such as cardiac luxation and pneumopericardium, and can also help detect the source of bleeding (Clancy et al., 2012; Shoar et al., 2021).

Treatment

Treatment options include conservative management for the clinically stable patient with minimal pericardial effusion, percutaneous pericardiocentesis, and surgical drainage through a subxiphoid incision or resuscitative thoracotomy in the emergency setting — the latter also provides exposure if other intrathoracic injuries are suspected.

The conservative approach is mainly reserved for patients who are in a hemodynamically stable condition, with evidence of stable effusion (Huang et al., 2018; Tanizaki et al., 2018). However, in hemodynamically unstable patients, only the drainage of fluid allows a normal ventricular filling and restoration of an adequate cardiac output (De Carlini & Maggiolini, 2017).

Conservative approach

Tanizaki reported his experience with 11 patients affected by blunt traumatic pericardial effusion. One patient with stable hemodynamics and a moderate pericardial effusion evidenced on FAST was successfully managed conservatively. However, of the remaining 10 patients in the series, 5 were in cardiopulmonary arrest and died despite surgical intervention, five had unstable hemodynamics and a large pericardial effusion was observed on FAST — two were managed with pericardiocentesis alone, one with both pericardiocentesis and pericardial window, one was managed with the pericardial window alone and one underwent median sternotomy because of unsuccessful pericardial drainage tube insertion (Tanizaki et al., 2018).

Huang reported a series of 30 patients with blunt pericardial effusion. Eleven patients had a systolic blood pressure less than 100 mmHg and an immediate surgical intervention was performed. In eight patients a cardiovascular lesion was identified and repaired. Nineteen patients had stable hemodynamics (systolic blood pressure >100 mmHg) and initially received conservative treatment. Seven of these patients received exclusively non-operative management since no increase in pericardial effusion was noted. In the remaining 12 patients, an indication for surgery was given due to the appearance of hypotension without response to medical treatment, tachycardia refractory to resuscitation, increasing pericardial effusion, signs of tamponade on echo (right ventricle collapse or global hypokinesia), increasing level of cardiac enzymes, any deterioration or clinical judgment. Cardiac rupture was identified and repaired through a sternotomy in 3 out of 12 cases. In the remaining 9 patients a pericardial effusion was evacuated through a subxiphoid pericardiotomy in 4 cases, a left minithoracotomy in 4 cases, and a sternotomy in one case (Huang et al., 2018).

The recent shift to more conservative treatment is explained by the theory that bleeding from blunt cardiac injury in patients who presented alive at the hospital could be temporarily stopped by blood clotting, while the low pressure in the right side of the heart might allow spontaneous closure of the injury. Hemodynamic instability can be managed by pericardial decompression, removing the requirement for surgical repair. Moreover, drainage is not always mandatory in stable patients. However, when a traumatic pericardial effusion is detected in hemodynamically unstable patients, prompt drainage remains essential (Tanizaki et al., 2018).

The same approach was shown in his series by Huang, who showed no remarkable predictor of cardiovascular injury that required surgical repair apart from systolic blood pressure below 100 mmHg (patients with systolic blood pressure >100 mmHg 16% vs those with systolic blood pressure <100 mmHg: 73%). Additionally, it was postulated that cardiovascular injury does not always result in unstable hemodynamics. The initial amount of blunt pericardial effusion could not reflect cardiac rupture or ongoing hemorrhage and much attention should be paid to pericardial effusion and stable hemodynamics. These patients should be hospitalized in the Intensive Care Unit (ICU) with hemodynamic monitoring, enzymes (creatine phosphokinase-MB (CPK−MB) and Troponin I) evaluation every 8 h, and bedside echography every 8 h for evaluation of the effusion (Huang et al., 2018).

Witt et al. showed that following blunt trauma, in patients with normal hemodynamics and without ECG changes, found to have minimal to small amounts of pericardial fluid, observation is appropriate. Likewise, when a moderate to large amount of fluid is present, in the case of normal hemodynamics, simple observation of the patient is a reasonable option. In the setting of abnormal hemodynamics or concerning clinical findings such as new ECG changes, regardless of the amount of fluid, operative drainage or exploration remains the gold standard (Marro et al., 2019).

According to these experiences the protocol for clinical management of blunt pericardial effusion proposed by Tanizaki appears to be reasonable (Tanizaki et al., 2018).

288 Chest Blunt Trauma

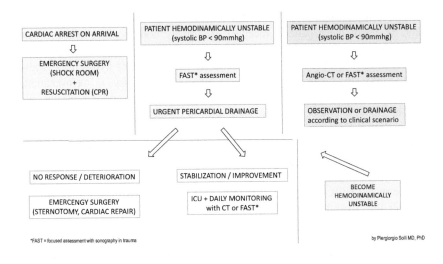

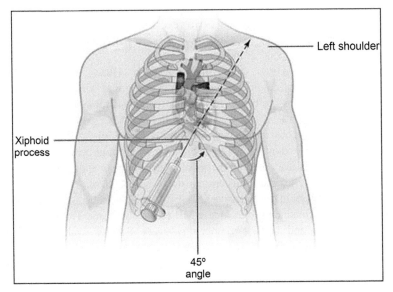

FIGURE 18.1 Anatomical landmarks and needle orientation for pericardiocentesis.

Pericardiocentesis

This is performed through a subxiphoid approach using a long 18−22 G needle attached to a syringe. With the needle directed towards the left shoulder, it is inserted between the xiphisternum and left costal margin, at a 30°−45° angle to skin (Fig. 18.1). Continual aspiration is performed as the needle passes through the layers, and once the pericardial fluid is aspirated, a pericardial drain with a 3-way tap is inserted into the pericardial space. A useful adjunct is attaching an ECG lead to the needle-ST elevation or ventricular ectopics signal contact with the ventricle (Nickson, 2024). If using an ultrasound-guided approach, the needle is inserted where the largest pocket of fluid is seen, this could be either subxiphoid or parasternal Figs. 18.1 and 18.2.

Discussion

Traumatic pericardial effusion is a significant concern in patients with chest injuries, as it is closely associated with cardiac injury. Most individuals with traumatic pericardial effusion die at the scene or during transportation due to cardiac tamponade. Those who survive to reach the hospital often have minor tears in the low-pressure chamber of the heart, with a blood clot temporarily stopping the bleeding, or the blood may decompress into the pleural cavity through a pericardial defect. Compared to penetrating chest injuries, cardiac injuries are more easily overlooked in cases of blunt trauma, especially when there are associated brain or abdominal injuries. Correct and timely diagnosis of cardiac injuries remains challenging but it is essential for a favorable outcome [AMO14].

Initial normal vital signs, ECG, and laboratory tests cannot exclude the presence of tamponade. The classic Beck's triad, consisting of muffled heart sounds, engorged jugular veins, and hypotension, is not highly sensitive or specific in patients with multiple traumatic injuries (Alerhand et al., 2022).

Patients with unexplained hypotension following chest trauma, in the absence of tension pneumothorax, abdominal or pelvic injuries, or spinal cord injuries, should raise suspicion of cardiac tamponade. Additionally, those with hypotension disproportionate to the estimated blood loss or who exhibit inadequate responses to fluid resuscitation should also be suspected of having a cardiac cause of shock or tamponade (Huang et al., 2010).

FAST is a rapid diagnostic tool used to identify intraperitoneal or pericardial fluid in hemodynamically unstable patients. Rapid cardiac ultrasound can be performed within seconds and repeated frequently, making it the first-line screening tool for suspected cardiac trauma. However, bleeding from cardiac lesions may not accumulate in sufficient amounts initially or become apparent after aggressive fluid resuscitation. In these cases, CT plays a crucial role. CT can be performed quickly (within minutes) and provides detailed information about pericardial effusion, solid organ injuries, retroperitoneal hematoma, and free intraperitoneal air. It guides decision-making regarding conservative management or the need for surgery. For these reasons, CT has become the gold standard for evaluating trauma patients with relatively stable vital signs and can be used to confirm or exclude traumatic pericardial effusion after an initial survey with FAST (Huang et al., 2010; Witt et al., 2017).

The management approach for blunt pericardial effusion varies. Some authors believe that surgical drainage is necessary regardless of the patient's clinical condition. However, others suggest that stable patients can be

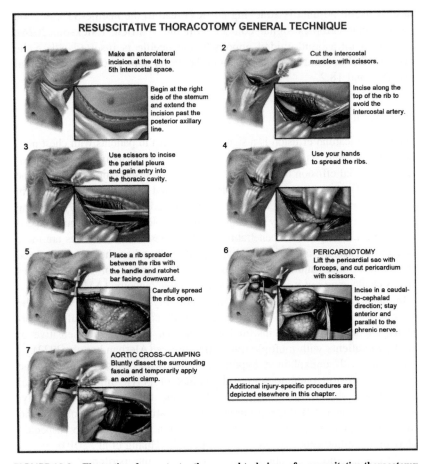

FIGURE 18.2 Illustration demonstrates the general technique of a resuscitative thoracotomy.

managed non-operatively, with a watchful waiting approach. The algorithm proposed in their series by Huang and Tanizaki demonstrated satisfactory results with this approach (Huang et al., 2018; Tanizaki et al., 2018).

Subxiphoid pericardiotomy is often preferred over percutaneous catheter drainage due to the frequent clotting of blood in the pericardium. Minor bleeding may stop after pericardial decompression, and sternotomy may not be required.

Full sternotomy is reserved for cases of persistent pericardial drainage. Sternotomy can be performed conveniently from a previous subxiphoid incision; moreover, it offers the best cardiac exposure, facilitates inspection, and allows for direct cardiac massage and repair. Thoracotomy offers limited exposure and is typically chosen in extreme circumstances when rapid intervention is necessary. Repairing cardiac injuries during initial exploratory thoracotomy for massive hemothorax can be challenging, and extending the incision may be considered. However, if visualization is limited and surgical access is difficult, immediate conversion to median sternotomy is recommended.

Concomitant cardiac repair or associated pericardial defects do not jeopardize survival, but the incidental discovery of pericardial effusion at the time of surgery is associated with higher mortality rates (Brathwaite et al., 1990).

1. https://ejtcm.gumed.edu.pl/articles/66
2. https://www.escardio.org/Journals/E-Journal-of-Cardiology-Practice/Volume-15/Pericardiocentesis-in-cardiac-tamponade-indications-and-practical-aspects
3. https://litfl.com/pericardiocentesis
4. https://www.emcurious.com/blog-1/2014/9/19/40b1x9dgskw47m1dvveu77bjjs0bvd

References

Alerhand, S., Adrian, R. J., Long, B., & Avila, J. (2022). Pericardial tamponade: A comprehensive emergency medicine and echocardiography review. *American Journal of Emergency Medicine*, *58*, 159–174. Available from https://doi.org/10.1016/j.ajem.2022.05.001, http://www.journals.elsevier.com/american-journal-of-emergency-medicine/.

ATLS (2018).

Brathwaite, C. E. M., Rodriguez, A., Turney, S. Z., Dunham, C. M., & Cowley, R. A. (1990). Blunt traumatic cardiac rupture: A 5-year experience. *Annals of Surgery*, *212*(6), 701–704. Available from https://doi.org/10.1097/00000658-199012000-00008, http://journals.lww.com/annalsofsurgery/pages/default.aspx.

Clancy, K., Velopulos, C., Bilaniuk, J. W., Collier, B., Crowley, W., Kurek., Lui, F., Nayduch, D., Sangosanya., Tucker, B., & Haut, E. R. (2012). Eastern Association for the Surgery of Trauma. Screening for blunt cardiac injury: An Eastern Association for the Surgery of Trauma practice management guideline. *Journal of Trauma and Acute Care Surgery*, *73*, 301–306. Available from https://doi.org/10.1097/TA.0b013e318270193a, PMID: 23114485.

De Carlini, C. C., & Maggiolini, S. (2017). Pericardiocentesis in cardiac tamponade: indications and practical aspects. *E-Journal of Cardiology Practice*, *15*(19). Available from https://www.escardio.org/Journals/E-Journal-of-Cardiology-Practice/Volume-15/Pericardiocentesis-in-cardiac-tamponade-indications-and-practical-aspects. Accessed 19.12.24.

Dogrul, B. N., Kiliccalan, I., Asci, E. S., & Peker, S. C. (2020). Blunt trauma related chest wall and pulmonary injuries: An overview. *Chinese Journal of Traumatology – English Edition*, *23*(3), 125–138. Available from https://doi.org/10.1016/j.cjtee.2020.04.003, http://www.elsevier.com/wps/find/journaldescription.cws_home/714951/description#description.

Huang, J. F., Hsieh, F. J., Fu, C. Y., & Liao, C. H. (2018). Non-operative management is feasible for selected blunt trauma patients with pericardial effusion. *Injury*, *49*(1), 20–26. Available from https://doi.org/10.1016/j.injury.2017.11.034, http://www.elsevier.com/locate/injury.

Huang, Y. K., Lu, M. S., Liu, K. S., Liu, E. H., Chu, J. J., Tsai, F. C., & Lin, P. J. (2010). Traumatic pericardial effusion: Impact of diagnostic and surgical approaches. *Resuscitation*, *81*(12), 1682–1686. Available from https://doi.org/10.1016/j.resuscitation.2010.06.026.

Lasek, J., & Jadczuk, E. (2020). Thoracic trauma – principals of surgical management. *Eur J Transl Clin Med*, *3*(1), 66–73.

Marro, A., Chan, V., Haas, B., & Ditkofsky, N. (2019). Blunt chest trauma: Classification and management. *Emergency Radiology*, *26*(5), 557–566. Available from https://doi.org/10.1007/s10140-019-01705-z, http://link.springer.de/link/service/journals/10140/index.htm.

Mattox, K. L., Moore, & Feliciano, D. V. (n.d.). *Trauma*. Chapter 26.

Nickson, C. (2024). Pericardiocentesis. Life in the Fast Lane. Available at: https://litfl.com/pericardiocentesis/. Accessed 19.12.24.

Shoar, S., Hosseini, F. S., Naderan., Khavandi, S., Tabibzadeh, E., Khavandi, S., & Shoar, N. (2021). Cardiac injury following blunt chest trauma: Diagnosis, management, and uncertainty. *International Journal of Burns and Trauma*, *11*(2).

Tanizaki, S., Nishida, S., Maeda, S., & Ishida, H. (2018). Non-surgical management in hemodynamically unstable blunt traumatic pericardial effusion: A feasible option for treatment. *American Journal of Emergency Medicine*, *36*(9), 1655–1658. Available from https://doi.org/10.1016/j.ajem.2018.06.066, http://www.journals.elsevier.com/american-journal-of-emergency-medicine/.

Witt, C. E., Linnau, K. F., Maier, R. V., Rivara, F. P., Vavilala, M. S., Bulger, E. M., & Arbabi, S. (2017). Management of pericardial fluid in blunt trauma: Variability in practice and predictors of operative outcome in patients with computed tomography evidence of pericardial fluid. *Journal of Trauma and Acute Care Surgery*, *82*(4), 733–741. Available from https://doi.org/10.1097/TA.0000000000001386, http://journals.lww.com/jtrauma.

Chapter 19

Airway and esophageal injuries

Sahar Hasanzade[1] and Marco Scarci[1,2]
[1]Department of Thoracic Surgery, Imperial College NHS Healthcare Trust, London, United Kingdom,
[2]National Heart and Lung Institute, Imperial College, London, United Kingdom

Airway and esophageal injuries

Introduction

Esophageal injuries are a rare and potentially life-threatening condition. They have a high morbidity and mortality rate because of the potential complications such as mediastinitis or sepsis which can cause a delay in patient diagnosis and their recovery.

Clinical presentation

Esophageal trauma is nonspecific and could range from feeling chest pain which can radiate to shoulders and back followed by dysphagia, odynophagia, hoarseness, regurgitation, hematemesis, hemoptysis to severe signs of infection, shock, and respiratory distress (Mubang et al., 2023).

Mechanisms of trauma

Penetrating trauma is the most common and is caused by objects penetrating the esophageal wall causing catastrophic cellular damage and can pierce into deeper and adjacent structures. This derives from gunshot wounds, stab wounds, and other foreign objects that can penetrate through the wall (Fig. 19.1).

Chemical injury is the second most common cause resulting from ingesting toxic substances which would cause detrimental cellular dysfunction by perforating the lining of the esophagus; this would happen via acidic or alkaline elements.

Blunt trauma is caused by sudden deceleration injuries at high speeds impacting the chest or abdomen leading to esophageal rupture or laceration. These could result from falls, physical abuse, car accidents, and other types of crushing and compressing trauma.

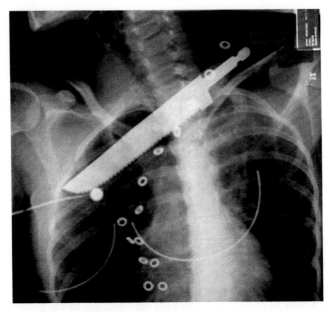

FIGURE 19.1 Esophageal penetrating trauma. Chest x-ray reveals stabbing via breadknife at the back of the chest causing a right-sided pneumothorax.

Iatrogenic trauma occurs because of surgical or medical interventions causing incidental injury, such as endoscopic procedures, esophageal dilation, surgical outcomes, and other injuries related to intubation. Thermal Injury appears from ingestion of extremely hot substances that could burn the esophageal tissue this would develop by consuming hot food, narcotics, and use of catheter ablation more commonly in the left atria (Clinic, n.d.) (Fig. 19.2).

Diagnosis

Obtaining a contrast-enhanced esophagography scan via contrast agents provides effective imaging revealing sites and extent of perforations. Clinically the most common diagnostic tool used is CT scans with contrast which detects air or fluid collections within the mediastinum and the structures (Fig. 19.3).

For direct visualization, a flexible esophagoscopy which uses air insufflation is used in penetrating trauma cases (Kassem & Wallen, 2024) (Fig. 19.4).

Treatment

Since esophageal trauma is so scarce their management is often based on clinical cases and series, a lot of which include discussions on nontraumatic etiologies of esophageal injury or perforation. Subsequently, common management plans for esophageal injuries are extrapolated and applied to both

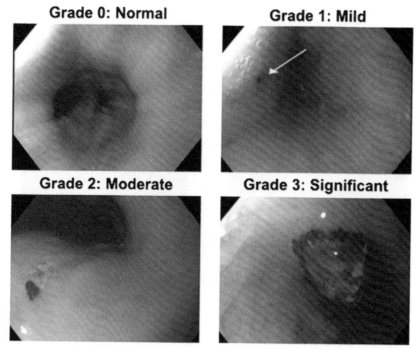

FIGURE 19.2 Esophageal thermal trauma via endoscopy. Classification of thermal trauma. *From Hanvesakul, R., Momin, A., Gee, M. J., & Marrinan, M. T. (2005). A role for video assisted thoracoscopy in stable penetrating chest trauma. Emergency Medicine Journal, 22(5), 386−387. https://doi.org/10.1136/emj.2003.014076.*

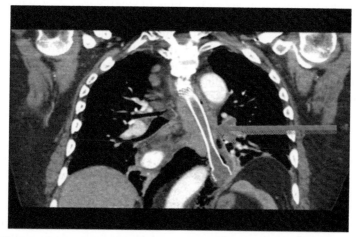

FIGURE 19.3 Computed tomography. CT image illustrating esophageal stent in situ. *From King, W. D., & Dickinson, M. C. (2015). Oesophageal injury. BJA Education, 15(5), 265−270. https://doi.org/10.1093/bjaceaccp/mku039.*

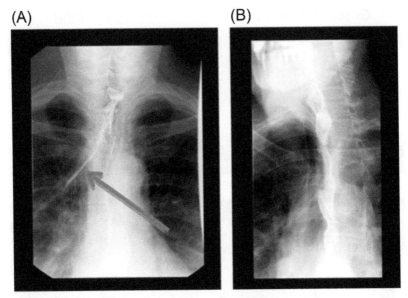

FIGURE 19.4 Example of postero anterior (PA) chest Gastrografin studies.

penetrating and blunt trauma. The location of injury plays a crucial role in determining the management (Hospital, 2024).

Generally, an immediate and prompt diagnosis and plan of esophageal trauma expedites patient recovery and rehabilitation. The consequences could lead to life-threatening mediastinitis followed by severe organ dysfunction sepsis and septic shock, resulting in delayed healing and interfering with surgical interventions. The lack of a protective serosal layer makes the esophagus vulnerable to injuries and more difficult to repair surgically than other tissues of the gastrointestinal tract. Therefore, a definite closure and drainage are ideal, however in rare cases other morbid and life-threatening injuries could take precedence over consequent conventional management. In such circumstances, meticulous assessment of risks and benefits should be addressed considering the overall patient clinical case (Medlineplus.gov, 2017).

Nonsurgical treatment of esophageal trauma is convenient for patients who are homeostatically stable with some symptoms or some who are too fragile to endure major surgery; this leads to a delay in diagnosis and better management. A standard approach must be met for consideration of conventional treatment, these include; (1) contained leak is when contrast via esophagram oozes the wall and reappears without leaving a trace of fluid or inflammation, (2) moderate symptoms, and (3) evidence of sepsis. Patients who cannot undergo surgical intervention are treated with a chest drain via thoracotomy performed at the bedside (Liao et al., 2022).

Operative management

Esophageal trauma is a surgical emergency that requires quick and efficient preoperative testing. The degree of tissue inflammation aids in determining the best approach between a surgical repair or drainage. If surgery is delayed it may still be effective if there's minimal contamination since the amount of contamination influences the risk of complications, therefore all factors must be considered during the initial treatment plan.

The location of the injury to the esophagus impacts the surgical technique and repair methods. Trauma to the cervical esophagus requires a left-sided incision on the neck made between the edge of the sternocleidomastoid muscle and the trachea and carotid sheath. This enables the identification and protection of the recurrent laryngeal nerve and membranous tissue of the trachea, while mobilizing the esophagus via the blunt technique (Sudarshan & Cassivi, 2019).

A right posterolateral thoracotomy is a prerequisite for thoracic esophageal injuries. This requires proper positioning for surgical visibility and repair. When entering the thoracic cavity it is crucial to preserve the intercostal muscle bundle since it can serve as a vascular tissue coverage for esophageal incision as well as a repair flap.

A left posterolateral thoracotomy provides easy access to the distal third of the thoracic esophagus, as does access via the abdomen. An upper midline laparotomy starting from the umbilicus to the xiphoid process, facilitates better access to the abdominal portion of the esophagus. In addition, a fundoplication and patches for the omentum and diaphragm would aid in reinforcing support to the site of injury. Local drainage as well as gastrostomy would be as impactful. For other esophageal perforation, less than 5 cm such as Boerhaave syndrome could be treated with a minimally invasive stent placement. All anatomical structures surrounding the site of injury must be thoroughly debrided and repaired with local tissue and muscular flaps (Brian et al., 2023).

The T Tube diversion is ideal for patients who cannot undergo primary conventional repair. It is done by placing the tube into the wound with prolonged suction to produce a controlled fistula anastomosis for a period to facilitate the formation of a passage. Once the patient is secure a gastrostomy tube will be utilized for pressure relief and a jejunostomy tube will be implemented and used for nutritional support (Gill, 2015).

An esophagectomy is a rare practice used and may be established when there's failure of the initial repair or when the diagnosis is delayed. Clinical data from retrospective studies advise that esophagectomy may be the desired method especially when performing the Ivor-Lewis approach. This is accomplished by resectioning the esophagus and making two incisions on the abdomen and thorax done by highly skilled surgeons (Mayoclinic, 2018) (Fig. 19.5).

The gastric "pull up" method is a simpler and more common repair that is chosen depending on the extent of injury and the attainability of healthy

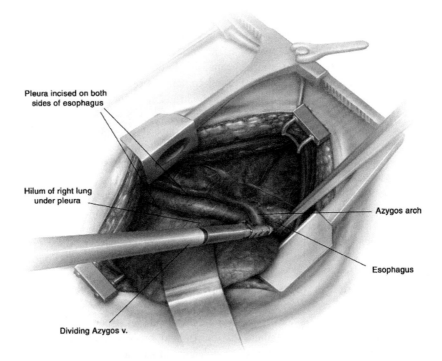

FIGURE 19.5 Technique of open Ivor-Lewis esophagectomy. *From Reed, C. E. (2009). Technique of open Ivor Lewis esophagectomy. Operative techniques. Thoracic and Cardiovascular Surgery, 14(3), 160–175. https://doi.org/10.1053/j.optechstcvs.2009.06.001.*

tissues. This provides a one-way path to the stomach. If the gastric "pull up" method fails and an esophageal resection occurs, the next approach would be the Roux-en-Y jejunal technique which creates an egg-sized pouch connecting the stomach to the small intestine.

Complications

Esophageal injuries may lead to several complications, these are more common injuries:

mediastinitis, abscess formation, and other infections within the mediastinum which may be life-threatening and require urgent treatment such as systemic infection sepsis.

Esophageal strictures and tracheoesophageal fistula could lead to aspiration of food and fluids resulting in intermittent pneumonia or choking. Aspiration pneumonia and hemorrhage lead to hemoptysis and hematemesis. Chronic persistent pain in the chest or throat. Altered esophageal function affects the movements of peristalsis and narrowing due to strictures or scar formation leading to potential blockage. Esophageal trauma could also cause psychological issues including depression and anxiety (S, 2013).

Airway injuries

This is a life-threatening case that can be obtained from various types of trauma to the chest, neck, and head. The severity of these injuries can range from minor mucosal abrasions to critical ruptures and extensive damage to nearby tissues. It is vital to also take into consideration that severe injuries adjunct to this type of injury could hinder the diagnosis resulting in premature death. Common contributing factors are asphyxiation due to impediment of the airway, ventilator failure, or pneumothorax tension. Furthermore, chronic complications involve frequent pulmonary infections or airway stenosis. Treatment and support are dependent on a prompt multidisciplinary method incorporating clinical assessments, advanced imaging, and endoscopic procedures to prevent further respiratory impediments and restore normal respirational purposes (Prokakis et al., 2014).

Mechanism

Airway trauma can occur from many mechanisms each resulting in specific and challenging diagnosis and management. Blunt trauma occurs less frequently in 1%–2% of trauma cases affecting the airway, often resulting from high-speed blows from motor vehicle accidents where forces can cause fractures, contusions, and lacerations. In contrast, penetrating trauma accounts for a significant proportion of airway injuries with 20%–25% of deaths up until the age of 40 being caused by this type. This injury is an outcome of sharp objects such as knives, bullets, glass, and sticks requiring surgical intervention. Chemical injuries are caused by exposure to corrosive elements resulting in burns and inflammation via ingestion or skin penetration. Thermal injuries are initiated by inhalation of gases and toxins which damage the airway lining causing deeper inflammation and obstruction (Fig. 19.6).

Iatrogenic trauma may arise from medical interventions such as endotracheal intubation, tracheostomy, laryngoscopy, bronchoscopy, mechanical ventilatory procedures, and thoracic and neck surgeries (Santiago-Rosado, 2024) (Fig. 19.7).

Clinical presentation

These injuries may vary on the type, location, and severity of damage. The most prevalent symptom is subcutaneous emphysema observed in 87% of patients accompanied by inflammation and edema in the neck. Occurrences of respiratory distress, and pneumothorax in 17%–70% of cases, complicating the clinical presentation while causing difficulties in lung re-expansion following

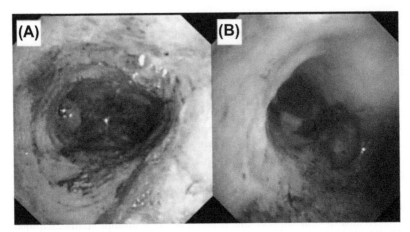

FIGURE 19.6 Bronchoscopic view. Showing visible inhalation injury with soot deposition 18 hours postexposure. *From Bronchoscopic findings of Patient B, eighteen hours after inhalation. (2024). https://www.researchgate.net/figure/Bronchoscopic-findings-of-Patient-B-eighteen-hours-after-inhalation-injury-The-trachea_fig2_258062948.*

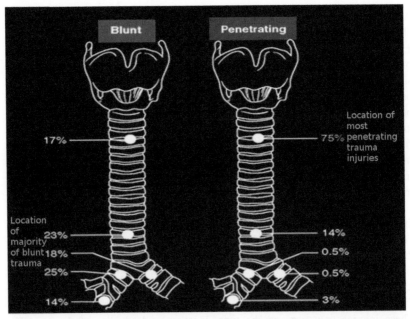

FIGURE 19.7 Location of tracheal injury. *From Locations of Tracheal Injury Contributed by O. Chaigasame. (2023). http://Nih.gov. https://www.ncbi.nlm.nih.gov/books/NBK547677/figure/article-30413.image.f4/.*

tube thoracostomy may occur. Other signs involve sharp and sore chest pain, signs of laryngeal damage via hoarseness, hemoptysis which reveals the severity of the injury, hematemesis, signs of infection, hypoxia, hypercapnia, and visible injuries. Understanding the types of airway trauma is essential for timely intervention and improved patient outcomes (Cakmak, 2022).

Evaluation

A chest x-ray reveals a subcutaneous emphysema which is the most common finding, followed by pneumothorax and air within the mediastinum. To distinguish a cervical tracheal tear it would be identified on a neck x-ray via a lateral scan exposed through signs in the scan such as elevation of the hyoid bone, abnormality of the trachea, change in tracheal location, presence of an inflated cuff surrounding the edge of the tracheal wall. In less common cases a bronchial tear also known as the 'fallen lung' would show as a collapsed lung pointing down to the diaphragm rather than inwards. Having said that, it is said that up to 10%– 20% of airway injured cases are asymptomatic (Fig. 19.8).

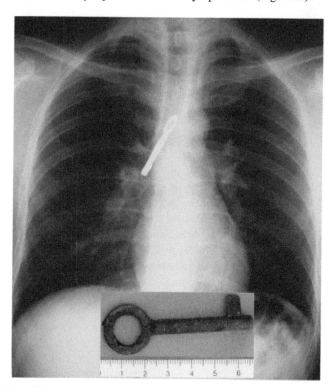

FIGURE 19.8 Chest X-ray. *Modified from Delage, A., & Marquette, C. -H. (2010). Airway foreign bodies: Clinical presentation, diagnosis and treatment. In Interventional pulmonology (pp. 135–148). European Respiratory Society (ERS). https://doi.org/10.1183/1025448x.00991009.*

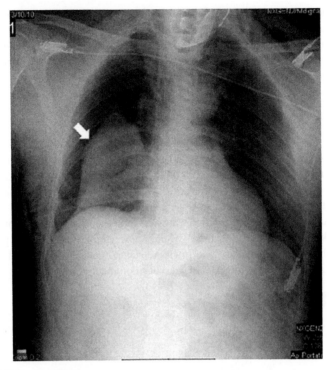

FIGURE 19.9 Chest X-ray.

Laryngeal injuries are best visualized by CT scans which are famous for detecting most abnormalities. Recent studies have proven that the most effective radiography device is CT with 3D imaging because of its 94%−100% precision. Modernized bronchoscopy also involves CT scans that produce 3D imaging of the airways. To confirm the diagnosis a bronchoscopy would be the gold standard to detect the site and extent of trauma. Endotracheal intubation generally uses a flexible bronchoscope, however in critical crisis, a rigid bronchoscope is utilized, however, failure of intubation would require a tracheostomy. An emergency thoracotomy is vital in such circumstances (Loren & Primack, 2019) (Fig. 19.9).

Management

The main priority in managing airway injuries is securing the airway to restore and maintain ventilation while also preventing infection and complications. Surgical intervention includes repairing or reconstructing organ structures after undergoing significant damage that cannot be rectified via medical therapy.

Vascular injuries to the cervical region of the trachea require access to the cricoid sheath which is achieved by a fine cut known as the collar incision along the sternocleidomastoid muscle. Which can be extended to creating a higher sternotomy to obtain views for the tracheas posterior wall injury. For the lower tracheal injuries including the carina and right main bronchus to be assessed would require a thoracotomy on the right side by the 4th intercostal space. In severe cases, a median sternotomy is made for trans pericardial access for better exploration. Left thoracotomy is achieved when injuries occur by the left main bronchus. Penetrating trauma to the anterior side of the trachea with no more than 2 rings of cartilage is injured and needs a tracheostomy, this is also used for complex injuries to adjacent tissues such as the recurrent laryngeal nerve (Giorgia et al., 2024).

Pulmonary resection is imperative for more complicated and permanent damage to the airways, its neighboring vessels, and parenchyma. To abstain from a pneumonectomy a cardiopulmonary bypass facilitates stability and maintains a better condition for hemodynamically stable patients. In addition in cases of blunt trauma to airways, a VV-ECMO would be utilized to aid in pulmonary ventilation in acute situations (Chiarelli et al., 2016).

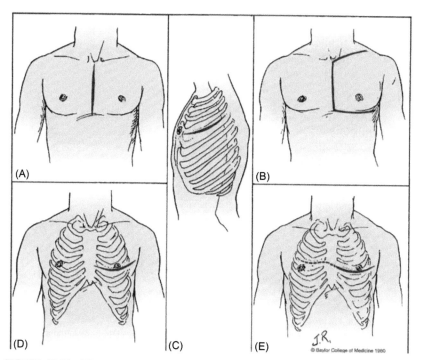

FIGURE 19.10 Thoracotomy principles and techniques. *Modified from Trauma thoracotomy: Principles and techniques. [online](n.d.). Thoracic Key. https://thoracickey.com/trauma-thoracotomy-principles-and-techniques/.*

Absorbable sutures are used for minor traumatic tears and lacerations for closure. In major injuries, debridement is essential for infection prevention followed by trimming of the affected sides to allow for clean anastomosis, and to avoid the formation of granulation tissue. Anastomosis with free tension is needed after resection by stitching the chin to the chest. Other techniques to decrease tension include pulmonary ligament division. Foreign bodies suspected would be initiated and excised by endoscopic removal. Additionally, trauma patients are mobilized and stabilized in a supine position for convenient access to enable other therapeutic interventions. This includes video-assisted thoracic surgery (VATS), endotracheal intubation, intercostal tube thoracotomy, and digital thoracotomy (John et al., 2024) (Fig. 19.10).

Complications

Prompt surgical management immediately after injury has over 90% of successful cases. Major morbidity is an outcome of a long duration in the ICU or delays in management. Airway obstruction and stenosis occur after a long period as a complication as well as vocal changes and issues and struggles with swallowing due to injuries to the larynx. Other impediments include hemorrhage, fistulas leading to further damage, persistent phlegm production, and chronic coughs (Cook, 2017).

References

Brian, C., Tucker, W. D., & Burns, B. (2023). *Thoracotomy.* https://www.ncbi.nlm.nih.gov/books/NBK557600/.

Cakmak, G. (2022). *Airway management in penetrating thoracic trauma.* https://pmc.ncbi.nlm.nih.gov/articles/PMC10156565/.

Chiarelli, M., Martino, G, & Giuseppe, V. (2016). Urgent pulmonary lobectomy for blunt chest trauma: Report of three cases without mortality. *Journal of Thoracic Disease, 8*(7). Available from https://jtd.amegroups.org/article/view/7879/7431.

Clinic Cleveland. (n.d.). *Esophageal rupture: Causes, symptoms and treatment.* https://my.clevelandclinic.org/health/diseases/24284-esophageal-rupture.

Cook, T. M. (2017). *Strategies for the prevention of airway complications — a narrative review.*https://associationofanaesthetists-publications.onlinelibrary.wiley.com/doi/full/10.1111/anae.14123.

Gill, R. C. (2015). https://pmc.ncbi.nlm.nih.gov/articles/PMC5067295/#:~:text = A%20T%2Dtube%20placed%20in,to%20iatrogenic%20or%20blunt%20trauma.

Giorgia, C., Stefano, M., & Andrea, S. (2024). *Thoracic trauma.* https://pmc.ncbi.nlm.nih.gov/articles/PMC11417912/.

Hospital, B. (2024). *Esophageal injury and trauma.* https://www.childrenshospital.org/conditions/esophageal-injury-and-trauma.

John, D., Gustavo, R., & Alex, P. (2024). *Video assisted thoracoscopic surgery (VATS) in trauma.* https://vats.amegroups.org/article/view/8812/html.

Kassem, M., & Wallen, J. (2024). https://www.ncbi.nlm.nih.gov/books/NBK532298/.

Liao, F., Zhu, Z., Pan, X., Li, B., Zhu, Y., Chen, Y., & Shu, X. (2022). Safety and efficacy of nonoperative treatment in esophageal perforation caused by foreign bodies. *Clinical and Translational Gastroenterology*, *13*(1)e00451. Available from https://doi.org/10.14309/ctg.0000000000000451.

Loren, K., & Primack, S. L. (2019). Thoracic trauma. https://www.ncbi.nlm.nih.gov/books/NBK553870/.

Mayoclinic. (2018). *Esophagectomy*. https://www.mayoclinic.org/tests-procedures/esophagectomy/about/pac-20385084#:~:text = Esophagectomy%20is%20a%20surgical%20procedure, treatment%20for%20advanced%20esophageal%20cancer.

Medlineplus.gov. (2017).*Esophageal perforation: MedlinePlus medical encyclopedia* https://medlineplus.gov/ency/article/000231.htm#:~:text = When%20there%20is%20a%20hole, injury%20during%20a%20medical%20procedure.

Mubang, R., Sigmon, D., & Stawicki, S. (2023). https://www.ncbi.nlm.nih.gov/books/NBK470161/.

Prokakis, C., Koletsis, E. N., Dedeilias, P., Fligou, F., Filos, K. Dougenis, D. (2014). Airway trauma: A review on epidemiology, mechanisms of injury, diagnosis and treatment. *Journal of Cardiothoracic Surgery, 30*(9), 117. https://pubmed.ncbi.nlm.nih.gov/24980209/.

Santiago-Rosado, L. (2024). *Tracheal trauma* https://www.ncbi.nlm.nih.gov/books/NBK500015/#:~:text = Tracheal%20trauma%20may%20result%20from,and%20treatment%20worsen%20the%20prognosis.

Sharma, S. (2013) *Management of complications of esophagectomy*. https://www.ncbi.nlm.nih.gov/pmc/articles/PMC3693150/.

Sudarshan, M., Cassivi, S. D. (2019). *Management of traumatic esophageal injuries.* https://pmc.ncbi.nlm.nih.gov/articles/PMC6389559/.

Index

Note: Page numbers followed by "*f*" and "*t*" refer to figures and tables, respectively.

A

Abbreviated Injury Scale, 56, 63–64
Abdominal organ herniation, 266
Acute lung injury (ALI), 8
Acute respiratory distress syndrome (ARDS), 8–12, 9f, 126
  adjunct therapies, 11
  diagnostic, 9–10
  differential diagnosis, 10
  epidemiology/pathophysiology, 8–9
  recovery and long-term results, 11–12
  research/future therapy options, 12
  risk factors, 9
  therapy, 10–11
Additive manufacturing. *See* 3D printing (3D-P) technology
Advanced Trauma Life Support (ATLS), 19, 245
AECC. *See* American-European Consensus Conference (AECC)
Airway and esophageal injuries, 293–299
  clinical presentation, 293
  complications, 298–299
  diagnosis, 294, 295f, 296f
  mechanisms of trauma, 293–294, 294f, 295f
  operative management, 297–298
  treatment, 294–296
Airway injuries, 299–304
  clinical presentation, 299–301
  complications, 304
  evaluation, 301–302
  management, 302–304
  mechanism, 299
Airway management, tracheobronchial injury and, 20–23
ALI. *See* Acute lung injury (ALI)
American-European Consensus Conference (AECC), 8–9
American Society of Emergency Radiology (ASER), 60–62
Analgesia in blunt chest trauma, 27–28
Anesthesia for surgical treatment, 26–27
Anesthesiological aspects in chest trauma patient, 19
  analgesia in blunt chest trauma, 27–28
  anesthesia for surgical treatment, 26–27
  cardiac tamponade, 24–25
  erector spinae plane (ESP) block, 30–31
  extracorporeal membrane oxygenation (ECMO), 25–26
  hemothorax, 24
  initial management of chest blunt trauma, 19–20
  intercostal nerve (ICN) block, 31
  open pneumothorax, 24
  regional anesthesia, 29
  serratus anterior plane (SAP) block, 30
  tension pneumothorax, 23
  thoracic epidural analgesia, 29
  thoracic paravertebral (TPV) block, 29–30
  tracheobronchial injury (TI) and airway management, 20–23
Angular malunion, 138
Antero-lateral thoracotomy, 112–113, 114f
AO Foundation/Orthopaedic Trauma Association (AO/OTA) fracture classification system, 56–57
ARDS. *See* Acute respiratory distress syndrome (ARDS)
ASER. *See* American Society of Emergency Radiology (ASER)
ATLS. *See* Advanced Trauma Life Support (ATLS)

B

Beck's triad, 285
Bleeding, 125
Blood loss and surgery, 249–250

307

Blunt chest trauma, analgesia in, 27–28
Blunt diaphragmatic trauma, 260–261
Blunt dissection technique
 steps for insertion of large-bore chest tubes using, 239
 steps for insertion of small-bore chest tubes using, 239

C

C-ABCDE approach, 19–20
Cardiac tamponade, 24–25
Chest drains, 233
 components, 234–235
 for pneumothorax, 246–248
 potential pitfalls and complications, 241
 for "small" haemothorax, 250–251
Chest Injuries International Database, 62
Chest trauma score (CTS), 46
Chest tube insertion, 238–240
 apical/anterior chest tube insertion, 239–240
 indications of, 235–237
 large-bore chest tubes insertion using blunt dissection technique, 239
 patient positioning and landmarks for insertion, 238
 precautions to, 237
 small-bore chest tubes insertion using Seldinger technique, 239
Chest tube management, 252–253
 prophylactic antibiotics, 252–253
 suction versus no suction, 252
Chest tube placement, 245
 chest drain for "small" haemothorax, 250–251
 indications, 245–246
 positive pressure ventilation (PPV), 248–249
 retained haemothorax, management of, 251–252
 volume of blood loss and surgery, 249–250
Chest tubes
 for haemothorax, 249
 for pneumothorax, 246
 removal, 240–241
Chest wall injuries, posttraumatic, video assisted thoracic surgery for, 176
Chest wall injury scoring systems and surgical stabilization, 48–49
Chest Wall Injury Society (CWIS), 38–39, 41–42, 49, 57–60, 58t, 88
 CWIS/ASER working group, 60–62, 61t

Chest wall Organ Injury Scale, 42–45, 44t
Chest X-ray, 262, 263t
Chronic pain and irritation, 131
Complex rib malunion, 139
Computed tomography (CT), 189–191, 189f, 190f, 222, 263–264, 264t
Costal cartilage, 64–65
Costal margin, injuries to, 65–66
CT. *See* Computed tomography (CT)
CTS. *See* Chest trauma score (CTS)
Current chest wall scoring systems, 42–48, 43t
CWIS. *See* Chest Wall Injury Society (CWIS)

D

Delayed hemothorax (DHTX), 215, 216t
 clinical and radiological predictive factors of, 215–220
 complications of, 221
 definition, epidemiology, and classification of, 213–215
 diagnostic and therapeutic role of VATS in, 222–225
 Quebec clinical decision rule, 220, 221t
 uniportal VATS in blunt chest trauma, 224–225
DHTX. *See* Delayed hemothorax (DHTX)
Diaphragm
 anatomy of, 258–259
 physiology of, 259
Diaphragmatic injuries, 174, 260–261
 blunt diaphragmatic trauma, 260–261
 penetrating trauma, 261
Diaphragmatic rupture
 clinical presentation and, 261–262
  acute presentation, 261
  delayed presentation, 261–262
  physical examination for, 262t
Diaphragmatic rupture and herniation following blunt trauma, 257
 commonly associated injuries, 265–267
  abdominal organ herniation, 266
  gastric perforation, 266
  liver laceration, 266
  pneumothorax and hemothorax, 266
  rib fractures, 266
  sepsis, 267
  spleen rupture, 266
 diagnostic imaging, 262–267
  chest X-ray, 262, 263t
  computed tomography scan, 263–264, 264t

ultrasound, 263, 263t
mechanisms of repair, 267–277
  laparoscopic approach, 268–270, 268t
  left-sided diaphragmatic laparoscopic repair, 273
  procedure, 271
  right-sided diaphragmatic laparoscopic repair, 272–273
mesh, 277
minimally invasive approaches, 273–275
  robotic transthoracic approach, 274–275
  video-assisted thoracoscopic surgery (VATS), 273–274
open approach, 275–277
  laparotomy, 275–276
  thoracotomy, 276–277
primary repair, 277

E

Early procedure-related complications, 125–127
  bleeding, 125
  neuromuscular weakness, 127
  pleural effusion, 125–126
  pneumothorax, 126
  pulmonary complications, 125
  ventilatory failure, 126
  wound infection, 126–127
Eastern Association for the Surgery of Trauma, 90
ECMO. See Extracorporeal membrane oxygenation (ECMO)
E-FAST. See Extended focused assessment with sonography for trauma (E-FAST)
Emergency room (ER), 20
Empyema, posttraumatic, video assisted thoracic surgery for, 173–174
ER. See Emergency room (ER)
Erector spinae plane (ESP) block, 30–31
ESP block. See Erector spinae plane (ESP) block
Essential emergency airway equipment, 23t
Extended focused assessment with sonography for trauma (E-FAST), 20
Extracorporeal Life Support Organization database, 25–26
Extracorporeal membrane oxygenation (ECMO), 25–26

F

FAST. See Focused assessment by sonography in trauma (FAST)

FixCon trial, 98
Flail chest, 4–5, 56
  adjunctive management, 6
  conservative management, 5–7
  diagnostic and definition, 4–5
  non-invasive/ventilation, 6
  pain management, 5
  rib stabilization, indications for, 6–7
  therapy, 5
Flail chest patients, management in, 72–73
Focused assessment by sonography in trauma (FAST), 187, 285–286, 289
Forced vital capacity (FVC), 38–39
Fracture nonunion or malunion, 130
FVC. See Forced vital capacity (FVC)

G

Gastric perforation, 266
Gram-positive organisms, 128

H

Haemothorax
  retained haemothorax, management of, 251–252
  video assisted thoracic surgery for posttraumatic haemothorax, 172
Hardware infection, 128–129
  mechanical hardware failure, 129–130
Hemopneumothorax, 24
Hemothorax, 24, 184–185, 192, 266
  massive, 191–192
  retained, 192–193
Hernias, posttraumatic, video assisted thoracic surgery for, 174–175
Heterotopic ossifications (HO) in rib malunion, 138
HO. See Heterotopic ossifications (HO)
Hydropneumothorax, 181
  epidemiology, 183–184
  historical background, 181–182
  imaging investigations, 187–191
    computed tomography, 189–191, 189f, 190f
    plain radiography, 187
    ultrasonography, 187–188
    ultrasonography versus radiography, 188–189
  management, 191–194
    hemothorax, 192
    massive hemothorax, 191–192
    pneumothorax, 193–194
    retained hemothorax, 192–193

Index

Hydropneumothorax (*Continued*)
  pathophysiology, 184–186
    hemothorax, 184–185
    pneumothorax, 185–186
  primary evaluation, 186
  prognosis, 195–196
  terminology, 182–183
Hypovolemia, 26–27

I

ICN block. *See* Intercostal nerve (ICN) block
ICU. *See* Intensive care unit (ICU)
Inframammary thoracotomy, 114, 116*f*
Initial management of chest blunt trauma, 19–20
Injury Severity Score (ISS), 1, 56, 123
Intensive care unit (ICU), 6, 26–27
Intercostal nerve (ICN) block, 31
Internal rib fixation, 105–106
Intrathoracic rib fixation, 106–110
ISS. *See* Injury Severity Score (ISS)
Ivor-Lewis esophagectomy, 297, 298*f*

L

Laparoscopic approach, 268–270, 268*t*
  left-sided diaphragmatic laparoscopic repair, 273
  right-sided diaphragmatic laparoscopic repair, 272–273
Laparoscopy, 258
Laparotomy, 257, 275–276
Large-bore chest tubes (LBCTs), 233–235, 237
  insertion using blunt dissection technique, 239
Laryngotracheal trauma, 22*f*
LBCTs. *See* Large-bore chest tubes (LBCTs)
Left-sided diaphragmatic laparoscopic repair, 273
Length of hospital stay (LOS), 6
Levosimendan, 12
Liver laceration, 266
Long-term complications, 130–131
  chronic pain and irritation, 131
  fracture nonunion or malunion, 130
LOS. *See* Length of hospital stay (LOS)
Lower rib subluxation. *See* Slipping rib syndrome (SRS)
Lung contusion, 1–4, 8
  diagnostic, 3–4
  epidemiology/pathophysiology, 1–2
  lung laceration, 2–3, 3*f*
  therapy, 4
Lung herniation, 174–175

M

Malunion. *See* Rib malunion
Massive hemothorax, 191–192
Mechanical hardware failure, 129–130
Mesh, 277
Minimal invasive surgical rib fixation, 104–105
Montelukast, 12
Muscle-sparing axillary thoracotomy, 116*f*
Muscle-sparing lateral thoracotomy, 115*f*
Muscle-sparing postero-lateral thoracotomy, 115*f*

N

National Institute for Health and Care Excellence (NICE), 88
Negative predictive value (NPV), 37–38
Neuromuscular weakness, 127
New Injury Severity Score (NISS), 123
NICE. *See* National Institute for Health and Care Excellence (NICE)
NISS. *See* New Injury Severity Score (NISS)
Non-flail chest patients
  management in, 76–79
  rib fixation in, 82
Nonpenetrating chest trauma. *See* Video assisted thoracic surgery (VATS)
Non-steroidal anti-inflammatory drugs, 28
NPV. *See* Negative predictive value (NPV)

O

Occult pneumothorax, 247–248
OIS. *See* Organ Injury Scale (OIS)
Open pneumothorax, 24, 247
Open reduction and internal fixation (ORIF), 123–124
Operative rib open reduction and internal fixation (ORIF), 90
Organ Injury Scale (OIS), 42–45, 56
ORIF. *See* Open reduction and internal fixation (ORIF); Operative rib open reduction and internal fixation (ORIF)
Orthopedic Trauma Association (OTA), 41
OTA. *See* Orthopedic Trauma Association (OTA)

P

PCS. *See* Pulmonary Contusion Score (PCS)
Pectoral thoracotomy, 115, 117*f*

PEEP. *See* Positive end-expiratory pressure (PEEP)
Pericardiocentesis, 288–289
Plain radiography, 187
Pleural effusion, 125–126
Pneumothorax, 126, 185–187, 193–194, 266
Positive end-expiratory pressure (PEEP), 10, 26–27
Positive predictive value (PPV), 37–38
Positive pressure ventilation (PPV), 248–249
Posterior thoracotomy, 115, 117f
Postero-lateral thoracotomy, 112, 113f
Postoperative complications, 123
    early procedure-related complications, 125–127
        bleeding, 125
        neuromuscular weakness, 127
        pleural effusion, 125–126
        pneumothorax, 126
        pulmonary complications, 125
        ventilatory failure, 126
        wound infection, 126–127
    hardware infection and failure, 127–130
        hardware infection, 128–129
        mechanical hardware failure, 129–130
    long-term complications, 130–131
        chronic pain and irritation, 131
        fracture nonunion or malunion, 130
Posttraumatic chest wall injuries, video assisted thoracic surgery for, 176
Posttraumatic empyema, video assisted thoracic surgery for, 173–174
Posttraumatic haemothorax, video assisted thoracic surgery for, 172
Posttraumatic hernias, video assisted thoracic surgery for, 174–175
PPV. *See* Positive predictive value (PPV); Positive pressure ventilation (PPV)
Primary repair versus mesh, 277
Prophylactic antibiotics, 252–253
Pulmonary complications, 125
Pulmonary Contusion Score (PCS), 1

Q

QALY. *See* Quality Adjusted Life Years (QALY)
Quality Adjusted Life Years (QALY), 124
Quebec clinical decision rule, 220, 221t

R

radiography, ultrasonography versus, 188–189

Radiography, ultrasonography versus, 188–189
Receiver operating characteristic (ROC) curve, 38–39, 39f
Regional anesthesia, 29
Respiratory failure, 4–5. *See also* Acute respiratory distress syndrome (ARDS) sequential clinical assessment of respiratory function (SCARF) score, 47–48
Retained haemothorax, 192–193
    management of, 251–252
RFS. *See* Rib fracture score (RFS)
Rib fixation
    early, 87
        benefits and risks of, 95f
        cost-effectiveness, 98–99
        elderly versus young, 93
        evidence synthesis, 88–90, 89t
        multi-disciplinary approach, 96–97
        patient selection and allocation challenges, 96
        quality of life, 97–99
        surgical approach, indications for, 93–94
        surgical intervention, movement toward, 90–93, 91t
        surgical intervention, timing to, 94–96
    modern approach to, 103
        full VATS approach, 106–107
        internal rib fixation, 105–106
        intrathoracic rib fixation, 106–110
        minimal invasive surgical rib fixation, 104–105
        partial VATS approach, 107–110
    surgical approaches to, 112–119
        modified open approaches, 113–116
        standard open approach, 112–113
        tunnel-based open approaches, 116–118
        video-assisted thoracoscopic hybrid approaches, 118–119
Rib fracture fixation, 3D-printing technology in, 110–112
    customized implants, surgical guides and templates, minimally invasive SSRF, 111
    future directions and challenges, 112
    preoperative planning with 3D printing models, 110–111
    training and education, potential for, 111
Rib fracture nonunion, management of, 80–82, 81f
Rib fractures, 266
Rib fractures and blunt chest wall injury, 55

312 Index

Rib fractures and blunt chest wall injury (*Continued*)
  AO Foundation/Orthopaedic Trauma Association (AO/OTA) fracture classification system, 56–57
  Chest Injuries International Database, 62
  Chest Wall Injury Society (CWIS), 57–60, 58*t*
  costal cartilage, 64–65
  costal margin, injuries to, 65–66
  CWIS/ASER working group, 60–62, 61*t*
  Sheffield Multiple Rib Fractures Study (SMuRFS), 64
Rib fracture score (RFS), 46, 123–124
Rib fractures management algorithm, 71, 73*f*
  flail chest patients, management in, 72–73
  non-flail chest patients
    management in, 76–79
    rib fixation in, 82
Rib malunion, 137. *See also* Slipping rib syndrome (SRS)
  associated complications, 149–150
  definition, 137
  diagnostic approach, 141–143
  differential diagnosis, 142–143
  incidence, 139–140
  prevention strategies, 150–151
  prognosis, 150
  risk factors, 140
  surgical treatment, 144–149
    indications for surgical intervention, 144–145
    surgical options, 145–149
  symptom, 140–141
  treatment options, 143–144
  types, 137–139
RibScore, 46–47
Rib stabilization, indications for, 6–7
Rib tip syndrome. *See* Slipping rib syndrome (SRS)
Right-sided diaphragmatic laparoscopic repair, 272–273
Robotic transthoracic approach, 274–275
ROC curve. *See* Receiver operating characteristic (ROC) curve
Rotational malunion, 138

S

SAP block. *See* Serratus anterior plane (SAP) block
SBCTs. *See* Small-bore chest tubes (SBCTs)
SCARF score. *See* Sequential clinical assessment of respiratory function (SCARF) score

Scoring systems, chest wall
  chest wall Organ Injury Scale, 42–45, 44*t*
  current, 42–48, 43*t*
  general properties of, 37–40
  ideal characteristics of, 40–41
  thoracic chest wall taxonomy, 41–42
  thoracic trauma severity score (TTSS), 45
Sepsis, 267
Sequential clinical assessment of respiratory function (SCARF) score, 47–48
Serratus anterior plane (SAP) block, 30
SF. *See* Sternal fracture (SF)
Sheffield classification of injuries, 66*f*
Sheffield Multiple Rib Fractures Study (SMuRFS), 64
Shortening malunion, 138
Simple pneumothorax, 247
Slipping rib syndrome (SRS), 153
  definition, 153–154
  demographics, 155
  diagnostic approach flow chart, 156–160
  diagnostic examinations, 159–160
  differential diagnosis, 159
  etiology, 155
  forms, 154
  history, 154
  incidence, 155
  indications for surgical intervention, 162
  main symptom, 156
  pathogenesis, 155–156
  patient monitoring, 165
  physical examination, 157–158
  prevention strategies, 165–166
  surgical options, 162–164
  surgical treatment, 161–164
  treatment options flow chart, 160–161
  types, 154
Small-bore chest tubes (SBCTs), 233–234, 237, 239
  insertion using Seldinger technique, 239
"Small" haemothorax, chest drain for, 250–251
SMuRFS. *See* Sheffield Multiple Rib Fractures Study (SMuRFS)
Spleen rupture, 266
Spontaneous simple pneumothorax, 248
SRS. *See* Slipping rib syndrome (SRS)
SSRF. *See* Subsequent segmental rib resection and fixation (SSRF); Surgical stabilization of rib fractures (SSRF)
Sternal fracture (SF), 203
  clinical presentation and diagnosis, 204–206

epidemiology, 203
management, 206–209
mechanism of injury and implications, 203–204
surgical management of, 208t
Subsequent segmental rib resection and fixation (SSRF), 145
Suction versus no suction, 252
Surgical stabilization of rib fractures (SSRF), 48, 50, 87
Surgical treatment, anesthesia for, 26–27

T

TDI. *See* Traumatic diaphragmatic injuries (TDI)
TEA. *See* Thoracic epidural analgesia (TEA)
Tension pneumothorax, 23, 246
Thoracic chest wall taxonomy, 41–42
Thoracic epidural analgesia (TEA), 29
Thoracic paravertebral (TPV) block, 29–30
Thoracic Trauma Score (TTS), 1
Thoracic trauma severity score (TTSS), 45
Thoracotomy, 258, 276–277
　antero-lateral, 112–113, 114f
　inframammary thoracotomy, 114, 116f
　muscle-sparing axillary thoracotomy, 116f
　muscle-sparing lateral thoracotomy, 115f
　muscle-sparing postero-lateral thoracotomy, 115f
　pectoral thoracotomy, 115, 117f
　posterior thoracotomy, 115, 117f
　postero-lateral, 112, 113f
3D printing (3D-P) technology, 110
　in rib fracture fixation, 110–112
　　customized implants, surgical guides and templates, minimally invasive SSRF, 111
　　future directions and challenges, 112
　　preoperative planning with 3D printing models, 110–111
　　training and education, potential for, 111
TI. *See* Tracheobronchial injury (TI)
TISS. *See* Trauma Injury Severity Score (TISS)
TPV block. *See* Thoracic paravertebral (TPV) block
Tracheobronchial injury (TI) and airway management, 20–23
Translation malunion, 138
Trauma Injury Severity Score (TISS), 1, 123
Traumatic diaphragmatic injuries (TDI), 267

Traumatic pericardial effusion and cardiac tamponade, 283
　clinical findings, 284–285
　conservative approach, 286–287
　discussion, 289–291
　investigations, 285–286
　pericardiocentesis, 288–289
　treatment, 286
TTS. *See* Thoracic Trauma Score (TTS)
TTSS. *See* Thoracic trauma severity score (TTSS)

U

Ultrasonography, 187–188
Ultrasound, 263, 263t
Uniportal VATS in blunt chest trauma, 224–225

V

VATS. *See* Video assisted thoracic surgery (VATS)
Ventilatory failure, 126
Video assisted thoracic surgery (VATS), 146–147, 273–274
　for delayed hemothorax, 213
　　clinical and radiological predictive factors, 215–220
　　complications, 221
　　definition, epidemiology, and classification, 213–215
　　diagnostic and therapeutic role of VATS, 222–225
　　Quebec clinical decision rule, 220, 221t
　　uniportal VATS in blunt chest trauma, 224–225
　full, 106–107
　in nonpenetrating chest trauma, 169
　　benefits, features, and timing of, 170–172
　　development of VATS surgery and its use in chest trauma, 169–170
　　for posttraumatic chest wall injuries, 176
　　for posttraumatic empyema, 173–174
　　for posttraumatic haemothorax, 172
　　for posttraumatic hernias, 174–175
　partial, 107–110
Video-assisted thoracoscopic hybrid approaches, 118–119

W

Wound infection, 126–127

Printed and bound by CPI Group (UK) Ltd, Croydon, CR0 4YY
05/06/2025
01894334-0001